AF388962

Fluid/Structure Interactions

Fluid/Structure Interactions

Editor

Yuriy Semenov

MDPI • Basel • Beijing • Wuhan • Barcelona • Belgrade • Manchester • Tokyo • Cluj • Tianjin

Editor
Yuriy Semenov
University College London
UK

Editorial Office
MDPI
St. Alban-Anlage 66
4052 Basel, Switzerland

This is a reprint of articles from the Special Issue published online in the open access journal *Journal of Marine Science and Engineering* (ISSN 2077-1312) (available at: https://www.mdpi.com/journal/jmse/special_issues/fluid_structure).

For citation purposes, cite each article independently as indicated on the article page online and as indicated below:

LastName, A.A.; LastName, B.B.; LastName, C.C. Article Title. *Journal Name* **Year**, *Volume Number*, Page Range.

ISBN 978-3-0365-3250-9 (Hbk)
ISBN 978-3-0365-3251-6 (PDF)

Contents

About the Editor

Yuriy Semenov works at the Institute of Hydromechanics of the National Academy of Science of Ukraine. In 2012–20, he was with University College London as a Senior Research Associate. From 1984–2006 he was with the Institute of Technical Mechanics of the National Space Agency of Ukraine. In 2000–02, he took leave to work at Osaka University on the development of a mathematical model describing the 'rotating choke': a cavitation instability first observed in a cavitating turbopump. In 2006–08, he was a Visiting Researcher at the Norwegian Centre of Excellence for Ship and Ocean Structures. In 2009–11, he worked as Research Professor at the School of Naval Architecture and Ocean Engineering, University of Ulsan in South Korea.

Journal of
Marine Science and Engineering

Editorial

Fluid/Structure Interactions

Yuriy A. Semenov

Institute of Hydromechanics of the National Academy of Sciences of Ukraine, 8/4 Marii Kapnist Street, 03680 Kiev, Ukraine; semenov@nas.gov.ua; Tel.: +38-044-456-4313

Citation: Semenov, Y.A. Fluid/Structure Interactions. *J. Mar. Sci. Eng.* **2022**, *10*, 159. https://doi.org/10.3390/jmse10020159

Received: 18 January 2022
Accepted: 21 January 2022
Published: 26 January 2022

Publisher's Note: MDPI stays neutral with regard to jurisdictional claims in published maps and institutional affiliations.

This Special Issue contains 12 papers devoted to fluid/structure interaction (FSI) problems. This is a fundamental subject of fluid dynamics, in which a lot has already been done in the theory, experiments, and computational techniques. However, every day further developments in marine engineering, bioengineering, and renewable energy call for addressing new problems and arising phenomena. That is why it would be helpful to explain in this preface basic motivations for publishing this issue.

FSI can be defined as a joint motion of a deformable structure with an internal or surrounding fluid flow. The structure and the fluid the governing equations are different, which calls for introducing an interface on which consistent boundary conditions for both the liquid and the solid regions are formulated. This is the main feature of FSI problems. In some cases, the liquid causes such a small deformation of the structure that it does not affect the motion of the liquid. These fluid–structure interaction systems are called weakly coupled systems, or one-way interaction, i.e., the liquid deforms the structure shape, but the deformation is so small that it does not affect the flow significantly. In contrast to this, a strong interaction, or two-way interaction, occurs when the deformation of the structure is great enough to change the flow characteristics, which, in its turn, may affect the structure deformation. Two-way interaction problems are complicated due to nonlinearity since the shape of the interface is unknown in advance and has to be determined as part of the solution of the problem using boundary conditions at the interface, which come from joining the solutions for the solid and the liquid parts.

The interest in FSI problems is very great due to their practical importance. Recent developments in engineering have led to advanced formulations of FSI problems. Some of them could not be formulated several years ago. These problems require progress in both numerical algorithms and mathematical apparatus and advanced computational techniques, such as parallel computations.

In this issue, we have tried to collect different FSI problems, new mathematical and numerical approaches, new numerical techniques and, of course, new results, which can provide an insight into FSI processes.

The issue opens with the paper by Ren et al. [1] on vortex induced vibration (VIV) of risers in oil/gas offshore production systems. Caused by ocean currents, vortices are periodically generated on the sides of the riser and manifest themselves as a periodic excitation force. When the frequency of the periodic force approaches one of the natural frequencies of the riser, an increase in the vibration amplitude can be expected.

The second paper by Qiu, Ren and Li [2] presents a study on FSI during the water impact of a lifeboat in free-fall from a ship into a rough see. The paper not only reviews recent developments in the water-entry theory, but also presents a way to account for the boat elasticity and friction forces. The computational results are verified by experiment.

The paper by Ni et al. [3] on ice–ship interaction is very remarkable. Activities in the Arctic Regions have rapidly increased in the last decade due to ice melting caused by the climate change and opening new routes for transportation. This paper presents a numerical simulation of a ship moving in level ice with a crack propagation process including radial and circular cracks. The computations are based on a one-way CFD-DEM coupling method.

A two-way fluid–structure interaction (FSI) approach is presented in the paper by Malazi et al. [4] considering three-dimensional FSI in the case of an elastic beam. They developed a two-way FSI coupling method and demonstrated its high efficiency. The method may be employed in various engineering fields, such as mechanical, civil, and ocean engineering. Chie and Wahab [5] investigated the flow field around an intake structure and derived engineering design criteria to mitigate the seawater intake operation impact on marine life.

Typically, pumps and turbines work with high loads on the impeller. In the case of flow unsteadiness, this may cause a response of the pump structure, which leads to vibration and even dangerous resonance phenomena. Mao et al. [6] investigated another type of instability of reversible pump/turbines caused by a transition from the turbine to the pump mode of operation. They studied the axial hydraulic force and determined the deformation of the support bracket and the main shaft, which contribute to the resultant force on the crown and the band.

The interesting study Liu et al. [7] related to the fish industry deals with the design parameters of a trawl net and its effect on the hydrodynamic characteristics. Based on the experiments conducted, it was found that the codend drag force oscillation mainly included a high-frequency and a low-frequency oscillation. The low-frequency oscillation of the drag force included a strong wave oscillation and a weak wave oscillation set up alternately.

A study on the vibration characteristics of a marine centrifugal pump unit and different excitation sources is presented by Dai et al. [8]. They developed a computational code coupling the fluid and the structural dynamics of the pump unit and studied the vibration characteristics caused by fluid excitation and electromagnetic excitation. The agreement between the calculated and the test results is quite impressive.

A two-way FSI coupling approach for the vp1304 marine propeller is presented by Masoomi and Mosavi [9]. They developed a code that predicts the pressure and stress distributions with a low-cost and high-precision approach. They pointed out that an important factor for the coupling approach is the rotational rate interrelated between two solution domains. The propeller strength was assessed by considering the blades' stress and strain for different load conditions.

The paper Liu et al. [10] deals with a numerical modal analysis of a prototype Francis pump turbine runner. They employed an acoustic–structure coupling method. The effect of the energy loss on the chamber wall on the natural modes of the runner was studied by the absorption boundary. The results show that the constraint condition (especially the rotating shaft) has significant impacts on various modes of the runner.

The case of one way fluid/structure interaction is presented by Marty et al. [11]. They studied the interaction of a coastal current with submerged components of floating wind turbines, risers and mooring lines for floating units taking into account surface roughness caused by mussel's colonies. The authors found two realistic shapes caused by mussel's colonization and presented tests of those shapes in a hydrodynamic tank.

Flow detachment conditions in the presence of surface tension are discussed in Savchenko et al. [12] for a case study of cavity flow past a wedge with rounded edges. The authors analyze the Brillouin–Villat criterion of flow detachment and its applicability to flows with surface tension. It was found that the Brillouin–Villat criterion has a limited applicability, especially for small Froude numbers and small edge radii. For moderate Weber numbers, surface tension slightly decreases the cavity size and the drag force. As the Weber number decreases further, the velocity at the point of cavity detachment increases, and the angle of flow detachment changes in such a manner that the wetting part of the edge becomes larger. The tendency seems to be towards wetting the whole of the edge and making the flow attached to the wedge.

We expect that this Special Issue will be helpful to researchers in different fields of hydrodynamics and will contribute to the study of fluid/structure interaction problems.

Funding: This research received no external funding.

Conflicts of Interest: The author declares no conflict of interest.

References

1. Ren, H.; Zhang, M.; Cheng, J.; Cao, P.; Xu, Y.; Fu, S.; Liu, C. Experimental Investigation on Vortex-Induced Vibration of a Flexible Pipe under Higher Mode in an Oscillatory Flow. *J. Mar. Sci. Eng.* **2020**, *8*, 408. [CrossRef]
2. Qiu, S.; Ren, H.; Li, H. Computational Model for Simulation of Lifeboat Free-Fall during Its Launching from Ship in Rough Seas. *J. Mar. Sci. Eng.* **2020**, *8*, 631. [CrossRef]
3. Ni, B.-Y.; Chen, Z.-W.; Zhong, K.; Li, X.-A.; Xue, Y.-Z. Numerical Simulation of a Polar Ship Moving in Level Ice Based on a One-Way Coupling Method. *J. Mar. Sci. Eng.* **2020**, *8*, 692. [CrossRef]
4. Malazi, M.T.; Eren, E.T.; Luo, J.; Mi, S.; Temir, G. Three-Dimensional Fluid—Structure Interaction Case Study on Elastic Beam. *J. Mar. Sci. Eng.* **2020**, *8*, 714. [CrossRef]
5. Chie, L.H.; Wahab, A.K.A. Derivation of Engineering Design Criteria for Flow Field Around Intake Structure: A Numerical Simulation Study. *J. Mar. Sci. Eng.* **2020**, *8*, 827. [CrossRef]
6. Mao, Z.; Tao, R.; Chen, F.; Bi, H.; Cao, J.; Luo, J.; Fan, H.; Wang, Z. Investigation of the Starting-Up Axial Hydraulic Force and Structure Characteristics of Pump Turbine in Pump Mode. *J. Mar. Sci. Eng.* **2021**, *9*, 158. [CrossRef]
7. Liu, W.; Tang, H.; You, X.; Dong, S.; Xu, L.; Hu, F. Effect of Cutting Ratio and Catch on Drag Characteristics and Fluttering Motions of Midwater Trawl Codend. *J. Mar. Sci. Eng.* **2021**, *9*, 256. [CrossRef]
8. Dai, C.; Zhang, Y.; Pan, Q.; Dong, L.; Liu, H. Study on Vibration Characteristics of Marine Centrifugal Pump Unit Excited by Different Excitation Sources. *J. Mar. Sci. Eng.* **2021**, *9*, 274. [CrossRef]
9. Masoomi, M.; Mosavi, A. The One-Way FSI Method Based on RANS-FEM for the Open Water Test of a Marine Propeller at the Different Loading Conditions. *J. Mar. Sci. Eng.* **2021**, *9*, 351. [CrossRef]
10. Liu, D.; Xia, X.; Yang, J.; Wang, Z. Effect of Boundary Conditions on Fluid–Structure Coupled Modal Analysis of Runners. *J. Mar. Sci. Eng.* **2021**, *9*, 434. [CrossRef]
11. Marty, A.; Schoefs, F.; Soulard, T.; Berhault, C.; Facq, J.-V.; Gaurier, B.; Germain, G. Effect of Roughness of Mussels on Cylinder Forces from a Realistic Shape Modelling. *J. Mar. Sci. Eng.* **2021**, *9*, 598. [CrossRef]
12. Savchenko, Y.N.; Savchenko, G.Y.; Semenov, Y.A. Cavity Detachment from a Wedge with Rounded Edges and the Surface Tension Effect. *J. Mar. Sci. Eng.* **2021**, *9*, 1253. [CrossRef]

Journal of
Marine Science and Engineering

Article

Experimental Investigation on Vortex-Induced Vibration of a Flexible Pipe under Higher Mode in an Oscillatory Flow

Haojie Ren [1,2], Mengmeng Zhang [1,2,*], Jingyun Cheng [3], Peimin Cao [3], Yuwang Xu [1,2], Shixiao Fu [1,2] and Chang Liu [4]

[1]   State Key Laboratory of Ocean Engineering, Shanghai Jiao Tong University, Shanghai 200240, China; rhaojie@163.com (H.R.); xuyuwang@sjtu.edu.cn (Y.X.); shixiao.fu@sjtu.edu.cn (S.F.)
[2]   Collaborative Innovation Center for Advanced Ship and Deep-Sea Exploration, Shanghai 200240, China
[3]   SBM Offshore, Houston, TX 77077, USA; jingyun.cheng@sbmoffshore.com (J.C.); peimin.cao@sbmoffshore.com (P.C.)
[4]   Department of Mechanical Engineering, Johns Hopkins University, Baltimore, MD 21218, USA; changliu0520@163.com
*   Correspondence: 15145029174@163.com

Received: 5 May 2020; Accepted: 2 June 2020; Published: 4 June 2020

Abstract: Different from the previous studies of the vortex-induced vibration (VIV) dominated by first mode of flexible pipe in an oscillatory flow, the features of a higher mode dominated are experimentally investigated in the ocean basin. The flexible pipe is forced to harmonically oscillate with different combinations of a period and amplitude. The design dominant mode consists of first and second modes under the maximum reduced velocity (V_R) of approximately 5.5 with a KC number ranging from 22 to 165. The VIV responses between only the excited first mode and the excited higher mode are compared and studied using displacement reconstruction and wavelet transform methods. The discrepancies of spatial and temporal response between smaller and larger KC numbers (KC = 56 and 121) are first observed. The strong alternate mode dominance and lock-in phenomena occur in the case of larger KC numbers, while they cannot be observed in the case of smaller KC numbers under higher modes. The VIV dominant frequency in the in-line (IL) direction is found to be always triple the oscillatory flow frequency and not twice that in the cross flow (CF) direction. The dominant frequency in the CF direction can be predicted by the Strouhal law, and the Strouhal number is approximately 0.18 under V_R = 5.5, which is not affected by the excited mode. Moreover, differences of response motion trajectory are also revealed in this paper. The present work improves the basic understanding of vessel motion induced VIV and provides helpful references for developing prediction methods of VIV in an oscillatory flow.

Keywords: vortex-induced vibration; higher mode; flexible pipe; oscillatory flow; motion trajectory; lock-in; dominant frequency; time-varying

1. Introduction

The riser, serving as the only channel to connect the seabed wellhead to the top floating vessel, is the weakest part of the entire oil and gas development system. As natural gas and oil production moves into deep and ultra-deep-water sea, risers are becoming increasingly slender. Under the action of ocean currents, vortices are periodically generated and alternately shed from the sides of these very slender risers, resulting in corresponding periodic excitation force. When the frequency of this force is near one of the natural frequencies of the flexible riser, a significant vibration will occur. This is termed Vortex-induced Vibration (VIV) [1], which has been proven to be the main reason for the fatigue damage of risers. Owing to the complexity of the ocean environment and stronger coupling interaction between fluids and structures, the observation of dynamic behaviors of VIV under different environments is the first step to solve this problem. Therefore, researchers in both industry and academia have carried out a lot work to investigate the VIV response performance and to improve the understanding of the mechanism behind VIV.

In last five decades, the studies, as shown in Table 1, have mainly focused on VIV features in a steady flow. A stationary rigid cylinder was first towed in the tank. The vortex-induced force, response frequency, and wake patterns were preliminarily investigated and found to be a function of Reynolds numbers [1–5]. Furthermore, self-oscillation tests of a rigid cylinder in steady flow were conducted [6–14]. The amplitude and frequency of the VIV response versus reduced velocity were studied. Three branches—the initial, the upper, and the lower branches—were defined, and many other features were revealed. These works enrich the basic understanding of vortex-induced vibration. However, the aforementioned cylinder model is rigid and quite different from the real riser. To get closer to the real riser, flexible pipe models were adopted and experimented in steady flows, such as uniform flow, linear shear flow, and stepped flow [15–23]. Higher-order and multi-mode responses, traveling wave and time-sharing features are further discussed. These particular phenomena greatly improve our understanding of VIV. Based on the above efforts, semi-empirical methods were established. Corresponding software was formed and widely used in industry, such as SHEAR7 [24], VIVA [25], and VIVANA [26]. However, observation under the steady flow field cannot fully reflect the vortex-induced vibration characteristics of the vertical pipe under the real ocean environment and limits development of VIV prediction methods.

Table 1. Summary of experimental studies on the Vortex-induced Vibration (VIV) features.

Author	Year	Current	Model Type	Experiment Type	Vibration Mode
Schewe et al., Williamson et al., Hallam et al., Achenbach et al.	1983,1988,1977,1981	Uniform flow	Rigid cylinder	Stationary towing	/
Williamson et al., Govardhan et al., Sarpkaya, Bearman, Feng, Griffin et al., Parkinson	2006, 2004, 2008, 2000, 2004, 1984, 1963, 1982, 1989	Uniform flow	Rigid cylinder	Self-oscillation	/
Lie et al., Frank etal., Trim et al.	2006, 2004, 2005	Linear shear flow	Flexible pipe	Flexible pipe	Multi-mode
Fu et al., Ren et al., Vandiver, Song et al.	2011, 2019, 1985, 2016, 2017	Uniform flow	Flexible pipe	Flexible pipe	Multi-mode
Chaplin et al.	2005	Stepped flow	Flexible pipe	Flexible pipe	Multi-mode
Wang et al., Pesce et al., Cunff et al., Pereira et al.,	2015, 2017, 2017, 2005, 2013	/	Flexible pipe	VMI-VIV	Multi-mode
Fu et al., Wang et al.	2014, 2015	Oscillatory flow	Flexible pipe	Flexible pipe	1st

In a real sea state, risers inevitably encounter the action of wave and vessel motions. This wave and wave-induced periodic motion of a platform always result in a relatively equivalent oscillatory flow around risers. Recently, experimental studies found that the oscillatory flow induced by vessel motion can also excite VIV at the sag-bend of steel catenary risers [27,28], which is the so-called Vessel Motion-induced VIV (VMI-VIV) [29]. VMI-VIV can cause serious fatigue damage to risers [30]. Similar experiments with the steel catenary riser are conducted by Cunff et al. (2005) and Pereira et al. (2013) [31,32]. Moreover, the experiment of a free-hanging riser under vessel motion was also carried out [33,34]. VMI-VIV has also been observed. The multimode participation, travelling wave, amplitude modulation and time-sharing phenomenon are directly found in these studies. To better understand the VMI-VIV, Fu et al. (2014) conducted a flexible riser model test in an oscillatory flow by forcing the model to oscillate in still water with different periods and amplitudes [35,36]. A VIV development process for a flexible riser in an oscillatory flow was first proposed, including the building-up, lock-in, and dying-out phases. However, only the first mode was excited in their experiments, which is different from the excited higher mode in real environments. This limitation prevents this experiment from providing a more basic understanding of VMI-VIV. Therefore, the features of VIV under a higher mode in an oscillatory flow need further study.

The remainder of this paper is organized as follows. Section 2 presents the VIV model tests of a flexible pipe under a higher mode in an equivalent oscillatory flow. The dominant mode is designed to be the first and second modes. The maximum reduced velocity, located in the significant vibration region of VIV, is selected as approximately 5.5. The KC numbers vary from 22 to 165 through forcing the flexible pipe to harmonically oscillate under various combinations of amplitude and period. Section 3 introduces the basic theory of data preprocessing, the displacement reconstruction method, and the time–frequency analysis method. Based on these methods, the differences of VIV features between only the first mode and the excited higher mode are compared and studied in Section 4. The investigation conclusions are then summarized in Section 5.

2. Model Test

2.1. Experimental Setup

To simulate an equivalent oscillatory flow, the same test apparatus made by Fu et al. (2014) [35] was adopted and forced a flexible pipe to oscillate in harmonic motions under various combinations of amplitudes and periods in an ocean basin at Shanghai Jiao Tong University. The whole experimental apparatus mounted under the bottom of the towing carriage primarily contains two horizontal and vertical tracks as shown in Figure 1. Two force sensors are placed at two ends of the pipe model through universal joints. A pretension force of 500 N was applied to the flexible pipe by a tensioner, which was connected to the force sensor and fixed to a side of the vertical tracks. The vertical track was used to adjust submerged depth of the model, and the horizontal tracks drove the pipe to oscillate. Two endplates were used to reduce the disturbance of the supporting frame to the equivalent oscillatory flow field.

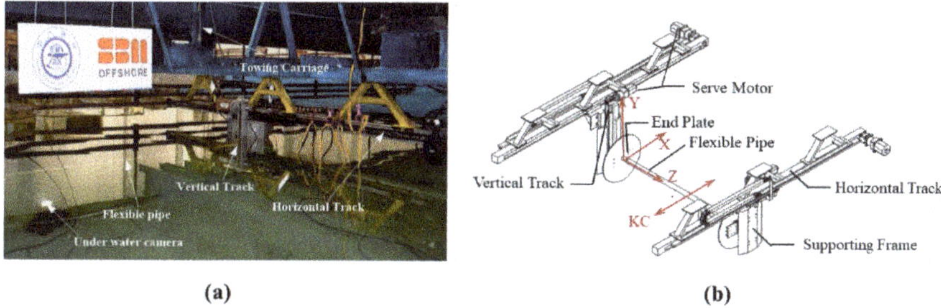

Figure 1. Overview and sketch of the whole experimental setup: ((**a**). the overview photo; (**b**). the schematic drawing).

The flexible pipe was made up of a polypropylene random pipe that is filled with copper cable inside. Silicone gel was placed between different layers to prevent relative slippage. The details of the flexible pipe are listed in Table 2. The bending stiffness is 46.433 N·m^2 and the damping ratio is 2.53%. The 1st and 2nd order eigen frequency of the flexible pipe in still water can be calculated by which can be calculated by Equation (1).

$$f_{n0} = \frac{n}{2L} \sqrt{\frac{F_{T_0}}{m} + \frac{n^2\pi^2}{L^2} \cdot \frac{EI}{m}}, \quad m = \overline{m} + \frac{1}{4}\pi D^2 \rho C_m, \tag{1}$$

where F_{T_0} is the pretension force of 500 N, m is the mass of bare pipe per unit length in still water, and L and D are the length and diameter of flexible pipe, respectively. EI is the bending stiffness of the pipe, and ρ is the density of water, ρ = 1000 kg/m^3. The added mass coefficient is chosen as C_m = 1.0. Notably, the added mass may deviate from the value of 1.0 in still water, and the forced oscillation periods will change with the KC number at the same maximum reduced velocity. n is the mode number.

Table 2. Parameters of the flexible pipe model.

Item	Value
Model length L (m)	4
Outer diameter D (mm)	29
Mass of flexible pipe in the air $\overline{m}$ (kg/m)	1.529
Mass ratio of flexible pipe (m^*)	2.3
Bending stiffness EI (N·m^2)	46.43
Tensile stiffness EA (N)	1.528×10^6
Pre-tension F_{T0} (N)	500
Damping ratio ζ	2.53%
Calculated first natural frequency f_{10} in still water (Hz)	1.90
Calculated second natural frequency f_{20} in still water (Hz)	4.08

Four groups of Fiber Bragg Grating (FBG) strain sensors were installed on the surface of the flexible pipe to measure the strain responses in both the CF and IL directions. Each of the FBG groups (CF_a, CF_c, IL_b and IL_d) had ten measurement points along the pipe separated by 0.36 m, as shown in the schematic diagram in Figure 2.

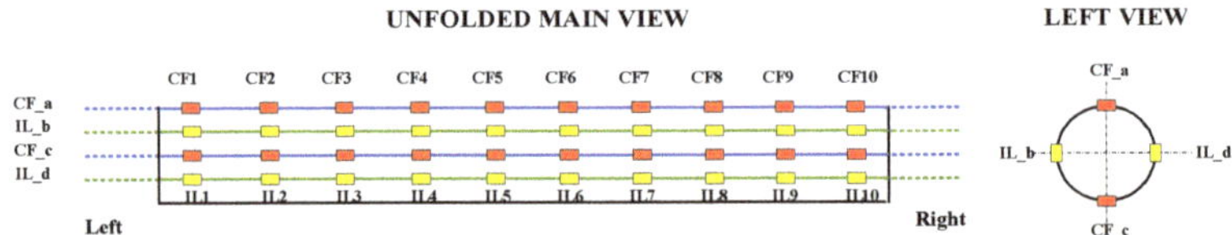

Figure 2. The Fiber Bragg Grating (FBG) strain sensor instrumentation along the flexible pipe model.

2.2. Test Arrangement

For convenience of description, the coordinate system is defined as O-XYZ as shown in Figure 1b. The origin is at the end point of the test model. The Z axis is along the length of pipe. The X axis is the in-line flow direction, and the Y axis is the cross-flow direction. In our model test, the instantaneous displacement $X(t)$ and velocity $U(t)$ of forced harmonic motions in the horizontal direction can be expressed as:

$$X(t) = A_m \sin\left(\frac{2\pi}{T}t\right) \tag{2}$$

$$U(t) = U_m \cos\left(\frac{2\pi}{T}t\right), \ U_m = A_m \frac{2\pi}{T} \tag{3}$$

where A_m and T are oscillation amplitude and period, respectively. U_m is the amplitude of the forced motion velocity.

The key parameters determining the VIV features of the flexible pipe under oscillatory flow are the KC and maximum reduced velocity V_R [37]. The KC number is defined as follows in Equation (4).

$$KC = \frac{2\pi \cdot A_m}{D}, \tag{4}$$

In our experiments, the excited dominant mode is designed to be the 1st and 2nd modes. The corresponding maximum reduced velocity V_{R1} and V_{R2} is expressed as Equations (5) and (6), respectively.

$$V_{R1} = \frac{2\pi \cdot A_m}{T \cdot f_{10} \cdot D} \tag{5}$$

$$V_{R2} = \frac{2\pi \cdot A_m}{T \cdot f_{20} \cdot D} \tag{6}$$

To investigate the VIV performance of the flexible pipe under higher modes in an oscillatory flow, the test cases are divided into two groups with different designed vibration modes under the same maximum reduced velocities. The comparison group and the group with higher modes used the f_{10} (f_{10} = 1.90Hz) and f_{20} (f_{20} = 4.08Hz) to design test cases, respectively. All test cases are listed in Table 3. The maximum reduced velocities are approximately 5.5. The KC number ranges from 22 to 160. The corresponding maximum Reynolds number (Re_{max}) is also listed in this table, which can be calculated by Equation (7):

$$Re_{max} = \frac{U_m D}{v}, \tag{7}$$

where U_m is the forced motion velocity amplitude, and v is the kinematic viscosity coefficient. In our experiment, the ambient temperature is maintained near 15 °C, and v is, therefore, approximately 1.14×10^{-6} m^2 s^{-1}.

Table 3. Test Matrix.

Case No.	V_R	Mode	A_m(m)	KC	Re_{max}
1–8	5.6	1st	0.10–0.76	22–165	7860
9–15	5.2	2nd	0.22–0.76	48–165	15,696

3. Data Analysis Procedures

3.1. Preprocessing

In the model test, the measured strain in both the IL and CF directions included three components: the initial axial strain caused by pretension, the varying axial strain caused by varying tension, and the bending strain resulted from hydrodynamic forces. Therefore, the pure VIV strain at position z, $\varepsilon_{CF}(z,t)$, can be calculated by:

$$\varepsilon_{CF}(z,t) = [\varepsilon_{CF\_a}(z,t) - \varepsilon_{CF\_c}(z,t)]/2 \tag{8}$$

where $\varepsilon_{CF\_a}(z,t)$ and $\varepsilon_{CF\_c}(z,t)$ are the original strain time histories at position z sampled by the CF_a and CF_c measurement points, respectively.

At the same time, the bending strain $\varepsilon_{IL}(z,t)$ in the IL direction can be calculated by Equation (9).

$$\varepsilon_{IL}(z,t) = [\varepsilon_{IL\_b}(z,t) - \varepsilon_{IL\_d}(z,t)]/2 \tag{9}$$

where $\varepsilon_{IL\_b}(z,t)$ and $\varepsilon_{IL\_d}(z,t)$ are the original strain time histories at position z sampled by the IL_b and IL_d measurement points, respectively.

Then, a band-pass filter was utilized to eliminate the higher frequency noise in the IL direction and to remove corresponding higher frequency noise and effects of pendulum motion caused by forced motion in the CF direction. The cutoff frequencies of the band-pass filter were 0 Hz and 15 Hz for the IL direction and 2.5f_o (f_o is the forced oscillation frequency) and 15 Hz for the CF direction.

3.2. Displacement Reconstruction

The displacement reconstruction method is a basic tool for the VIV study [15,21,38]. According to the Euler–Bernoulli beam theory, the VIV displacement response of a flexible pipe under an external load can be expressed as the sum of the modal shapes multiplied by the generalized coordinate values at each step. Taking the response in the CF direction as an example, the VIV displacement response $y(z,t)$ can be expressed as:

$$y(z,t) = \sum_{i=1}^{n} p_i(t)\varphi_i(z), \quad z \in [0,L] \tag{10}$$

where $p_i(t)$ is the ith generalized coordinate displacement value at time t, and $\varphi_i(z)$ is the displacement at position z in the ith modal shape.

Based on an assumption of small deformation, the curvature $\kappa(z,t)$ can be expressed as:

$$\kappa(z,t) = \frac{\partial^2 y(z,t)}{\partial z^2} = \sum_{i=1}^{n} p_i(t)\varphi_i''(z), \quad z \in [0,L] \tag{11}$$

where φ_i'' is the ith modal shape of the curvature. According to the geometric relationship between the curvature and strain, the strain can be calculated by:

$$\varepsilon(z,t) = \kappa(z,t)R = R\sum_{i=1}^{n} p_i(t)\varphi_i''(z), \quad z \in [0,L] \tag{12}$$

where R denotes the radius of the flexible pipe model at position z.

In the flexible pipe model, the modal shape of the displacement is sinusoidal since the two ends of the beam are hinged boundary using universal joints, and thus can be expressed as:

$$\varphi_i(z) = \sin\frac{i\pi z}{L}, \quad i = 1, 2... \tag{13}$$

From Equation (11), the modal shape of the curvature is also sinusoidal. After obtaining the modal shapes of the displacement and curvature, the generalized coordinates can be obtained from Equation (12). Then, the VIV displacement response in the CF direction can be calculated by Equation (10). Using the same method, the displacement response of the IL direction can also be obtained.

3.3. Time–Frequency Analysis

In an oscillatory flow, the shedding frequency changes with the periodic oscillation velocity of the model, which can be written as:

$$f_{st}(t) = \frac{St \cdot |U(t)|}{D} \tag{14}$$

where St refers to the Strouhal number (typically, $St = 0.2$), and f_{st} is the vortex shedding frequency.

Under the effect of a periodically varying shedding frequency, the VIV characteristics in the CF direction vary with time. This time-varying feature has been reported to be the major difference between the VIV in steady flow and that in an oscillatory flow [35]. To investigate this time-varying feature under higher modes, the wavelet transform is introduced to analyze the time–frequency distribution of VIV in an oscillatory flow. The continuous wavelet transform equation is expressed as:

$$WT_f(a,\tau) = \langle f(t), \psi_{a,\tau}(t) \rangle = a^{-1/2} \int_{-\infty}^{+\infty} f(t)\psi*(\frac{t-\tau}{a})dt \tag{15}$$

where $WT_f(a,\tau)$ is the coefficient of the time-domain signal $f(t)$ after the wavelet transform representing the frequency variation at that time scale, the parameter a is the scale factor, τ is the shift factor, and $\psi(t)$ is the mother wavelet. In this paper, the Morlet wavelet equation is employed as the mother wavelet, and this wavelet can be defined as:

$$\psi(t) = Ce^{-t^2/2}\cos(5t) \tag{16}$$

where C is the wavelet transform coefficient.

4. Results and Discussions

4.1. Spatial and Temporal Distributions of VIV Responses

Based on the method of displacement reconstruction described in Section 3.2, the displacement modal weights and the displacement response can be obtained. Figure 3 shows the root mean square

(RMS) values of the displacement weights (p_{rms}) in the CF direction in the cases of KC = 56 and 121 under V_{R2} = 5.2. It shows that the second mode dominates the VIV response in all cases. The dominant vibration mode is expected as designed. Moreover, the first mode vibration is also obvious, except for second order modes, especially for larger KC numbers (KC = 121). This indicates that the multimode will participate in vibrations.

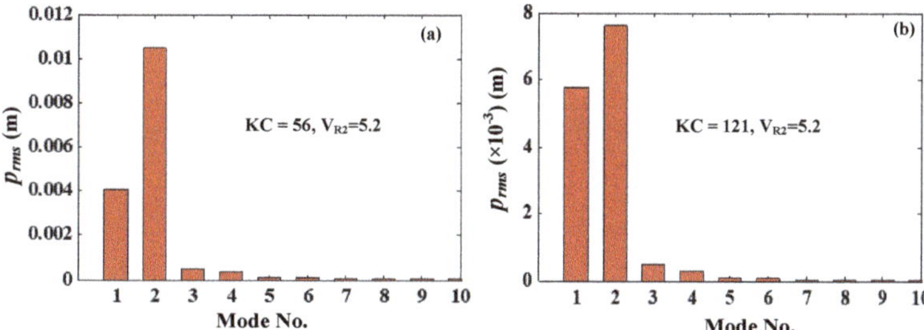

Figure 3. Root mean square values of the displacement modal weights in the cross flow (CF) direction for different KC numbers under V_{R2} = 5.2.

Figure 4 shows the corresponding RMS value distribution of the displacement response along the flexible pipe in both the CF and IL directions. The blue solid line and the red dashed line represent the displacement response for KC = 56 and 121 under V_{R2} = 5.2, respectively. Figure 4a shows that the RMS value of the VIV response (Y_{RMS}) for KC = 121 is smaller than that for KC = 56, except for the node of the second mode shape (Z/L = 0.5). Meanwhile, the value in the IL direction (X_{RMS}) for KC = 121 is correspondingly less than that for KC = 56 as shown in Figure 4b. This means that the drag force acting on the flexible pipe in the IL direction for KC = 121 is weaker than that for KC = 56. The results are consistent with the belief that VIV will enlarge the incoming flow area and increase drag in the IL direction [17,18].

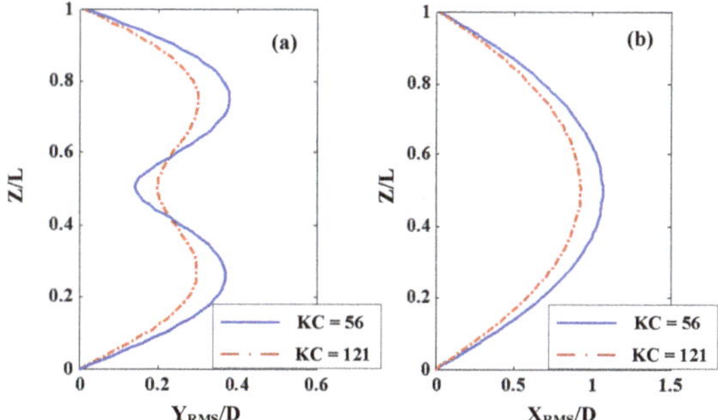

Figure 4. Distribution of displacement response along the flexible pipe in the CF and in-line (IL) directions for different KC numbers under V_{R2} = 5.2: (**a**) The VIV displacement response in the CF direction; (**b**) The VIV displacement response in the IL direction.

To verify the displacement reconstruction method, the strain was recalculated based on the reconstructed displacement response. Through Equation (12), the curvature can be obtained by the second order difference of the displacement shape multiplying the generalized coordinate displacement value. Then, the strain values are further recalculated by Equation (13). Figures 5 and 6 illustrate the time history of measured and calculated strain in both the CF and IL direction for KC = 56 and 121 under V_{R2} = 5.2, respectively. The blue solid line and red dashed line respectively represent measured and calculated strains. Figure 7 presents the distribution of root mean squares of measured and calculated strains along the flexible pipe in both the CF and IL directions in the case of KC = 56 and 121 under V_{R2} = 5.2. The blue circles and the red dashed lines represent the measured and calculated values, respectively. It reveals that these calculated values are in good agreement with the measured one. This consistency demonstrates the validity of the displacement reconstruction method for the flexible pipe in an oscillatory flow.

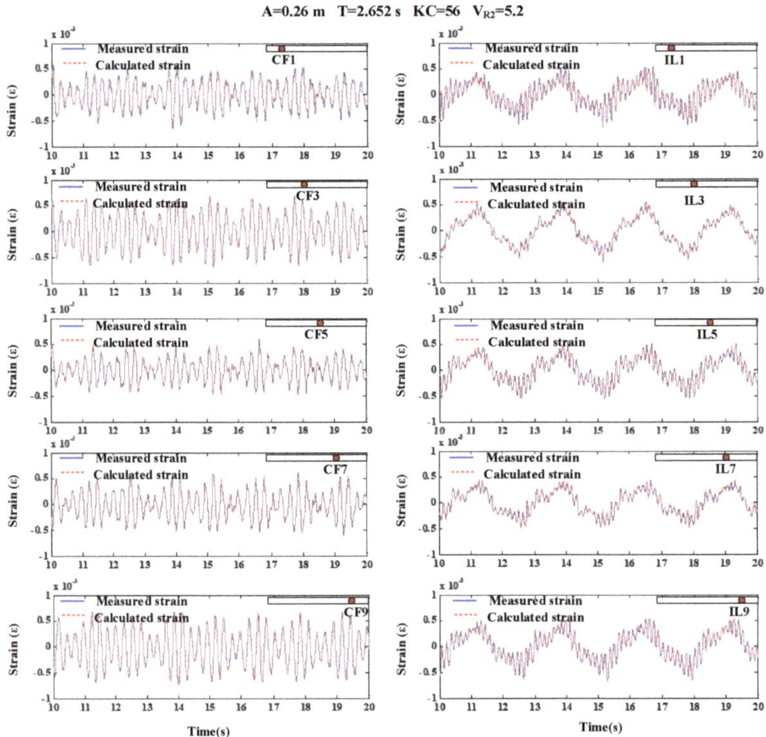

Figure 5. Time history of measured and calculated strain at different gauge points in both the CF and IL directions for KC = 56 and V_{R2} = 5.2.

In oscillatory flows, the VIV responses have distinctive time-varying features [35,38,39]. Figures 8 and 9 show the spatial and temporal distribution of the VIV response for KC = 56 and 121 under V_{R2} = 5.2, respectively. Each figure has two subfigures: Subfigure (a) is the time history of forced motion velocity, and Subfigure (b) presents spatial–temporal distribution of VIV response. Under a smaller KC number (KC = 56), the displacement response is always dominated by the second mode over time. Steady standing waves can be clearly seen in the whole forced motion as shown in Figure 8b. Different from the results of KC = 56, the first mode and the second mode alternately dominate the VIV response in the case of KC = 121, as shown in Figure 9b. The travelling wave can be observed in the transition region of two modes. The standing wave occurs in the dominated region of the second mode. Thus, VIV under higher modes has the characteristic of time-sharing dominance for each mode for larger KC number. The reason for this can be attributed to the fact that the vortex shedding does not immediately change with forced motion velocity altering [40]. Under KC = 56, the oscillating period is shorter, and the vortex shedding causing the second mode vibration will cover it, resulting in the first mode vibration. As the KC number increases and the correspondingly oscillating period becomes longer, the above effects gradually weaken. Time-sharing of the two modes appears in the case of a larger KC number. More insightful experiments and Computational Fluid Dynamics (CFD) simulations will be carried out in the near future work to further investigate this interesting phenomenon.

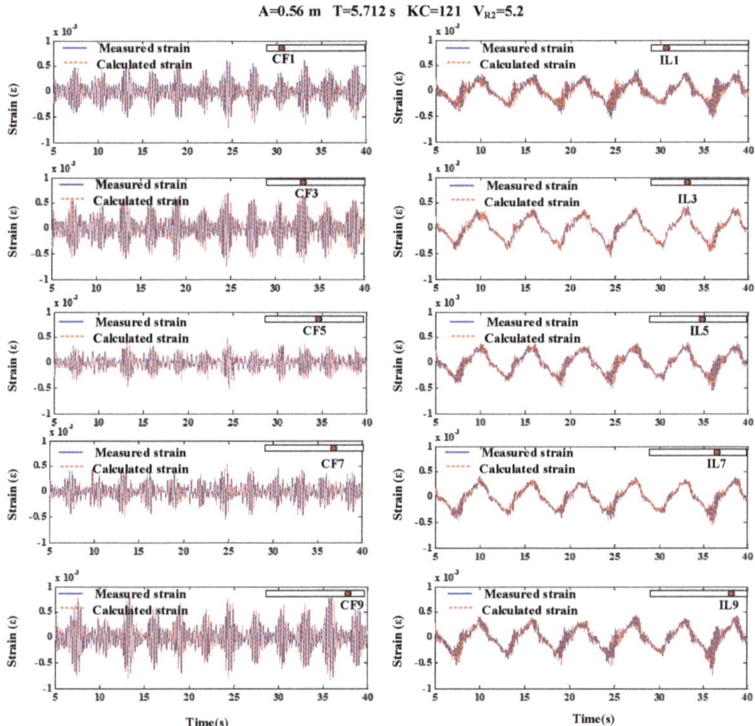

Figure 6. Time history of measured and calculated strain at different gauge points in both the CF and IL directions for KC = 121 and V_{R2} = 5.2.

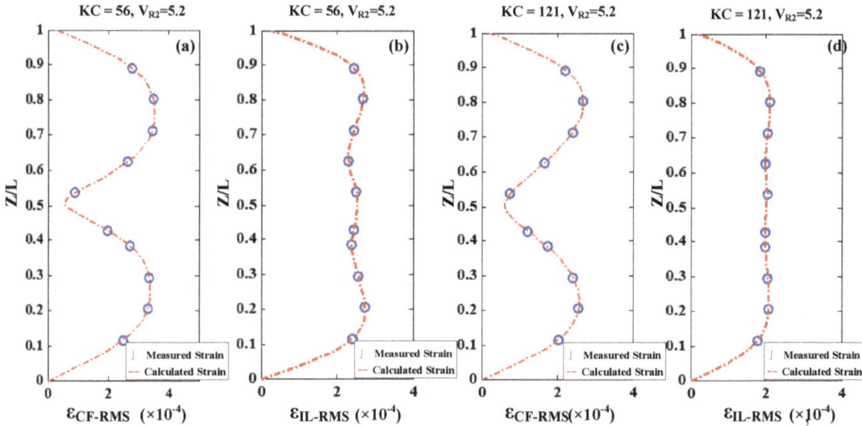

Figure 7. Distributions of the measured and calculated strains in the CF and IL directions for different KC numbers.

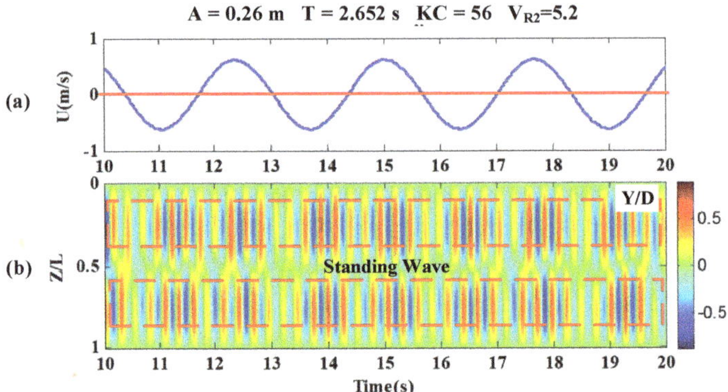

Figure 8. Spatial and temporal distribution of the VIV response in the case of KC = 56 for V_{R2} = 5.2.

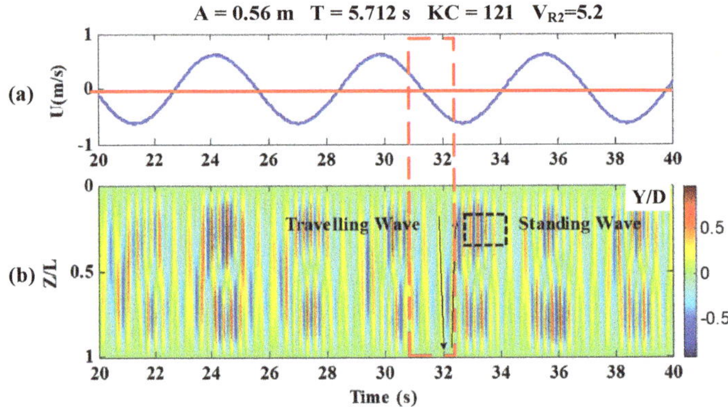

Figure 9. The spatial and temporal distribution of the VIV response in the case of KC = 121 for V_{R2} = 5.2.

4.2. Time-Varying Features of VIV Responses

To further reveal the VIV response features under the higher mode (second mode), the VIV responses near the antinode of second modal shape (Z/L = 0.3) were selected for investigation. Figures 11 and 13 respectively present the time history of the VIV response and time–frequency distributions at Z/L = 0.3 for KC = 56 and 121 under V_{R2} = 5.2. Figures 10 and 12 present the corresponding results at Z/L = 0.5 for KC = 56 and 121 under V_{R1} = 5.6. Each figure above has four subfigures. Subfigure (a) shows the time history of the forced motion velocity. Subfigure (b) is the time history of VIV displacement response. Subfigure (c) shows the wavelet analysis of the VIV displacement response. The depth of the color indicates the concentration level of the VIV response components. Subfigure (d) indicates the calculated time-varying shedding frequency (f_{st}), time varying natural frequency (f_1, f_2), and VIV response dominant frequency (f_{domi}). The black dashed line and purple solid line, respectively, represents VIV response dominant frequency and vortex shedding frequency. The blue dot dashed line and red dot line are the first- and second-order eigen frequency, respectively. Different from Equation (1), f_1, f_2 here considered the measured time-varying tension $F_T(t)$, which is calculated by:

$$f_n(t) = \frac{n}{2L} \sqrt{\frac{F_T(t)}{m} + \frac{n^2\pi^2}{L^2} \cdot \frac{EI}{m}}, \quad m = \overline{m} + \frac{1}{4}\pi D^2 \rho C_m \tag{17}$$

while we still assumed the added mass coefficient C_m to be equal to 1 the same as Equation (1).

Under smaller KC number (KC = 56), similar relatively mild amplitude modulation was seen in both Figures 10b and 11b. The maximum VIV response of the flexible pipe can be reached 0.65 D and 0.74 D for V_{R1} = 5.6 and V_{R2} = 5.2, respectively. The VIV response dominant frequencies are respectively always lock-in the first and second natural frequency as presented in Figures 10d and 11d. Although the vortex shedding frequency varies with time, the dominant frequency is always near the natural eigen frequency of the dominant mode for smaller KC number. There is no mode transition phenomenon in case of smaller KC number under higher mode. This is a distinctive feature of VIV response of flexible under higher mode in an oscillatory flow under smaller KC number.

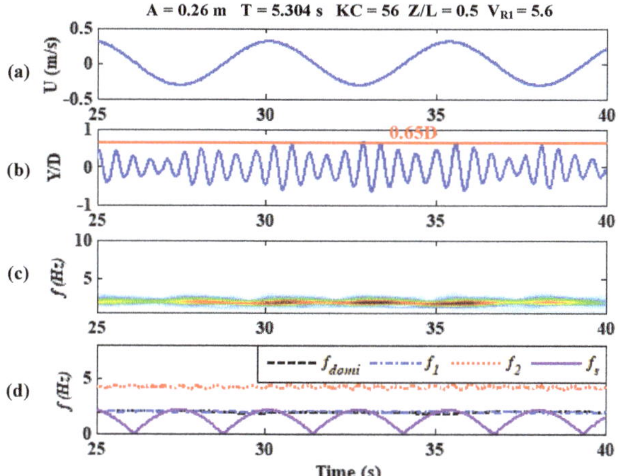

Figure 10. The time history of the VIV response and time–frequency distribution at Z/L = 0.5 for V_{R1} = 5.6 and KC = 56: (**a**) The forced motion velocity; (**b**) The VIV displacement response; (**c**) The time–frequency distribution of the VIV response; (**d**) The time-varying frequencies.

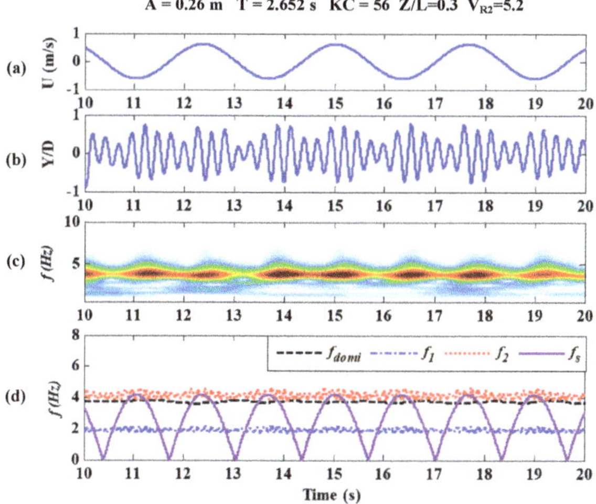

Figure 11. The time history of the VIV response and time–frequency distribution at Z/L = 0.3 for V_{R2} = 5.2 and KC = 56: (**a**) The forced motion velocity; (**b**) The VIV displacement response; (**c**) The time–frequency distribution of the VIV response; (**d**) The time-varying frequencies.

As KC number increases to 121, amplitude modulation became stronger as shown in Figures 12b and 13b. When V_{R1} = 5.6, the VIV developing process including building up, lock-in and dying out can be clearly witnessed in Figure 12b, which were also found by Fu et al. (2014) [35]. However, these three phases cannot be directly seen in Figure 13b and the VIV response was more disordered when V_{R2} = 5.2 and KC = 121. These differences are caused by co-participation and intermittent dominance of multiple modes under higher mode vibration for larger KC number. Figure 14 gives the time histories of the first two modal weights. The cyan dashed lines were the envelope of response obtained by the Hilbert transform. The similar VIV development process under V_{R1} = 5.6 can be easily found in the second modal weight under V_{R2} = 5.2 as shown in Figure 14d, while it is difficult to identify this process in the first modal weight as shown in Figure 14c. Thus, the secondary vibration modes confuse the development process of the dominant vibration mode and lead to more chaotic features of VIV.

In the oscillatory flow, the forced motion contains continuous acceleration and deceleration phases. The second VIV modal response in acceleration and deceleration stages has an obvious asymmetrical characteristic as shown in Figure 14d. The amplitude value in the acceleration phase at half a maximum forced motion velocity (I, III, V) was nearly 0.10 D, while approximately 0.25 D in deceleration phase at points (II, IV, VI). The latter response in the second modal space is larger than the former one. This is termed as "hysteresis" [35]. Different from the VIV response dominated by the first mode, the interesting hysteresis phenomenon did not appear in the total VIV displacement or the first modal response under the higher vibration mode for the larger KC number. The "hysteresis" only occurs in the dominant response modal space.

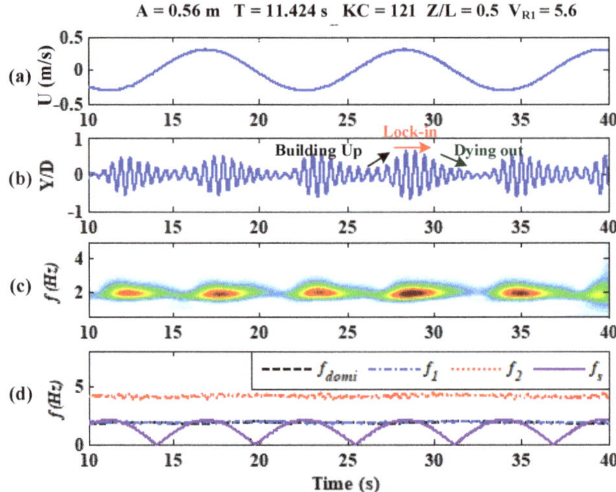

Figure 12. The time history of the VIV response and time–frequency distribution at Z/L = 0.5 for V_{R1} = 5.6 and KC = 121: (**a**) The forced motion velocity; (**b**) The VIV displacement response; (**c**) The time–frequency distribution of the VIV response; (**d**) The time-varying frequencies.

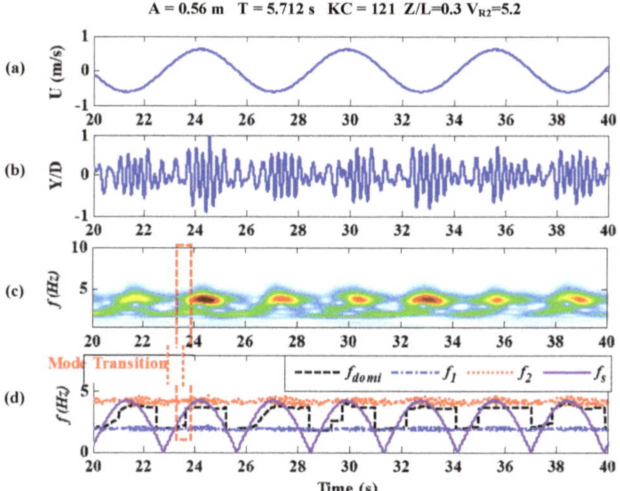

Figure 13. The time history of the VIV response and time–frequency distribution at Z/L = 0.3 for V_{R2} = 5.2 and KC = 121: (**a**)The forced motion velocity; (**b**) The VIV displacement response; (**c**) The time–frequency distribution of VIV response; (**d**) The time-varying frequencies.

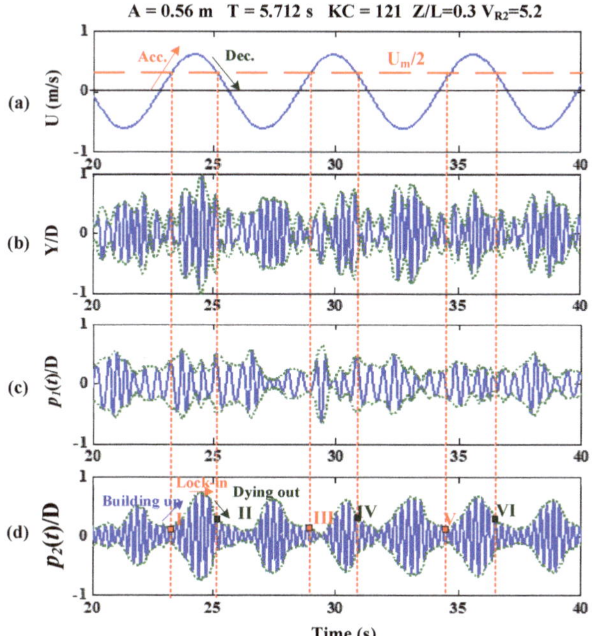

Figure 14. The time history of the VIV response at Z/L = 0.3 and the response of the first and second modal spaces for V_{R2} = 5.2 and KC = 121: (**a**) The forced motion velocity; (**b**) The VIV displacement response; (**c**) The time history of first modal weight; (**d**) The time history of second modal weight.

Beyond that, the "mode transition" and the alternate mode lock-in are prominent for the larger KC number (KC = 121) under V_{R2} = 5.2 as presented in Figure 13c,d. This is absent in the VIV response dominant by only the first vibration mode as illustrated in Figure 12c,d, and it is also not found under the smaller KC number (KC = 56) under the same maximum reduced velocity of the higher mode as presented in Figure 11c,d. With vortex shedding frequency increasing from zero to the maximum value, the dominant frequency first locked in the first natural frequency of approximately 1.95 Hz and jumped to lock in at the second natural frequency of nearly 3.8 Hz. When the vortex shedding frequency reached around the second natural frequency in the acceleration phase, the response dominant frequency started jumping from the first mode to the second mode. However, the frequency turning points of the lock-in mode decreased from the second to the first mode occurred when the vortex shedding frequency fell between 0 Hz to the first natural frequency and displayed some uncertainties. Thus, the higher mode locking establishment frequency and unlocking frequency are significantly different, which can also be attributed to the lag of vortex shedding as previously mentioned. Vortex shedding does not change immediately with forced motion velocity altering, and the locking time of the higher mode vibration becomes longer, accordingly. Although the interesting phenomenon of "mode transition" and the alternate mode were revealed above, the prediction of the critical transition frequency awaits to be determined by more extensive experiments and studies in the future.

4.3. Response Frequencies and Trajectories

For the convenience of investigating performances of response trajectory and frequency under higher mode vibrations, Figure 15 and Figure 17 first show the spatial distribution of response frequency in the IL and CF directions and the cross-section displacement trajectories at Z/L = 0.3, 0.5, 0.7 when KC = 56 and 121 under V_{R1} = 5.6 in one oscillation cycle, respectively, which is processed by MATLAB software. Figure 16 and Figure 18 present the above under V_{R2} = 5.2. The dominant frequency of VIV response in both the IL and CF direction are summarized in Table 4. Under the smaller KC number (KC = 56) with V_{R1} = 5.6, Figure 15a shows the dominant response frequency in the IL direction, the same as expected for the forced motion frequency at f = 0.188 Hz (f_o). The secondary contribution from the VIV response in the IL direction is dominated by f = 0.565 Hz ($3f_o$). The VIV response in the CF direction is dominated by f = 1.907 Hz ($10f_o$) as shown in Figure 15b. Comparing two spatial distributions of VIV response frequencies, the corresponding twice VIV dominant frequency in the CF direction is inconspicuously observed in Figure 15a. There is no clear relationship where the VIV dominant frequency in the IL direction is twice of that in the CF direction. This is inconsistent with the results of the frequency relationship found in steady flow reported by Blevins and Saunders (1977) [1].

With the design vibration mode increasing to 2 for KC = 56, removing the response corresponding to the forced motion frequency (f_o = 0.38 Hz) in the IL direction, the VIV response in the IL direction is dominated by f = 1.129 Hz ($3f_o$) and participated in an insignificant frequency contributor f = 1.892 Hz ($5f_o$) as shown in Figure 16a. The VIV response frequency in the CF direction is dominated by f = 3.785 Hz ($10f_o$) as shown in Figure 16b. f = 0.322 Hz ($9f_o$), and f = 4.548 Hz ($12f_o$) can also be seen, but is not obvious. The twice relationship of the VIV response dominant frequency between the CF and IL directions is also not easy to find under the higher mode in the case of a smaller KC number. Under the larger KC number (KC = 121), the frequency results are similar to those in the case with smaller KC numbers. The VIV response dominant frequency in the case of KC = 121 and V_{R1} = 5.6 is f = 0.267 Hz ($3f_o$) and f = 1.92 Hz ($22f_o$) for the IL and CF direction as shown in Figure 17a,b, respectively. The frequencies f = 0.53 Hz ($3f_o$) and f = 3.68 Hz ($21f_o$), respectively, dominates the VIV response in the IL and CF directions with KC = 121 under V_{R2} = 5.2 as presented in Figure 18a,b. Thus, it suggests that the traditional VIV response frequency relationship between the IL and CF directions cannot be directly used to predict the VIV response in an oscillatory flow. Beyond that, an interesting phenomenon is witnessed in that the dominant response frequency and other secondary frequencies all maintain the multiple relationship with the frequency of the forced motion. Similar phenomena are also found from stationary rigid cylinder experiments in an oscillatory flow [37]. Moreover, the VIV responses in the IL direction are always coincidentally dominated by $3f_o$. This provides a possible reference for a VIV prediction in the near future.

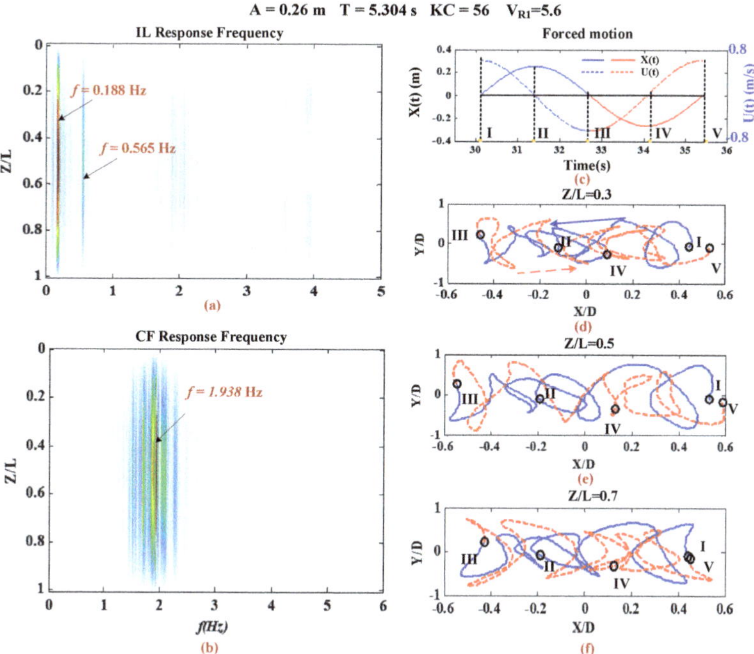

Figure 15. The spatial distribution of response frequency and displacement trajectories at different positions when KC = 56 and V_{R1} = 5.6.

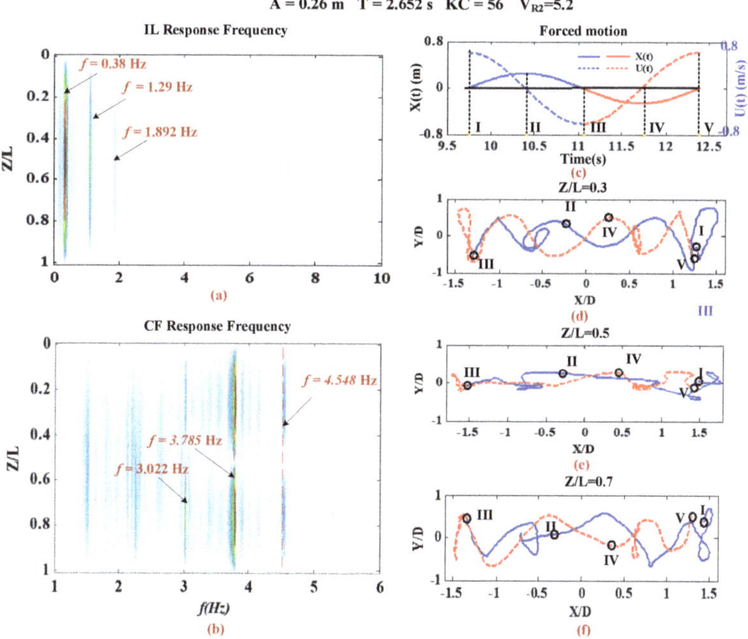

Figure 16. The spatial distribution of response frequency and displacement trajectories at different positions when KC = 56 and V_{R2} = 5.2.

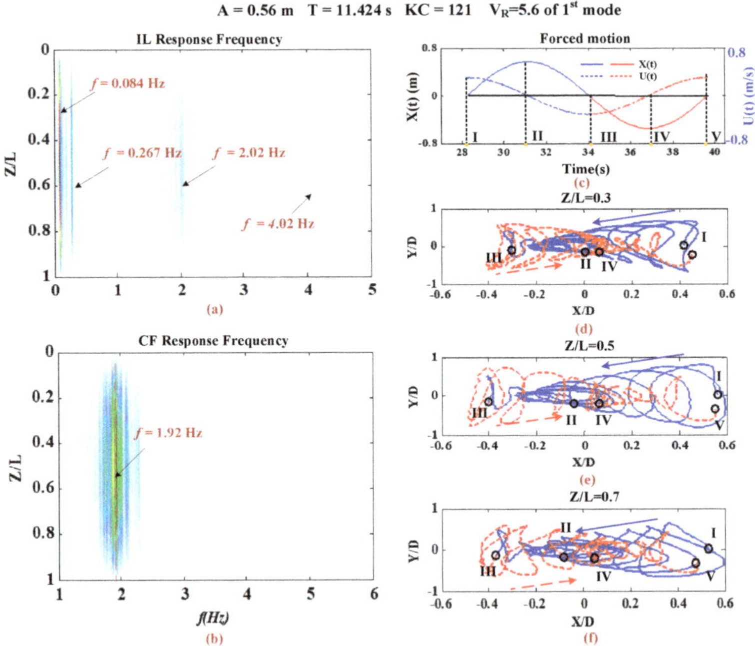

Figure 17. The spatial distribution of response frequency and displacement trajectories at different positions when KC = 121 and V_{R1} = 5.6.

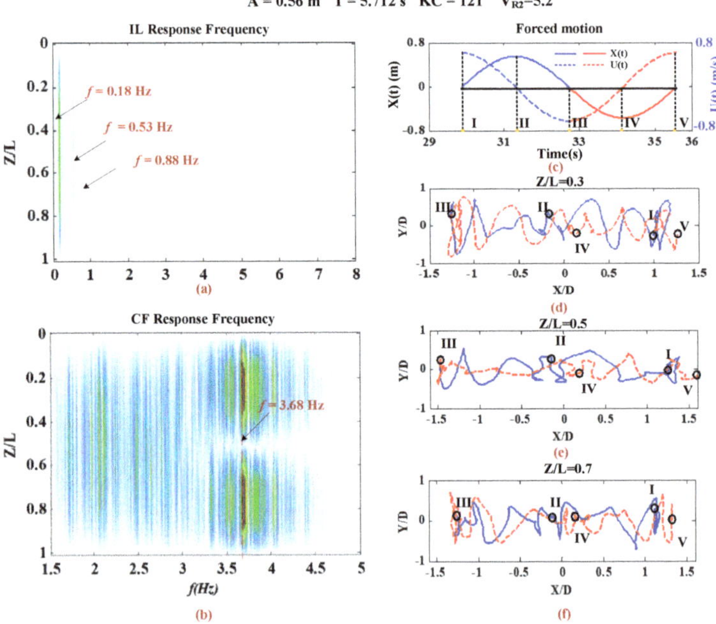

Figure 18. The spatial distribution of response frequency and displacement trajectories at different positions when KC = 121 and V_{R2} = 5.2.

Table 4. Summary of dominant frequency of VIV for different test cases.

V_R	KC	Dominant Mode	Forced Motion Frequency (f_o)	Dominant Frequency of VIV	Direction
V_{R1} = 5.6	56	1st	0.188 Hz	0.565 Hz ($3f_o$)	IL
V_{R1} = 5.6	56	1st	0.188 Hz	1.938 Hz ($10f_o$)	CF
V_{R2} = 5.2	56	2nd	0.38 Hz	1.290 Hz ($3f_o$)	IL
V_{R2} = 5.2	56	2nd	0.38 Hz	3.785 Hz ($10f_o$)	CF
V_{R1} = 5.6	121	1st	0.084 Hz	0.267 Hz ($3f_o$)	IL
V_{R1} = 5.6	121	1st	0.084 Hz	1.920 Hz ($22f_o$)	CF
V_{R2} = 5.2	121	2nd	0.18 Hz	0.530 Hz ($3f_o$)	IL
V_{R2} = 5.2	121	2nd	0.18 Hz	3.680 Hz ($21f_o$)	CF

Displacement trajectories can directly reveal the whole response process of a flexible pipe in an oscillatory flow. We firstly select five typical time points (I, II, III, IV, and V) to mark and to help us to describe the features of displacement trajectories at different acceleration or deceleration phases as shown in Subfigure (c) of Figures 15–18. The necessary comparisons of displacement trajectories mainly include the following two items:

(1) The performance of trajectories changes before and after reverse motion such as those in the stages I→II and II→III, III→IV, and IV→V. Under the smaller KC number of V_{R1} = 5.6, taking one pair stages (I→II and II→III) in half an oscillation cycle for observation, the trajectories from time I to II are not symmetrical to those from time II to III as shown in Figure 15d–f. The response in phase (I→II) is slightly larger than that in phase (II→III), which indicates that the response was suppressed in the reverse motion stages. The same results can be also witnessed in phases (III→IV) and (IV→V) as shown in Figure 15d.

One reason for this asymmetry may be attributed to the vortex shedding interaction, when the cylinder reverses its motion direction and encounters the previously shed vortices [6,36]. The vortex shedding interaction leads to a correspondingly different response before and after the reverse motion. Nevertheless, we cannot rule out the effects of vortex shedding lag. At point I, the forced motion starts to slow down, and stronger shedding vortices catch up with the flexible pipe, resulting in a larger response in the process from I to II. These effects will weaken over time. Thus, the response asymmetry and a larger response of I→II than that of II→III were observed. When the KC number increases to 121 under V_{R1} = 5.6, the asymmetric features in phase (I→II) and (II→III) of the trajectory become very apparent, as shown in Figure 17d–f.

With the dominated mode increasing to the second mode, similar results can also be found under the smaller KC number (KC = 56) as shown in Figure 16d–f. However, the aforementioned prominent features are not obvious in the case of the larger KC number (KC = 121) under V_R = 5.2 of the second mode as illustrated in Figure 18d–f. A multifrequency response and a multimode intermittent disrupt the features above.

(2) Differences of trajectory in acceleration and deceleration stages merit study. In the case of the smaller KC number when V_{R1} = 5.6, trajectory discrepancies are very obvious between acceleration stage II→III and deceleration stages III→IV as presented in Figure 15d–f. More significant results can be found when KC increases to 121 under V_{R1} = 5.6 as shown in Figure 17d–f. Under smaller KC numbers with V_{R1} = 5.6, the discrepancies of the cross-section trajectory are mainly derived from the response phase differences between the CF and IL directions, while response amplitude discrepancies contribute to the corresponding differences in trajectory under larger KC numbers with V_{R1} = 5.6. Moreover, under the higher mode V_{R2} = 5.2, although a distinctively inconsistent trajectory can directly observed in both smaller and larger KC numbers, as shown in Figures 16d–f and 18d–f, respectively, differences of trajectories can be all attributed to response phase differences. The response amplitude difference is slight and is a secondary reason for trajectory discrepancies in the case of VR = 5.2 of the second mode.

Beyond that, under V_{R1} = 5.6, the trajectory of the walking "8" and "o" shapes can be observed, especially for the larger KC number (KC = 121) as shown in Figure 17d–f. When the design dominated mode increased to the second mode, the quasi-regular track characteristics like the walking "8" and "o" shapes disappeared without any trace for smaller or larger KC numbers as presented in Figures 16d–f and 18d–f. Furthermore, some differences do exist in the trajectory along the length of the flexible pipe whether the higher order mode is dominated or only the first mode is involved. Subfigures (d–f) of Figures 15–18 display this well. It should be emphasized that the response amplitude and the response phase between the IL and CF directions are two essential factors for hydrodynamic force coefficients. The results above indicate that hydrodynamic coefficients before and after the reverse motion, acceleration, and deceleration stages will be different. A different trajectory along the flexible pipe means that a different cross section will have different hydrodynamic coefficients. Thus, hydrodynamic coefficients may display spatial and temporal varying features [39,41–43]. Moreover, the trajectory discrepancies between the higher mode dominated and only the first mode involved manifest that hydrodynamic coefficients under a response dominated by a higher mode will have certain novel and complex characteristics that need to be studied in future work.

4.4. General Discussions

Based on this detailed case study, features of the VIV response of a flexible pipe under oscillatory flow with the higher mode dominated are different from that with only the first mode dominated through comparing smaller and larger KC numbers under the same maximum-reduced velocity of different design dominated modes. Under the higher mode with the larger KC number, the mode transition and alternate mode lock-in are prominent. A travelling wave occurs in the mode transition stage and the standing wave appears in the lock-in one. However, the alternate mode lock-in phenomenon cannot occur, and the VIV response is always locked in the dominated mode for the smaller KC number with the higher mode. The standing wave is always seen in the spatial and temporal distributions of the VIV. The travelling wave characteristic is not found in the case of the smaller KC number under the higher mode. Different from only the first mode excited case, VIV processing, including building up, lock-in and dying out, and hysteresis, are clearly witnessed only in the dominated modal space. This indicates that the response of the secondary participate mode does not affect the VIV process or features in the dominated modal space.

The aforementioned detailed features provide us with a basic understanding of the VIV response under the higher mode in the case of different KC numbers. The VIV displacement amplitude of each case should be further summarized. Figure 19 shows the maximum root mean square VIV displacement versus the KC number under different dominated modes. The blue triangle and red circle represent the responses under $V_{R1} = 5.6$ and $V_{R2} = 5.2$, respectively. The decreasing tendency of response amplitude with KC number can be observed under two dominant designed modes. Under the small KC number, the interaction of vortex shedding before and after the reverse forced motion is stronger for the small KC number case, leading to a larger response [36]. In the case of the larger KC number, the maximum value of Y_{RMS}/D approaches towards a constant value asymptotically as the KC number increases. The value differences of VIV response amplitudes under two dominated modes are not large. This indicates that the effects of higher dominated mode (>2) in an oscillatory flow on VIV response amplitude may be slight.

In Section 4.3, the response frequency and trajectory under different dominated modes in an oscillatory flow are compared. Discrepancies of response trajectories are revealed between different dominated modes and may cause some unknown effects on hydrodynamic force. Future work should emphasize the investigation of hydrodynamic coefficients under higher modes. Moreover, response frequency always maintains a multiple relationship with the frequency of the forced motion whether is dominated by a higher mode or not. The response frequency calculation is an essential issue in VIV prediction. For the VIV of a rigid cylinder in an oscillatory flow, Sumer and Fredsøe (1988) [37] found the integral multiple relationship between the VIV dominant response frequency f_{domi} in the CF direction and the oscillatory flow frequency f_o. This result is consistent with our observation on the

VIV response of the flexible pipe in an oscillatory flow as previously described. It indicates that there are integral vortices shedding for each motion period. These vortex shedding pairs N are defined as:

$$N = \frac{f_{domi}}{f_0} \tag{18}$$

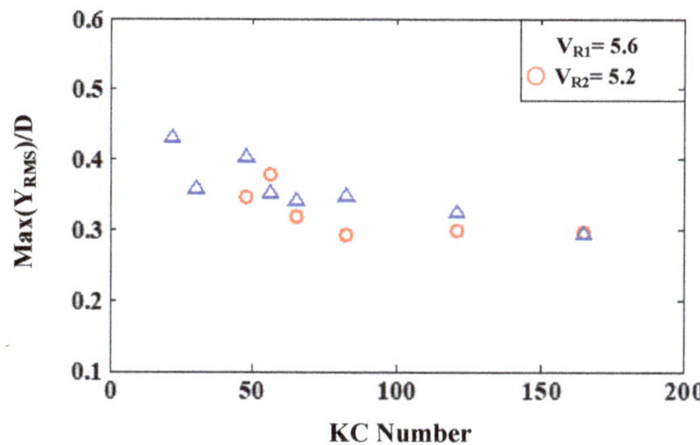

Figure 19. The maximum root mean square VIV displacement responses versus KC numbers under different dominated modes.

In a steady flow, the VIV dominant frequency is governed by the Strouhal number (St). The typical value of the St number is found to be approximately 0.2 through the stationary smooth cylinder in a steady flow [1]. To further investigate the effects of dominated modes on the St number and to extract the exact value of the St number, the relationship among the VIV dominant frequency, the number of vortex shedding pairs, the KC number under oscillatory flow, and the Strouhal number can be expressed as follows:

$$\frac{N}{KC} = \frac{f_{domi}}{f_0} \cdot \frac{D f_0}{U_m} = \frac{f_{domi} D}{U_m} = St$$
$$N = St \cdot KC \tag{19}$$

The aforementioned relationship was first introduced to investigate the St number of a smooth rigid cylinder in an oscillatory flow by Sumer and Fredsøe (1988) [37]. However, it can be also applied for the flexible pipe and has been used in the SCR and free-hanging riser model tests [27,33,34]. Through Equation (18), the dominant frequency of all cases is first summarized, and the vortices shedding pairs N are then calculated. The distributions of vortex shedding pairs versus KC numbers are presented in Figure 20. The vortex shedding pairs maintain a good linear relationship with the KC numbers. Based on Equation (19), the values of the St number under $V_{R1} = 5.6$ and $V_{R2} = 5.2$ are 0.18 and 0.174, respectively. These values are slightly lower than the typical value of 0.20, which is consistent with the results reported by Wang et al. (2015) [27]. Comparing St numbers under two different dominated modes in an oscillatory flow, the two values are very close, and the differences can be ignored. Thus, we can infer that the St number is controlled by the maximum reduced velocity and not related with the dominated modes in the CF direction.

Moreover, the dominant frequency of the VIV response in the IL direction is different from that in the CF direction as depicted in Figures 15a, 16a, 17a and 18a. The dominant frequency seems to always be triple the oscillatory flow frequency. Thus, the dominant frequency prediction of the VIV response in the IL direction can be directly calculate by this simple relationship. Although the VIV response dominant frequency in both the IL and CF directions can be preliminarily predicted, the time varying frequency features under larger KC numbers require a critical value to determine whether dominant lock-in frequency jump has occurred.

Future work should conduct a more systematic series of experiments. The lock-in frequency jump between multimode needs a critical value to be further determined. The hydrodynamic coefficients under different excited modes, KC numbers and maximum reduced velocity are also emphasized for future investigation.

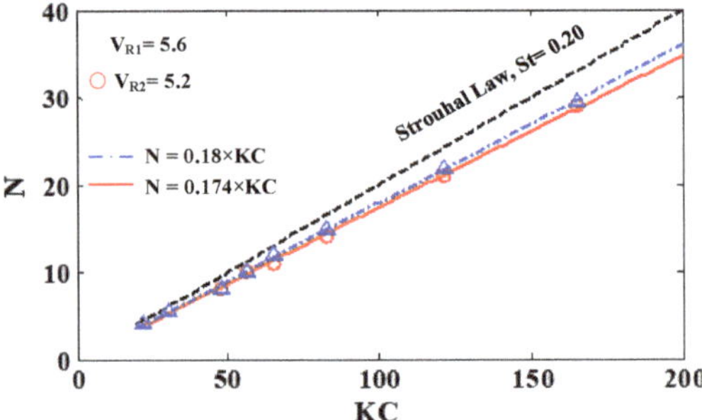

Figure 20. The summarized relationship between vortex-shedding pairs and KC numbers under different dominant modes.

5. Conclusions

In this paper, an experimental study on the VIV of a flexible pipe under a higher mode in an oscillatory flow was carried out with a V_R of approximately 5.5 in the case of KC numbers varying from 22 to 165. The dominant excited mode was designed for the first and second modes. The VIV displacement and time-varying frequency are reconstructed and identified through the modal superposition method and wavelet transform, respectively. Compared with only the first mode excited cases, the features of VIV were further investigated. The main conclusions are as follows:

1. Under the higher mode, a travelling wave is observed in the mode transition regions for larger KC numbers. An alternate mode lock-in occurs in the case of larger KC numbers, but does not occur for smaller KC ones. A distinctive feature for smaller KC number is that the VIV response of the flexible pipe always locks in the dominant mode.
2. The features of response motion trajectory under higher modes are different from those under only the first mode excited, especially for larger KC numbers. The walking "8" and "o" shapes of the motion trajectory are observed under only the first mode excited and disappear under the higher mode. The discrepancies of trajectory along the flexible pipe and in different time phases indicate that the hydrodynamic coefficient may exhibit spatial and temporal features.
3. The dominant frequency of the VIV is always kept triple that of the oscillatory flow frequency in the IL direction and maintains the Strouhal law in the CF direction. Under a V_R of approximately 5.5, the St number is equal to 0.18 and is not affected by the excited mode number.

Generally speaking, these findings will improve the basic understanding of VMI-VIV phenomena and provide a good reference for developing a VIV prediction method of a flexible pipe in an oscillatory flow.

Author Contributions: Conceptualization, H.R. and M.Z.; methodology, H.R.; validation, H.R., M.Z. and Y.X.; formal analysis, H.R.; investigation, H.R., M.Z., Y.X. and S.F.; resources, S.F., J.C. and P.C.; data curation, M.Z. and Y.X.; writing—original draft preparation, H.R. and C.L.; writing—review and editing, H.R. and C.L.; visualization, H.R.; supervision, M.Z., S.F., J.C., and P.C.; project administration, J.C. and P.C. All authors have read and agreed to the published version of the manuscript.

Funding: This research was funded by the National Natural Science Foundation of China, grant number: 51825903, 51909159; the Shanghai Science and Technology Program, grant number:19XD1402000, 19JC1412800, 19JC1412801; the Joint Funds of the National Natural Science Foundation of China, grant number: U19B2013; the Key Projects for Intergovernmental Cooperation in International Science, grant number: SQ2018YFE0125100; Ub-project of the Important National Science & Technology Specific Projects of China, grant number: 2016ZX05028-002-004.

Acknowledgments: The authors gratefully acknowledge the support from the National Natural Science Foundation of China (No. 51825903, 51909159), the Shanghai Science and Technology Program (No. 19XD1402000, 19JC1412800, 19JC1412801), the Joint Funds of the National Natural Science Foundation of China (No. U19B2013), the Key Projects for Intergovernmental Cooperation in International Science (No.SQ2018YFE0125100), Ub-project of the Important National Science & Technology Specific Projects of China (No. 2016ZX05028-002-004). Additionally, all authors would also like to acknowledge the permission and support from SBM Offshore to prepare and publish this work.

Conflicts of Interest: The authors declare no conflict of interest.

References

1. Blevins, R.D.; Saunders, H. *Flow-Induced Vibration*; Elsevier: Amsterdam, The Netherlands, 1977.
2. Schewe, G. On the force fluctuations acting on a circular cylinder in crossflow from subcritical up to transcritical Reynolds numbers. *J. Fluid Mech.* **1983**, *133*, 265–285. [CrossRef]
3. Williamson, C.H.K. Defining a universal and continuous Strouhal–Reynolds number relationship for the laminar vortex shedding of a circular cylinder. *Phys. Fluids* **1988**, *31*, 2742. [CrossRef]
4. Hallam, M.G. *Dynamics of Marine Structures: Methods of Calculating the Dynanic Response of Fixed Structures Sub*; Ciria underwater engineering group: London, UK, 1977.
5. Achenbach, E.; Heinecke, E. On vortex shedding from smooth and rough cylinders in the range of reynolds number multiplied by 10^3 to multiplied by 10^6. *J. Fluid Mech.* **1981**, *109*, 239–251. [CrossRef]
6. Williamson, C.H.K. Sinusoidal flow relative to circular cylinders. *J. Fluid Mech.* **2006**, *155*, 141–174. [CrossRef]
7. Williamson, C.H.K.; Govardhan, R. Vortex-induced vibrations. *Annu. Rev. Fluid Mech.* **2004**, *36*, 413–455. [CrossRef]
8. Williamson, C.H.K.; Govardhan, R. A brief review of recent results in vortex-induced vibrations. *J. Wind. Eng. Ind. Aerodyn.* **2008**, *96*, 713–735. [CrossRef]
9. Govardhan, R.; Williamson, C.H.K. Modes of vortex formation and frequency response of a freely vibrating cylinder. *J. Fluid Mech.* **2000**, *420*, 85–130. [CrossRef]
10. Sarpkaya, T. A critical review of the intrinsic nature of vortex-induced vibrations. *J. Fluids Struct.* **2004**, *19*, 389–447. [CrossRef]
11. Bearman, P.W. Vortex shedding from oscillating bluff bodies. *Annu. Rev. Fluid Mech.* **1984**, *16*, 195–222. [CrossRef]
12. Feng, C.C. The Measurement of Vortex Induced Effects in Flow past Stationary and Oscillating Circular and D-Section Cylinder. Master's Thesis, University of Britich Columbia, Vanccuver, BC, Canada, 1963.
13. Griffin, O.M.; Ramberg, S.E. Some recent studies of vortex shedding with application to marine tubulars and risers. *J. Energy Resour. Technol.* **1982**, *104*, 2–13. [CrossRef]
14. Parkinson, G. Phenomena and modelling of flow-induced vibrations of bluff bodies. *Prog. Aerosp. Sci.* **1989**, *26*, 169–224. [CrossRef]
15. Lie, H.; Kaasen, K.E. Modal analysis of measurements from a large-scale VIV model test of a riser in linearly sheared flow. *J. Fluids Struct.* **2006**, *22*, 557–575. [CrossRef]
16. Fu, S.; Ren, T.; Li, R.; Wang, X. Experimental Investigation on VIV of the Flexible Model under Full Scale Re Number. In Proceedings of the ASME 2011 30th International Conference on Ocean, Offshore and Arctic Engineering, Rotterdam, The Netherlands, 19–24 June 2011; pp. 43–50.
17. Ren, H.; Xu, Y.; Zhang, M.; Fu, S.; Meng, Y.; Huang, C. Distribution of drag coefficients along a flexible pipe with helical strakes in uniform flow. *Ocean Eng.* **2019**, *184*, 216–226. [CrossRef]
18. Vandiver, J.K. Drag coefficients of long flexible cylinders. In Proceedings of the 15th Annual Offshore Technology Conference, Houston, TX, USA, 1–4 May 1985.
19. Chaplin, J.R.; Bearman, P.W.; Huarte, F.J.H.; Pattenden, R.J. Laboratory measurements of vortex-induced vibrations of a vertical tension riser in a stepped current. *J. Fluids Struct.* **2005**, *21*, 3–24. [CrossRef]
20. Song, L.; Fu, S.; Dai, S.; Zhang, M.; Chen, Y. Distribution of drag force coefficient along a flexible riser undergoing VIV in sheared flow. *Ocean Eng.* **2016**, *126*, 1–11. [CrossRef]
21. Song, L.; Fu, S.; Li, M.; Gao, Y.; Ma, L. Tension and drag forces of flexible risers undergoing vortex-induced vibration. *China Ocean Eng.* **2017**, *31*, 1–10. [CrossRef]

22. Frank, W.R.; Tognarelli, M.A.; Slocum, S.T.; Campbell, R.B.; Balasubramanian, S. Flow-induced vibration of a long, flexible, straked cylinder in uniform and linearly sheared currents. In *Offshore Technology Conference*; Offshore Technology Conference: Houston, TX, USA, 2004; p. 8.

23. Trim, A.D.; Braaten, H.; Lie, H.; Tognarelli, M.A. Experimental investigation of vortex-induced vibration of long marine risers. *J. Fluids Struct.* **2005**, *21*, 335–361. [CrossRef]

24. Vandiver, J.K.; Li, L. *SHEAR7 V4.4 Program Theoretical Manual*; Massachusetts Institute of Technology: Cambridge, MA, USA, 2005.

25. Triantafyllou, M.; Triantafyllou, G.; Tein, Y.D.; Ambrose, B.D. Pragmatic Riser VIV Analysis. In Proceedings of the Offshore Technology Conference (OTC), Houston, TX, USA, 3–6 May 1999.

26. SINTEF Ocean. *VIVANA 4.12.2 Theory Manual*; Marintek: Trondheim, Norway, 2018.

27. Wang, J.; Fu, S.; Baarholm, R.; Jie, W.; Larsen, C.M. Out-of-plane vortex-induced vibration of a steel catenary riser caused by vessel motions. *Ocean Eng.* **2015**, *109*, 389–400. [CrossRef]

28. Wang, J.; Fu, S.; Larsen, C.M.; Baarholm, R.; Wu, J.; Lie, H. Dominant parameters for vortex-induced vibration of a steel catenary riser under vessel motion. *Ocean Eng.* **2017**, *136*, 260–271. [CrossRef]

29. Pesce, C.P.; Franzini, G.R.; Fujarra, A.L.C.; Gonçalves, R.T.; Salles, R.; Mendes, P. Further experimental investigations on vortex self-induced vibrations (vsiv) with a small-scale catenary riser model. In Proceedings of the ASME 2017 36th International Conference on Ocean, Offshore and Arctic Engineering, Rondheim, Norway, 25–30 June 2017.

30. Wang, J.; Fu, S.; Baarholm, R.; Jie, W.; Larsen, C.M. Fatigue damage of a steel catenary riser from vortex-induced vibration caused by vessel motions. *Mar. Struct.* **2014**, *39*, 131–156. [CrossRef]

31. Le Cunff, C.; Biolley, F.; Damy, G. Experimental and numerical study of heave-induced lateral motion (HILM). In Proceedings of the ASME 2005 24th International Conference on Offshore Mechanics and Arctic Engineering, Halkidiki, Greece, 12–17 June 2005; pp. 757–765.

32. Rateiro Pereira, F.; Gonçalves, R.T.; Pesce, C.P.; Fujarra, A.L.C.; Franzini, G.R.; Mendes, P. A model scale experimental investigation on vortex-self induced vibrations (vsiv) of catenary risers. In Proceedings of the ASME 2013 32nd International Conference on Ocean, Offshore and Arctic Engineering. Volume 7: CFD and VIV, Nantes, France, 9–14 June 2013. V007T08A029.

33. Wang, J.; Fu, S.; Wang, J.; Li, H.; Ong, M.C. Experimental investigation on vortex-induced vibration of a free-hanging riser under vessel motion and uniform current. *J. Offshore Mech. Arct. Eng.* **2017**, *139*. [CrossRef]

34. Wang, J.; Xiang, S.; Fu, S.; Cao, P.; Yang, J.; He, J. Experimental investigation on the dynamic responses of a free-hanging water intake riser under vessel motion. *Mar. Struct.* **2016**, *50*, 1–19. [CrossRef]

35. Fu, S.; Wang, J.; Baarholm, R.; Wu, J.; Larsen, C.M. Features of vortex-induced vibration in oscillatory flow. *J. Offshore Mech. Arct. Eng.* **2014**, *136*, 011801. [CrossRef]

36. Wang, J.; Fu, S.; Baarholm, R.; Wu, J.; Larsen, C.M. Fatigue damage induced by vortex-induced vibrations in oscillatory flow. *Mar. Struct.* **2015**, *40*, 73–91. [CrossRef]

37. Sumer, B.M.; Fredsøe, J. Transverse vibrations of an elastically mounted cylinder exposed to an oscillating flow. *J. Offshore Mech. Arct. Eng.* **1988**, *110*, 387–394. [CrossRef]

38. Ren, H.; Xu, Y.; Cheng, J.; Cao, P.; Zhang, M.; Fu, S.; Zhu, Z. Vortex-induced vibration of flexible pipe fitted with helical strakes in oscillatory flow. *Ocean Eng.* **2019**, *189*, 106274. [CrossRef]

39. Liu, C.; Fu, S.; Zhang, M.; Ren, H.; Xu, Y. Hydrodynamics of a flexible cylinder under modulated vortex-induced vibrations. *J. Fluids Struct.* **2020**, *94*, 102913. [CrossRef]

40. Fu, B.; Zou, L.; Wan, D. Numerical study of vortex-induced vibrations of a flexible cylinder in an oscillatory flow. *J. Fluids Struct.* **2018**, *77*, 170–181. [CrossRef]

41. Song, L.; Fu, S.; Cao, J.; Ma, L.; Wu, J. An investigation into the hydrodynamics of a flexible riser undergoing vortex-induced vibration. *J. Fluids Struct.* **2016**, *63*, 325–350. [CrossRef]

42. Mengmeng, Z.; Shixiao, F.; Leijian, S.; Jie, W.; Halvor, L.; Hanwen, H. Hydrodynamics of flexible pipe with staggered buoyancy elements undergoing vortex-induced vibrations. *J. Offshore Mech. Arct. Eng.* **2018**, *140*, 061805.

43. Liu, C.; Fu, S.; Zhang, M.; Ren, H. Time-varying hydrodynamics of a flexible riser under multi-frequency vortex-induced vibrations. *J. Fluids Struct.* **2018**, *80*, 217–244. [CrossRef]

Journal of
Marine Science
and Engineering

Article

Computational Model for Simulation of Lifeboat Free-Fall during Its Launching from Ship in Rough Seas

Shaoyang Qiu *, Hongxiang Ren * and Haijiang Li

Key Laboratory of Marine Dynamic Simulation and Control for Ministry of Communications,
Dalian Maritime University, Dalian 116026, China; hai@dlmu.edu.cn
* Correspondence: qsy@dlmu.edu.cn (S.Q.); dmu_rhx@dlmu.edu.cn (H.R.);
  Tel.: +86-1804-115-4892 (S.Q.); +86-1894-092-9166 (H.R.)

Received: 17 June 2020; Accepted: 12 August 2020; Published: 20 August 2020

Abstract: In order to improve the accuracy of the freefall of lifeboat motion simulation in a ship life-saving simulation training system, a mathematical model using the strip theory and Kane's method is established for the freefall of the lifeboat into the water from a ship. With the boat moving on a skid, the model of the ship's maneuvering mathematical group (MMG) is used to model the motion of the ship in the waves. Based on the formula of elasticity and friction theory, the forces of the skid acting on the boat are calculated. When the boat enters the water, according to the analytical solution theory of slamming, the slamming force of water entry is solved. The simulation experiments are carried out by the established model. The results of the numerical simulation are compared with the calculation results of the hydrodynamics software Star CCM+ at water entry under initial condition A in the paper. The position and velocity of the center of gravity of the boat, the angle, and velocity and acceleration of pitch calculated by the two methods are in good agreement. There is a little difference between the values of translation acceleration calculated by the two methods, which is acceptable. This shows that our numerical algorithm has good accuracy. A qualitative analysis is performed to find the safe point of water entry under the condition of different wave heights and two situations of a ship encountering waves. Finally, the model is applied to the ship life-saving training system. The model can meet the system requirements and improve the accuracy of the simulation.

Keywords: lifeboat; freefall; ship motion; Kane's method

---

1. Introduction

1.1. Motivation

The lifeboat is the main life-saving equipment onboard ships. When a shipwreck accident occurs, the crew onboard can quickly escape from the ship by the lifeboat. During the freefall of the lifeboat, the hull of the boat usually slams into the water at high speed, with huge instantaneous impact pressure. If the crews make a mistake when launching the lifeboat, it can cause serious damage to the hull structure and threaten the personal safety of the crew [1]. According to the International Convention on Standards of Training, Certification and Watchkeeping for Seafarers, 1978, as amended in 1995 (STCW 78/95), the crew must be trained and pass a lifeboat assessment before boarding [2]. Training is limited by factors such as time and costs. In recent years, virtual reality technology has developed rapidly, and has been used in training for marine life-saving [3,4]. In order to improve the immersion experience and the reality of the ship's life-saving training system, a mathematical model is established for the motion of the lifeboat's freefall during its launch from the ship.

1.2. Related Work

Re et al. [5–7] conducted a series of model experiments of a boat launched from a fixed platform. The main focus of the experimental evaluation was the performance of the boat in a range of different weather conditions. Hollyhead et al. [8] and Hwang et al. [9] performed experiments on launching boats from moving ships. They analyzed the motion parameters and the load of the boat. Due to the limitation of experimental conditions, they could only analyze parameters that affected the water entry motion of the boat and did their experiments under good environmental conditions.

In addition to the model experiment, some scholars used a computational method to analyze the freefall of the boat, which was mainly divided into two aspects. First, computational fluid dynamics (CFD) technology was used to numerically simulate the water entry of the lifeboat, and the slamming load of the hull was analyzed by the simulated results [10–12]. The second method was to establish a mathematical model for the freefall of the lifeboat for predicting the lifeboat motion attitude, and analyzing the risk of injury to the crew [13–15]. The mathematical model of the water entry of the lifeboat's freefall started from Karman's momentum theorem, solving the problem of water impact [16]. Boef [17,18] applied Karman's theory to the modeling of the lifeboat's motion, and divided the force of the lifeboat entering the water into gravity, buoyancy, drag, and slamming force. Arai et al. [19,20] simplified Boef's model. They divided the lifeboat's motion into four stages, sliding phase, rotation phase, freefall phase, and water entry phase, and analyzed the local acceleration of the bow, midship, and stern. The calculation results of local acceleration were close to the model experiment data. Khondoker et al. [21–23] applied Arai's model to analyze the parameters that affect the boat's water entry. Karim et al. [24,25] applied Arai's model and took the effect of regular waves into account when calculating speed of the boat entering the water. Raman-Nair and White [26] used multibody dynamics to analyze the entry of the boat from an offshore platform with simple movement. The rotation phase at skid exit is automatically modeled in this way. They regarded the skid as a slope, and the lifeboat as a cube on the slope and as a cylinder at water entry. When calculating the slamming force, they added the item of incident wave force in addition to the item of momentum theory, considering the effect of waves [27,28]. Dymarski and Dymarski [29] studied the model of a lifeboat released into the water from the stern of a ship. They also regarded the skid as a slope and the lifeboat as a cube on the slope, considering the reaction of the boat to the skid, but the motion of the ship was not considered. The detail of the algorithm was not given.

In summary, the current mathematical models for the freefall of the lifeboat do not take motion of ships in waves into account, and the contact force between the boat and skid is not modeled according to their actual structure. The slamming force is calculated according to momentum theory. All algorithms are not compared with real boat experiments or fluid dynamics software.

1.3. Our Contributions

This paper presents a computational model for the simulation of lifeboat freefall during its launching from a ship in rough seas. In order to consider the effect of the ship's motion on lifeboat motion, the maneuvering mathematical group (MMG) model is used to simulate the motion of the ship in the waves. In order to improve the calculation accuracy, the contact force between boat and skid is calculated based on strip theory, according to the actual structure of the boat and skid. When the boat enters the water, the slamming force is solved based on the theory of energy. The added mass of the boat is calculated according to its size and depth of water entry. Finally, the equations of motion of a freefall lifeboat are formulated using Kane's method.

A series of simulation experiments are carried out by using the computational model established in this paper. The results of simulation experiments are compared with results of the fluid dynamics software Star CCM+. They show that our numerical algorithm has good accuracy. A qualitative analysis is performed to find a safe point of water entry under the condition of different wave heights and two situations of a ship encountering waves.

Section 2 describes the mathematical model of the motion of boat on the skid. Section 3 describes the mathematical model of boat's water entry. The setup, results, and analysis of the simulation experiments are presented in Section 4. Section 5 describes an application of the mathematical model and Section 6 gives conclusions.

2. Motion of the Boat on the Skid

After the lifeboat is unhooked and released, the lifeboat slides down the skid away from the mother ship by its own gravity. When the center of gravity of the boat slides out of the lowest point of the skid, the boat begins to rotate owing to the vector of the gravity and the force of the skid not acting on the boat in a straight line.

2.1. Coordinate System

Figure 1 shows the central longitudinal section of the ship. There are three Cartesian coordinate systems: $oxyz$ is the inertial coordinate system, with unit vectors of the three axes N_1, N_2, N_3; the coordinate system of $o_0 x_0 y_0 z_0$ is fixed on the ship; o_0 is located at the center of gravity of the ship, the direction of $o_0 x_0$ points to the bow, the direction of $o_0 z_0$ points to the keel, and i_0, j_0, k_0 are the unit vectors of the axes. The coordinate system of $o_1 x_1 y_1 z_1$ is fixed on the skid and o_1 is located at the center of the upper end of the skid, with $o_1 x_1$ pointing down along the skid and $o_1 z_1$ pointing up the perpendicular to the slope of the skid; i_1, j_1, k_1 are the unit vectors of the three axes; ϕ is the angle between the plane of the skid and the horizontal plane. As shown in Figure 2, G_b is the center of gravity of the lifeboat, $o_2 x_2 y_2 z_2$ is the coordinate system attached to the boat, o_2 is located at G_b, and b_1, b_2, b_3 are the unit vectors for the three axes.

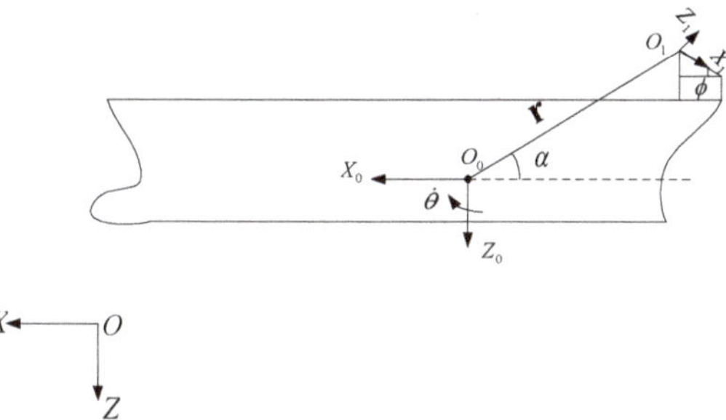

Figure 1. The three coordinate systems are the inertial coordinate system $oxyz$, the coordinate system $o_0 x_0 y_0 z_0$ fixed on the ship, and the coordinate system $o_1 x_1 y_1 z_1$ fixed on the skid.

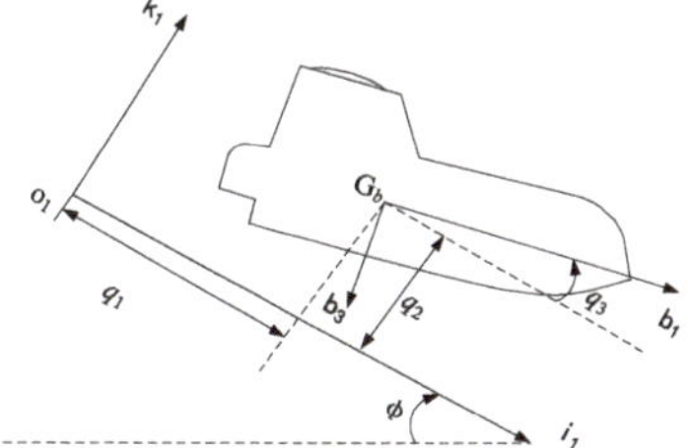

Figure 2. Generalized coordinates q_r of the boat in the coordinate system fixed on the skid and the coordinate system attached to the boat with the unit vectors b_r ($r = 1, 2, 3$).

2.2. Mathematical Model of Boat

The relative position of o_0 and o_1 remains unchanged in $o_0x_0y_0z_0$. The vector o_0o_1 is represented by r and a is the angle of r with the horizontal line. Based on the theorem for composited velocity and acceleration, the velocity and acceleration of o_1 are

$$\left\{ \begin{array}{l} v_{o_1} = v_{o_0} + \omega_{o_0} \times r \\ a_{o_1} = a_{o_0} + \left(\dot{\omega}_{o_0} \times r\right) + \omega_{o_0} \times \left(\omega_{o_0} \times r\right) \end{array} \right. , \tag{1}$$

where v_{o_0}, ω_{o_0} are the velocity and acceleration of the ship.

When considering only the ship's longitudinal motion, the ship's pitch angle is θ ; u and w are components of the ship's velocity in the direction of axis ox_0 and oz_0. The velocity and acceleration of o_1 are

$$\left\{ \begin{array}{l} v_{o_1} = \left(u - \dot{\theta}|r|sina\right)i_0 + \left(w + \dot{\theta}|r|cosa\right)k_0 \\ a_{o_1} = \left(\dot{u} + w\dot{\theta} - \ddot{\theta}|r|sina + \dot{\theta}^2|r|cosa\right)i_0 + \left(\dot{w} - u\dot{\theta} + \ddot{\theta}|r|cosa + \dot{\theta}^2|r|sina\right)k_0 \end{array} \right. . \tag{2}$$

The velocity and acceleration of G_b are

$$\left\{ \begin{array}{l} v_G = v_{o_1} + \left(\dot{q}_1 + q_2\dot{\phi}\right)i_1 + \left(\dot{q}_2 - q_1\dot{\phi}\right)k_1 \\ a_G = a_{o_1} + \left(\ddot{q}_1 + q_2\ddot{\phi} + 2\dot{q}_2\dot{\phi} - q_1\dot{\phi}^2\right)i_1 + \left(\ddot{q}_2 - q_1\ddot{\phi} - 2\dot{q}_1\dot{\phi} - q_2\dot{\phi}^2\right)k_1 \end{array} \right. . \tag{3}$$

The angular velocity of the rigid body is irrelevant to the choice of the base point, so $\dot{\phi} = \dot{\theta}$. The angular velocity and acceleration of the boat are

$$\left\{ \begin{array}{l} \omega_b = \left(\dot{\phi} - \dot{q}_3\right)N_2 \\ \alpha_b = \left(\ddot{\phi} - \ddot{q}_3\right)N_2 \end{array} \right. . \tag{4}$$

The lifeboat is divided into n cross sections with equal thickness along the length of the boat. The coordinates of the center of each section S_k, $(k = 1, \ldots, n)$ are expressed as $\left(x_{S_k}, 0, 0\right)$ in the coordinate system of the boat, and l_b is the length of boat. The thickness of each section is $t_s = l_b/n$; the velocity at the center of each cross section is

$$v_{S_k} = v_{o_1} + (\dot{q}_1 + q_2\dot{\phi} + \dot{\phi}x_{S_k}sinq_3 - \dot{q}_3x_{S_k}sinq_3)i_1 + \left(\dot{q}_2 - q_1\dot{\phi} - \dot{\phi}x_{S_k}cosq_3 + \dot{q}_3x_{S_k}cosq_3\right)k_1. \tag{5}$$

Generalized coordinates are q_r, generalized velocities are $u_r = \dot{q}_r (r = 1, 2, 3)$, partial velocities associated with points G_b, S_k and the angular velocity are written as follows [30]:

$$v_G^r = \left\{ \begin{array}{ll} i_1 & (r = 1) \\ k_1 & (r = 2) \\ 0 & (r = 3) \end{array} \right. , \tag{6}$$

$$v_{S_k}^r = \left\{ \begin{array}{ll} i_1 & (r = 1) \\ k_1 & (r = 2) \\ x_{S_k}cosq_3k_1 - x_{S_k}sinq_3i_1 & (r = 3) \end{array} \right. , \tag{7}$$

$$\omega_r^b = \left\{ \begin{array}{ll} 0 & (r = 1) \\ 0 & (r = 2) \\ -N_2 & (r = 3) \end{array} \right. . \tag{8}$$

The force analysis of the boat at the skid is shown in Figure 3. Forces acting on the boat are gravity G, supporting force F_n, and friction F_f.

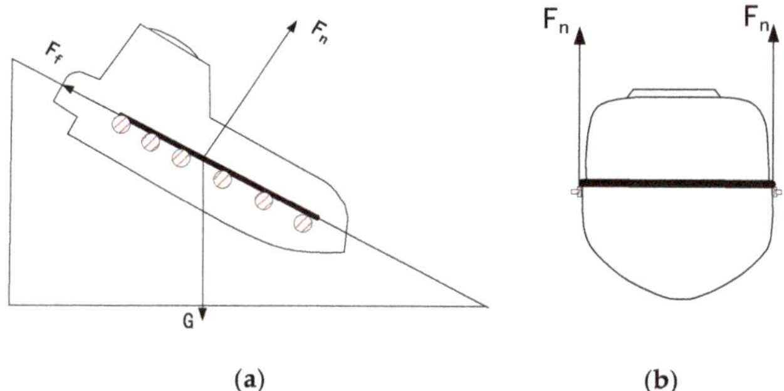

(a)   (b)

Figure 3. The side view (**a**) of forces and front view (**b**) of supporting force acting on the boat at the skid. The lifeboat is located on the contact rollers on both sides of the skid. The top point of the rollers is in contact with the boat and provides supporting force and friction.

The generalized force of gravity and inertia are, respectively,

$$F_G^r = v_G^r \cdot m_b g k, \tag{9}$$

$$F_*^r = v_G^r \cdot (-m_b a_G) + \omega^r \cdot (-I_2 \alpha_b), \tag{10}$$

where m_b is the mass of the boat, g is the acceleration of gravity, and I_2 is the moment of inertia around oy_2.

For the supporting force F_n, this paper regards the contact roller as a spring with stiffness coefficient k_r and damping coefficient c_r. The frictional contact force is modeled using a coefficient of μ_r. The coordinates of the center of each section S_k ($k = 1, \ldots, n$) are $\left(q_1 + x_{s_k} cosq_3, 0, q_2 + x_{s_k} sinq_3\right)$ in the boat coordinate system. We define them as follows:

$$\begin{cases} \alpha_{s_k} = q_1 + x_{s_k} cosq_3 \\ \beta_{s_k} = q_2 + x_{s_k} sinq_3 - z_p \\ \gamma_{s_k} = 1/2\left(\left|\beta_{s_k}\right| - \beta_{s_k}\right) \end{cases} . \tag{11}$$

The velocity of each section relative to its contacting point is $v_{s_k/s_{k'}}$. The positions of each contact roller are known; there are n_p contact rollers on each side of the skid. The coordinates of the top of the contact roller on each side of the axes of $o_1 x_1$ and $o_1 z_1$ at the skid coordinate system are $x_{p_i}, z_p\left(i = 1, \ldots, n_p\right)$ as $x_{p_i} - \frac{t_s}{2} cosq_3 < \alpha_{s_k} < x_{p_i} + \frac{t_s}{2} cosq_3$. The contacting force between S_k and contact rollers [31,32] is

$$F_{sk} = 2\left(k_r * \gamma_{s_k} - c_r\left(v_{s_k/s_{k'}} \cdot k_1\right) sign\left(\gamma_{s_k}\right)\right)k_1 - 2\left(\mu_r k_r \gamma_{s_k} sign\left(v_{s_k/s_{k'}} \cdot i_1\right)\right)i_1 \tag{12}$$

where

$$\begin{cases} v_{s_k/s_{k'}} \cdot k_1 = \dot{q}_2 - q1\dot{\phi} - \dot{\phi}X_{S_k} cosq_3 + \dot{q}_3 X_{S_k} cosq_3 + \dot{\phi}\alpha_{S_k} \\ v_{s_k/s_{k'}} \cdot i_1 = \dot{q}_1 - q2\dot{\phi} - \dot{\phi}X_{S_k} sinq_3 + \dot{q}_3 X_{S_k} sinq_3 \end{cases} . \tag{13}$$

The associated generalized force acting on the S_k by the skid is

$$F_{S_k}^r = v_{S_k}^r \cdot F_{S_k} \ (r = 1, 2, 3). \tag{14}$$

The combined generalized force of the skid that acted on the boat is $F_c^r = \sum F_{S_k}^r (r = 1, 2, 3)$. The equation of the boat motion is

$$F^r + F_*^r = 0 \ (r = 1, 2, 3), \tag{15}$$

where $F^r = F_c^r + F_G^r$. Finally, the fourth-order Runge-Kutta method is used to solve the differential equation.

2.3. Mathematical Model of Ship Motion

Based on the model of MMG [33], the coordinate system is as shown in Figure 1. The ship dynamic equation is

$$
\begin{cases}
(m + m_x)(\dot{u} - vr + qw) = X_H + X_{wave} \\
(m + m_y)(\dot{v} + ur - pw) = Y_H + Y_{wave} \\
(m + m_z)(\dot{w} + pv - qu) = Z_H + Z_{wave} \\
(I_{xx} + J_{xx})\dot{p} + (I_{zz} - I_{yy})qr = K_H + K_{wave} \\
(I_{yy} + J_{yy})\dot{q} + (I_{xx} - I_{zz})pr = M_H + M_{wave} \\
(I_{zz} + J_{zz})\dot{r} + (I_{yy} - I_{xx})pq = N_H + N_{wave}
\end{cases}
\tag{16}
$$

where the variables with subscript H are the forces and moments on the hull; the variables with subscript *wave* are wave forces. m is the mass of the ship; m_x, m_y and m_z are, respectively, added masses in the direction of the axes ox_0, oy_0 and oz_0 ; I_{xx}, I_{yy} and I_{zz} are the rotational moment of inertia around axes ox_0, oy_0 and oz_0; J_{xx}, J_{yy} and J_{zz} are the added moment of inertia around axes ox_0, oy_0, oz_0 ; u, v, w are velocities in the direction of axes ox_0, oy_0 and oz_0; and p, q, r are the angular velocities around axes ox_0, oy_0, oz_0. $m_z = m$; other added masses are calculated as follows:

$$
\begin{cases}
\frac{m_x}{m} = \frac{1}{100}\left(0.398 + 11.988c_b\left(1 + 3.73\frac{d}{B}\right) - 2.89\frac{c_b L_p}{B}\left(1 + 1.13\frac{d}{B}\right) + 0.175c_b\left(\frac{L_p}{B}\right)^2\left(1 + 0.541\frac{d}{B}\right) - 1.107\frac{d L_p}{B^2}\right) \\
\frac{m_y}{m} = 0.882 - 0.54c_b\left(1 - 1.6\frac{d}{B}\right) - 0.156(1 - 1.673c_b)\frac{L_p}{B} + 0.826\frac{d}{B}\frac{L_p}{B}\left(1 - 0.678\frac{d}{B}\right) - 0.638\frac{d}{B}\frac{L_p}{B}\left(1 - 0.669\frac{d}{B}\right)
\end{cases}
\tag{17}
$$

where c_b is the block coefficient, d is the draft, B is the ship width, and L_p is the length between perpendiculars. $I_{xx}, I_{yy}, I_{zz}, J_{xx}, J_{yy}$ and J_{zz} are calculated as follows:

$$
\begin{cases}
I_{xx} + J_{xx} = mB\left(0.3085 + 0.0227\frac{B}{d} - 0.0043\frac{L_p}{100}\right) \\
I_{yy} = J_{yy} = 0.83\frac{B}{2d}\left(0.25L_p c_p\right)^2 m \\
I_{zz} = \left(1 + c_b^{4.5}\right)m + \left(L_p^2 + B^{2.4}\right)/24 \\
J_{zz} = 0.01m\left(33L_p^2 - 76.85c_b(1 - 0.784c_b) + 3.43\frac{L_p}{B}(1 - 0.63c_b)\right)
\end{cases}
\tag{18}
$$

where c_p is the prismatic coefficient.

The kinematics equation is

$$
\begin{cases}
\dot{x} = \mu\cos\psi\cos\theta + v(\cos\psi\sin\theta\sin\varphi - \sin\psi\cos\varphi) + w(\cos\varphi\sin\theta\cos\psi + \sin\varphi\sin\psi) \\
\dot{y} = \mu\sin\psi\sin\theta + v(\sin\psi\sin\theta\sin\varphi + \cos\psi\cos\varphi) + w(\cos\varphi\sin\theta\cos\psi + \sin\varphi\cos\psi) \\
\dot{z} = -\mu\sin\theta + v\cos\theta\sin\varphi + w\cos\varphi\cos\theta \\
\dot{\varphi} = p + q\sin\varphi \tan\theta + r\cos\varphi \tan\theta \\
\dot{\theta} = q\cos\varphi - r\sin\varphi \\
\dot{\psi} = \frac{q\sin\varphi}{\cos\theta} + \frac{r\cos\varphi}{\cos\theta}
\end{cases}
\tag{19}
$$

where x, y, z and φ, θ, ψ are displacements and Euler angles relative to the inertial coordinate system, respectively.

The forces and moments on the hull are

$$
\begin{cases}
X_H = X_{uu}u^2 + X_{vv}v^2 + X_{vr}vr + X_{rr}r^2 \\
Y_H = Y_v v + Y_r r + Y_{r|r|}r|r| + Y_{vr}vr \\
Z_H = -Z_w w - Z_q q - Z_{\dot{q}}\dot{q} - Z_\theta \theta \\
K_H = -2K_p p - \Delta \cdot GM \cdot sin\varphi - Y_H z_H \\
M_H = -M_w w - M_{\dot{w}}\dot{w} - M_q q - M_\theta \theta \\
N_H = N_v v + N_r r + N_{r|r|}r|r| + N_{vr}vr + Y_H x_c
\end{cases}
, \tag{20}
$$

where $X_{uu}, X_{vv}, X_{vr}, X_{rr}, Y_v, Y_r, Y_{r|r|}, Y_{vr}, Z_w, Z_q, Z_{\dot{q}}, Z_\theta, K_p, M_w, M_{\dot{w}}, M_q, M_\theta, N_v, N_r, N_{r|r|}, N_{vr}$, and N_{vr} are hydrodynamic derivatives, which are easily calculated according to the ship parameters [34]; Δ is the ship displacement, GM is the metacentric height, z_H is the coordinate of action point of Y_H in the direction of oz_0, and x_c is the distance between the center of gravity and the center.

For the wave force, it is estimated, based on the Frude-Krenov assumption, that the hull is simplified to a box, and the six-degrees-of-freedom wave force and moment are as follows [34]:

$$
\begin{cases}
X_{wave} = 2\rho ga\dfrac{sin(\frac{kBsin(\chi)}{2})}{\frac{kBsin(\chi)}{2}}e^{-kd}Bdsin(k\frac{L}{2}cos(\chi))sin(\omega_e t) \\[3mm]
Y_{wave} = -2\rho ga\dfrac{sin(\frac{kLcos(\chi)}{2})}{\frac{kBcos(\chi)}{2}}e^{-kd}Ldsin(k\frac{B}{2}sin(\chi))sin(\omega_e t) \\[3mm]
Z_{wave} = \rho gak\dfrac{sin(\frac{kBsin(\chi)}{2})}{\frac{kBsin(\chi)}{2}}e^{-kd}BLd\dfrac{sin(\frac{kLcos(\chi)}{2})}{\frac{kBcos(\chi)}{2}}cos(\omega_e t) \\[3mm]
K_{wave} = \rho gasin(\chi)\dfrac{sin(\frac{kBsin(\chi)}{2})}{\frac{kBsin(\chi)}{2}}e^{-kd}d^2\dfrac{sin(\frac{kLcos(\chi)}{2})}{cos(\chi)}sin(\omega_e t) \\[3mm]
M_{wave} = \rho ga\dfrac{sin(\frac{kBsin(\chi)}{2})}{\frac{ksin(\chi)}{2}}e^{-kd}d\dfrac{2sin(\frac{kLcos(\chi)}{2})-kLcos(\chi)cos(\frac{kLcos(\chi)}{2})}{k^2cos^2(\chi)}sin(\omega_e t) \\[3mm]
N_{wave} = \rho ga\dfrac{sin(\frac{kBsin(\chi)}{2})}{\frac{ksin(\chi)}{2}}e^{-kd}d\dfrac{2sin(\frac{kLcos(\chi)}{2})-kLcos(\chi)cos(\frac{kLcos(\chi)}{2})}{k^2cos^2(\chi)}cos(\omega_e t)
\end{cases}
\tag{21}
$$

where a is the amplitude of the wave, k is the number of waves, ω_e is the encounter frequency, χ is the encounter angle, ρ is the water density, and L is the ship waterline length.

The motion parameters of the ship can be obtained by solving the differential Equations (16) and (19) using the fourth-order Runge-Kutta method.

3. Water Entry

The geometric orientation relationship of the inertial and boat coordinate system is represented by $\beta_1, \beta_2, \beta_3$. The transformation relationship between the two coordinate systems is as follows, where $[^N C^b]$ is transformation matrix:

$$
\begin{pmatrix} N_1 \\ N_2 \\ N_3 \end{pmatrix} = [^N C^b]\begin{pmatrix} b_1 \\ b_2 \\ b_3 \end{pmatrix}, \tag{22}
$$

$$
[^N C^b] = \begin{pmatrix}
c_2 c_3 & s_1 s_2 c_3 - s_3 c_1 & c_1 s_2 c_3 + s_3 s_1 \\
c_2 s_3 & s_1 s_2 s_3 + c_1 c_3 & c_1 s_2 s_3 - c_3 s_1 \\
-s_2 & s_1 c_2 & c_1 c_2
\end{pmatrix}, \tag{23}
$$

where $s_i = sin\beta_i, c_i = cos\beta_i, (i = 1, 2, 3)$.

The generalized coordinates are

$$
\begin{cases}
q_i^b = \beta_i \\
q_{3+i}^b = \overrightarrow{OO_b} \cdot b_i
\end{cases}
(i = 1, 2, 3). \tag{24}
$$

The generalized velocities are

$$\begin{cases} u_i^b = \omega^b \cdot b_i \\ u_{3+i}^b = v^b \cdot b_i \end{cases} (i = 1, 2, 3), \tag{25}$$

where ω^b, v^b are the velocity and angular velocity of the lifeboat, respectively.

The relationships between generalized coordinates and generalized velocities are

$$\begin{cases} \dot{q}_1^b = u_1^b + s_2/c_2\left(u_2^b s_1 + u_3^b c_1\right) \\ \dot{q}_2^b = u_2^b c_1 - u_3^b s_1 \\ \dot{q}_3^b = \left(u_2^b s_1 + u_3^b c_1\right)/c_2 \end{cases}, \tag{26}$$

$$\begin{cases} \dot{q}_4^b = u_4^b - u_2^b q_6^b + u_3^b q_5^b \\ \dot{q}_5^b = u_5^b + u_1^b q_6^b - u_3^b q_4^b \\ \dot{q}_6^b = u_6^b - u_1^b q_5^b + u_2^b q_4^b \end{cases}. \tag{27}$$

The acceleration and angular acceleration of the boat are

$$a^b = {}^N d\left(\omega^b\right)/dt = {}^b d\left(\omega^b\right)/dt + \omega^b \times \omega^b = \dot{u}_i^b b_i, \tag{28}$$

$$a^b = (\dot{u}_4^b + u_2^b u_6^b - u_3^b u_5^b)b_1 + (\dot{u}_5^b - u_1^b u_6^b + u_3^b u_4^b)b_2 + (\dot{u}_6^b + u_1^b u_5^b - u_2^b u_4^b)b_3. \tag{29}$$

The partial angular velocities and velocities of the boat are

$$\omega_r^b = \begin{cases} b_r & (r = 1, 2, 3) \\ 0 & (r = 4, 5, 6) \end{cases}, \tag{30}$$

$$v_r^b = \begin{cases} 0 & (r = 1, 2, 3) \\ b_{r-3} & (r = 4, 5, 6) \end{cases}. \tag{31}$$

The generalized inertial force is

$$F_r^{*b} = \omega_r^b \cdot T^* + v_r^b \cdot \left(-m_b a^b\right), \tag{32}$$

$$F_r^{*b} = \begin{cases} -[\dot{u}_1^b I_1 - u_2^b u_3^b (I_2 - I_3)] \\ -[\dot{u}_2^b I_2 - u_3^b u_1^b (I_3 - I_1)] \\ -[\dot{u}_3^b I_3 - u_1^b u_2^b (I_1 - I_2)] \\ -m_b(\dot{u}_4^b + u_2^b u_6^b - u_3^b u_5^b) \\ -m_b(\dot{u}_5^b - u_1^b u_6^b + u_3^b u_4^b) \\ -m_b(\dot{u}_6^b + u_1^b u_5^B - u_2^b u_4^b) \end{cases}, \tag{33}$$

where I_1, I_2, I_3 is the moment of inertia around ox_2, oy_2, oz_2.

The generalized force caused by the gravity of the boat is

$$F_r^{G/b} = m_b g N_3 \cdot v_r^b, \tag{34}$$

$$F_r^{W/b} = \begin{Bmatrix} 0 \\ 0 \\ 0 \\ m_b g(\sin - q_2) \\ -m_b g(\sin q_1 \cos q_2) \\ -m_b g(\sin q_1 \cos q_2) \end{Bmatrix}. \tag{35}$$

The generalized force caused by air drag is

$$F_r^{W/b} = F_b^W \cdot v_r^b, \tag{36}$$

$$F_b^W = -\frac{1}{2}\rho_w A_w C_w \left| v^{b/W} \right| v^{b/W}, \tag{37}$$

where $v^{b/W}$ is the vector difference between the lifeboat velocity and the wind velocity $v^{b/W} = v^b - v^W$, ρ_w is the air density, A_w is the projected area of the lifeboat in the plane perpendicular to $v^{b/W}$, and C_w is the coefficient of air drag.

The generalized force caused by fluid drag is

$$F_r^{D/b} = \sum_{k=1}^{n} F_r^{D/s_k} (r = 1, \cdots, 6), \tag{38}$$

$$F_r^{D/s_k} = F_{s_k}^D \cdot v_r^{s_k}, \tag{39}$$

where $F_{s_k}^D$ is the fluid drag acting on S_k and $v_r^{s_k}$ is the partial velocity of S_k. The velocity of S_k is as follows:

$$v^{s_k} = v_1^{s_k} b_1 + v_2^{s_k} b_2 + v_3^{s_k} b_3, \tag{40}$$

where $v_1^{s_k} = u_4, v_2^{s_k} = u_5 + x_{s_k} u_3, v_3^{s_k} = u_6 - x_{s_k} u_2$.

The partial velocities of S_k are

$$v_r^{s_k} = \begin{Bmatrix} 0 & (r = 1) \\ -x_{s_k} b_3 & (r = 2) \\ x_{s_k} b_2 & (r = 3) \\ b_{r-3} & (r = 4, 5, 6) \end{Bmatrix}. \tag{41}$$

The components of the fluid drag acting on S_k are as follows [22]:

$$F_{s_k}^D = \sum_{i=1}^{3} D_i^{s_k} b_i, \tag{42}$$

$$\begin{cases} D_1^{s_k} = -\frac{1}{2n}\rho_a A_1^{s_m} c_{D_1}^{s_m} \left| v_1^{s_m/R} \right| v_1^{s_m/R} \\ D_2^{s_k} = -\frac{1}{2}\rho_a A_2^{s_k} c_{D_2}^{s_k} \left| v_2^{s_k/R} \right| v_2^{s_k/R} \\ D_3^{s_k} = -\frac{1}{2}\rho_a A_3^{s_k} c_{D_3}^{s_k} \left| v_3^{s_k/R} \right| v_3^{s_k/R} \end{cases}, \tag{43}$$

where ρ_a is the density of seawater and $A_i^{s_k}$ is the area of s_k perpendicular to b_i below the water surface. The section s_m has maximum area $A_1^{s_m} = max(A_1^{s_k})(k = 1 \cdots n)$, $c_{D_i}^{s_k}$ is the fluid drag coefficient, and $v_i^{s_k/R}$ is the component of the velocity of the section s_k relative to the wave surface on the axis of b_i.

The generalized force caused by buoyancy is

$$F_r^{B/b} = \sum_{k=1}^{n} F_r^{B/s_k} (r = 1, \cdots, 6), \tag{44}$$

$$F_r^{B/s_k} = F_{s_k}^B \cdot v_r^{s_k}, \tag{45}$$

$$F_{S_k}^B = -\rho_a V_{S_k} g N_3,\tag{46}$$

where V_{S_k} is the volume of S_k in water and $F_{S_k}^B$ is the buoyancy acting on s_k.

The generalized force caused by the slamming force is

$$F_r^{p/b} = \sum_{k=1}^{n} F_r^{p/s_k} (r = 1, \cdots, 6),\tag{47}$$

$$F_r^{p/s_k} = F_{S_k}^p \cdot v_r^{s_k},\tag{48}$$

$$F_{S_k}^p = \sum_{i=1}^{3} p_i^{s_k} b_i,\tag{49}$$

where $F_{S_k}^p$ is the slamming force that acted on section s_k. Previous scholars adopted the momentum theory to calculate the slamming force. When an object enters the water, part of its initial momentum will be transferred to the surrounding water. Assuming that the momentum conversion process is irreversible, the slamming force acting on the boat can be calculated by the rate of its momentum change:

$$m_0 v_0 = (m_0 + m_a) v,\tag{50}$$

$$F^p = \frac{d}{dt}(m_0 v) = -\frac{d}{dt}(m_a v) = -\left(\frac{dm_a}{dt}v + \frac{dv}{dt}m_a\right),\tag{51}$$

where v_0 is the velocity of the object before entering the water and v is the velocity of the object after entering the water; m_0 is the mass of the object and m_a is the added mass of the object.

Another method is based on energy theory, which was derived by Wu [35]. It has more physical significance and is widely used [36–38]. The relationship between slamming force and added mass is

$$F^p = -\frac{dv}{dt}m_a - \frac{1}{2}\frac{dm_a}{dt}v.\tag{52}$$

In order to consider the effect of the wave on the boat, the incident wave force F^{in} is added into the slamming force:

$$F^{in} = \rho_a V_{sub} a^f,\tag{53}$$

where V_{sub} is the submerged volume and a^f is the fluid acceleration.

In this paper, the slamming force $F_{S_k}^p$ is calculated as the sum of energy theory and incident wave force for the first time. $p_i^{s_k}$ are the components of $F_{S_k}^p$ on the axis of b_i:

$$\begin{cases} p_1^{s_k} = \frac{1}{n}(\rho_a V_{s_h} a_1^{f/s_h} - (0.5\frac{dm_1^{s_h}}{dt}v_1^{s_h/R} + \frac{dv_1^{s_h/R}}{dt}m_1^{s_h})) \\ p_2^{s_k} = \rho_a V_{s_k} a_2^{f/s_k} - (0.5\frac{dm_2^{s_k}}{dt}v_2^{s_k/R} + \frac{dv_2^{s_k/R}}{dt}m_2^{s_k}) \\ p_3^{s_k} = \rho_a V_{s_k} a_3^{f/s_k} - (0.5\frac{dm_3^{s_k}}{dt}v_3^{s_k/R} + \frac{dv_3^{s_k/R}}{dt}m_3^{s_k}) \end{cases},\tag{54}$$

where a^{f/s_k} is the wave surface acceleration at section s_k at the coordinate system of the boat, a_i^{f/s_k} are the components of a^{f/s_k} at the coordinate system of the boat, $m_i^{s_k}$ are the added mass of the cross section in the direction of b_i, $m_1^{s_h}$ is the added mass of the boat in the direction of b_1, and $v_1^{s_h}$ is the velocity component of the midsection s_h between the bow and the center of gravity of the boat in the direction of b_1.

s_k have coordinates $(x_{s_k}^g, y_{s_k}^g, z_{s_k}^g)$ in an inertial coordinate system:

$$\begin{pmatrix} x_{s_k}^g \\ y_{s_k}^g \\ z_{s_k}^g \end{pmatrix} = \begin{bmatrix} {}^N C^b \end{bmatrix} \begin{pmatrix} q_4 + x_{s_k} \\ q_5 \\ q_6 \end{pmatrix}.\tag{55}$$

The depth of s_k is

$$h_{s_k} = C_{31} \cdot \left(q_4 + x_{s_k}\right) + C_{32} \cdot q_5 + C_{33} \cdot (q_6 - h_0) - \eta\left(x_{s_k}^g, t\right), \tag{56}$$

where C_{ii} are elements of the matrix of $[^N C^b]$, the subscript is its location in the matrix, h_0 is the vertical distance from the center of gravity to the keel, h_0 is 1.15 m in Figure 4, and $\eta(x, t)$ is a known function describing the wave surface.

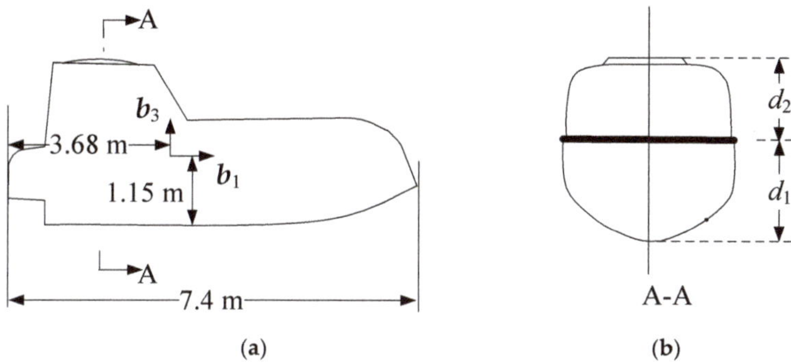

Figure 4. Lifeboat cross section: (a) longitudinal section in center plane, (b) A-A cross section.

The added mass is $m_2^{s_k} = m_3^{s_k}$ in this paper. $m_3^{s_k}$ depends on the depth and the instantaneous half width of s_k at the wave surface:

$$m_3^{s_k} = \begin{cases} \rho \pi C^2\left(x_{s_k}, h_{s_k}\right)/2 & h_{s_k} < d_1 \\ \rho \pi C^2\left(x_{s_k}, d_1\right)/2 & h_{s_k} \geq d_1 \end{cases}, \tag{57}$$

where $C(x_{s_k}, h_{s_k})$ is half the width of s_k at depth h_{s_k}. When the depth of s_k is greater than d_1, the added mass is calculated according to the half-width at depth d_1, as shown in Figure 4. We take the derivative of $m_3^{s_k}$ to time $\frac{dm_3^{s_k}}{dt} = \frac{dm_3^{s_k}}{dh_{s_k}} \frac{dh_{s_k}}{dt}$ with $\frac{dh_{s_k}}{dt} = \max(v_3^{s_k}/R, 0)$.

To calculate $m_1^{s_h}$, define α_{s_k}

$$\alpha_{s_k} = \begin{cases} 1 & h_{s_k}^g \geq 0 \\ 0 & h_{s_k}^g < 0 \end{cases}, \tag{58}$$

where $h_{s_k}^g = z_{s_k}^g - \eta(x_{s_k}^g, t)$. The length of water entry in the direction of b_1 is l_1:

$$l_1 = \sum_{k=1}^{n} \alpha_{s_k} \cdot t_s, \tag{59}$$

$$m_1^{s_h} = \begin{cases} m_{ax}\left(2(l_1/l_b)^3 - 4(l_1/l_b)^2 + 2.5(l_1/l_b)\right) & l_1 < l_b/2 \\ m_{ax}/2 & l_1 \geq l_b/2 \end{cases}, \tag{60}$$

where m_{ax} is the added mass of the boat in the direction of b_1 when the boat is completely in the water.

$$m_{ax} = \frac{k_m \pi \rho_a l_b (d_1 + d_2)^2}{6}, \tag{61}$$

where k_m is a coefficient depending on $\frac{l_b}{(d_1 + d_2)}$. We take the derivative of $m_1^{s_h}$ to time $\frac{dm_1^{s_h}}{dt} = \frac{dm_1^{s_h}}{dl_1} \frac{dl_1}{dt}$ with $\frac{dl_1}{dt} = \max(v_1^{s_h}/R, 0)$.

The acceleration of the section is

$$a^{s_k} = \sum_{i=1}^{3} a_i^{s_k} b_i,$$ (62)

where $a_1^{s_k} = \dot{u}_4 + z_1^{s_k}$, $a_2^{s_k} = \dot{u}_5 + x_{s_k}\dot{u}_3 + z_2^{s_k}$, $a_3^{s_k} = \dot{u}_6 - x_{s_k}\dot{u}_2 + z_3^{s_k}$, $z_1^{s_k} = u_2u_6 - u_3u_5 - x_{s_k}(u_2^2 + u_3^2)$, $z_2^{s_k} = -u_1u_6 + u_3u_4 + x_{s_k}u_1u_3$, $z_3^{s_k} = u_1u_5 + u_2u_4 + x_{s_k}u_1u_3$.

So

$$\dot{v}_1^{s_h} = \dot{u}_4^b + u_2^b u_6^b - u_3^b u_5^b - \frac{1}{2}L_f\left(u_2^2 + u_3^2\right),$$ (63)

where L_f is the distance from the center of gravity of the boat to the bow.

In summary, the equation of motion is

$$F_r^{*b} + F_r^{G/b} + F_r^{W/b} + F_r^{D/b} + F_r^{B/b} + F_r^{p/b} = 0 \ (r = 1,\cdots,6).$$ (64)

The motion parameters of the lifeboat can be solved by Equation (64) by the fourth-order Runge-Kutta method.

4. Results and Analysis

4.1. Experimental Setup and Implementation

The basic information of the boat, ship, and skid is shown in Tables 1–3. Air drag coefficient $C_w = 0.5$. Fluid drag coefficient $c_{D_i}^{s_k} = 1.2$. Figure 5 gives the structural dimension diagram of the ship and boat in the simulation experiment.

Table 1. Basic information of the boat.

Item (Unit)	Value
Length (m)	7.4
Width (m)	2.65
Draft (m)	0.74
Maximum height (m)	2.3
Maximum half width (m)	1.42
Distance between center of gravity and center (m)	0.02
I_2 (kg·m^2)	34,825
mass (kg)	6200
The projected areas of the boat at plane perpendicular to b_1 (m^2)	4.2
The projected areas of the boat at plane perpendicular to b_2 (m^2)	13.2
The projected areas of the boat at plane perpendicular to b_3 (m^2)	13.2

Table 2. Basic information of the ship.

Item (Unit)	Value
Length over all (m)	144.4
Waterline length (m)	133.55
Length between perpendiculars (m)	129
Width (m)	20.8
Draft (m)	4.4
Depth molded (m)	11.4
GM (m)	5.57
Distance between center of gravity and center (m)	2.3
Block coefficient	0.68
Water plane coefficient	0.83
Prismatic coefficient	0.693
Displacement (kg)	7,550,000

Table 3. Basic information of the skid.

Item (Unit)	Value
Vertical distance between the low end of skid and deck (m)	3.5
Inclination (°)	30
Length (m)	9
Sliding distance of the center of gravity (m)	4.96

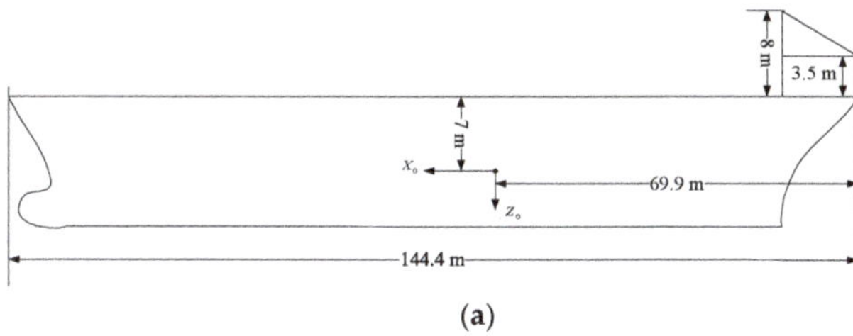
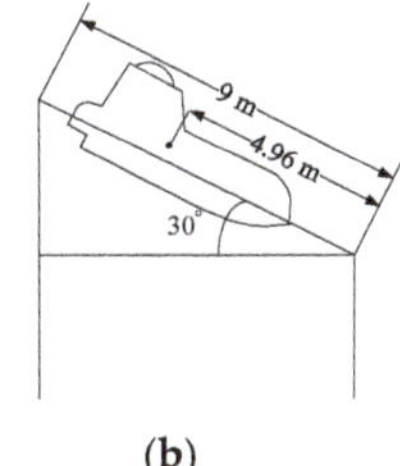

(a)   (b)

Figure 5. Structural dimension diagram of the ship and boat: (**a**) the size of the longitudinal section in the center plane of the ship and skid; (**b**) the inclination of the skid and the position of the boat at the skid.

The simulation experiment of the whole algorithm was realized by MATLAB. The flow of the algorithm consists of two stages that are separated by the time point of the boat out of the skid. In the first stage, the motion of the skid is calculated according to parameters of ship motion solved by Equations (16) and (19). The generalized contact force can be solved by Equation (14), the generalized force caused by gravity and inertia can be solved by Equations (9) and (10), and the generalized velocities and generalized coordinates of the boat can be solved by Equation (15). In the second stage, the generalized force caused by inertial, gravity, air drag, fluid drag, buoyancy, and slamming force can be solved by Equations (32), (34), (36), (38), (44) and (47); the generalized velocities of the boat can be solved by Equation (64); and the generalized coordinates of the boat can be solved by Equations (26) and (27). For the two stages, 20 lifeboat segments were used. For convenience of analysis and understanding, the data of the boat's translation and pitch are shown in the coordinates obtained by the inertial coordinate system, rotating 180° around the axis of oy.

The results of a simulation experiment by the numerical algorithm in this paper were compared with the results of Star CCM+ at the initial condition A, as shown in Table 4. The version of Star CCM+ used was 13.02. These computations were performed using an overlapping grid, as shown in Figure 6a, in order to provide accurate wave representation at any location irrespective of the lifeboat position. The behavior of two fluids (air and water) in the same continuum is modeled by using the model of volume of fluid (Figure 6b). Due to the existence of two fluids in different phases, the Euler multiphase model is activated and the gravity model is used to take into account the gravity effect of the two fluids. The initial condition of water entry is obtained by our numerical algorithm: the boat enters the water at the crest of the wave, the pitch angle of the boat when entering the water is 50.7° with horizontal, vertical, and rotational velocity of 5.9 m/s, −11 m/s, and 0.46 rad/s.

Table 4. Details of initial condition A.

Item (Unit)	Value
Wave height (m)	2
Period of wave (s)	8
Wind velocity (m/s)	$(0,0,0)$
Ship's initial position (m)	$(0,0,0)$
Ship's initial velocity (knot)	$(1,0,0)$
Situation of ship encountering waves (0: head the wave; 1: follow the wave)	1

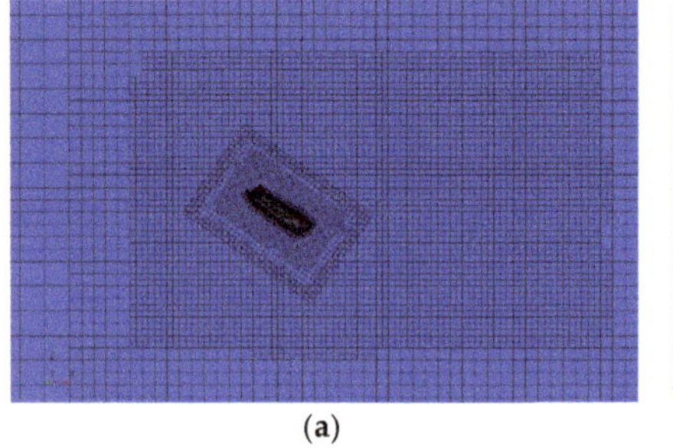
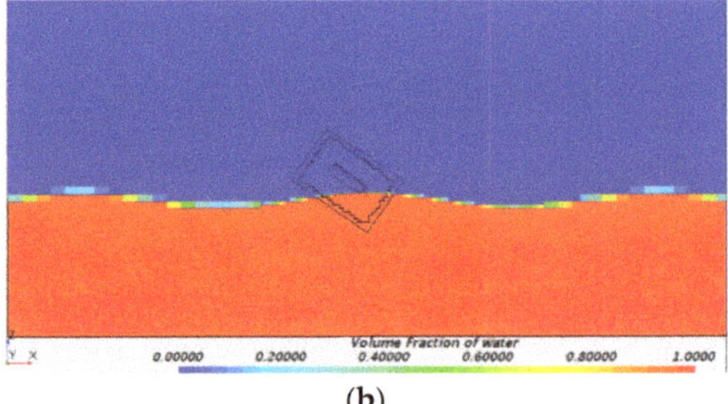

(a) **(b)**

Figure 6. Simulation experiment diagram of Star CCM+: (**a**) overlapping grid; (**b**) volume fraction of water.

4.2. Comparison and Analysis of Experimental Results

The red line is the result of the whole motion process calculated by the model in this paper. The red blue is the result of the water entry process calculated by Star CCM+.

Figure 7 shows the trajectory and pitch angle of the boat. The results show that the trajectory and pitch angle calculated by two methods are in good agreement. When the boat moves on the skid, the pitch angle begins to change with no obvious logic. The pitch angle begins to increase at the bottom of the skid. After the boat leaves the skid, the pitch angle of the boat gradually increases to about 53°. The pitch angle begin to decrease after increases in short time because the bow of the boat hits the water. The center of gravity of the boat is sinking, the bow of the boat is raised, resulting in the stern hitting the water. The buoyancy increases with the increase of the depth of the stern. Subsequently, the pitch angle of the boat begins to increase again, and the lifeboat begins to float. After it comes out of the water, the bow hits the water again.

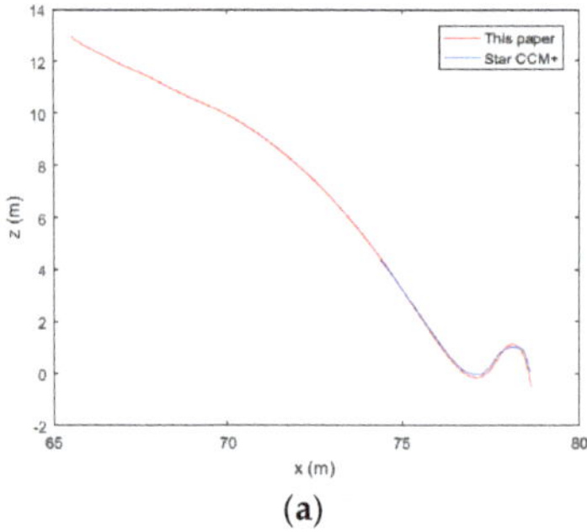
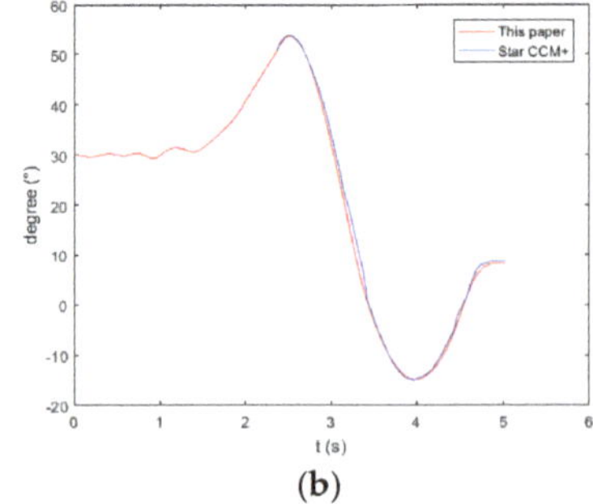

(a) **(b)**

Figure 7. Trajectory (**a**) and pitch (**b**) of the lifeboat freefall during its launch from the ship.

Figure 8 shows the ratio of the acceleration of the center of gravity to *g*. When the boat is on the skid, the acceleration value of the center of gravity of the boat begins to vibrate. When the boat is at the low end of the skid, the vibration frequency of the acceleration value increases and the amplitude decreases. When the boat leaves the skid, the horizontal acceleration of the boat is zero and the vertical acceleration is *g* before entering the water. The horizontal acceleration of the boat is negative in the initial stage of water entry because of the slamming force and the fluid drag force, and the acceleration becomes positive when the boat floats up because of its large buoyancy. The vertical acceleration of the boat is almost positive because of the upward force, and its negative value caused by gravity and fluid drag when the boat floats up. There is a little difference between the acceleration values calculated by the two methods, especially at the initial stage of water entry. The difference is acceptable. This phenomenon may be due to the fact that the deformation of the water caused by boat is not considered in this algorithm.

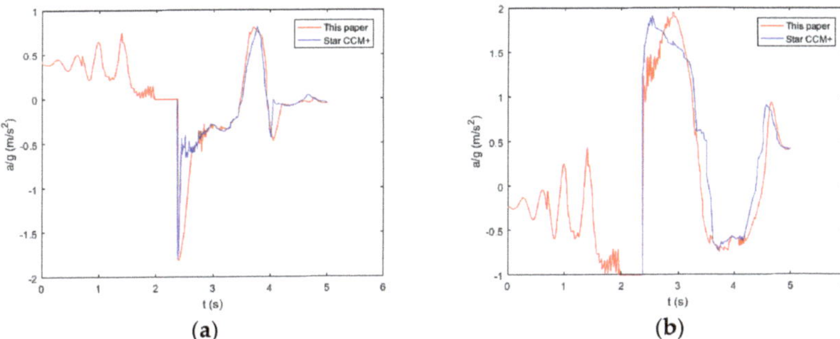

Figure 8. Acceleration of center of gravity of the lifeboat in the horizontal (**a**) and vertical (**b**) direction.

Figure 9 shows the velocity of the center of gravity of the boat. The horizontal velocity of the boat gradually begins to increase at the skid, and remains the same in the air before entering the water; it begins to decrease due to the negative acceleration at the beginning of the water entry phase, and begins to increase when the boat floats up. The value of vertical acceleration is decreasing because the boat moves down the skid. The value of vertical acceleration decreases at the same rate because of *g* in the air before entering the water. Its value gradually increases until the sign is positive because of the slamming force and the buoyancy at the water entry phase; its sign becomes negative when the boat falls into the water again. The velocity values calculated by the two methods are in good agreement.

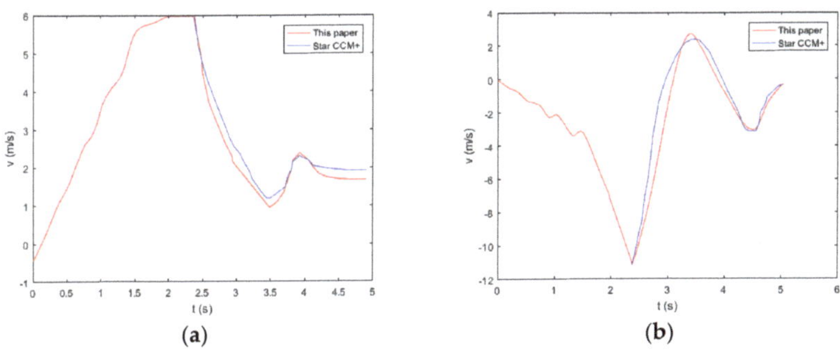

Figure 9. Velocity of center of gravity of the lifeboat in the horizontal (**a**) and vertical (**b**) direction.

Figure 10 shows the velocity and acceleration of pitch. When the boat is on the skid, the value of the velocity of the pitch begins to vibrate. When the boat leaves the skid, the velocity of the pitch remains the same before entering the water. The velocity of the pitch decreases to a negative value when the boat enters the water. Its sign is positive when the boat's bow begins to fall into the water again. The acceleration of the pitch has the same rule of change as the horizontal acceleration. The velocity and acceleration of pitch calculated by the two methods are in good agreement.

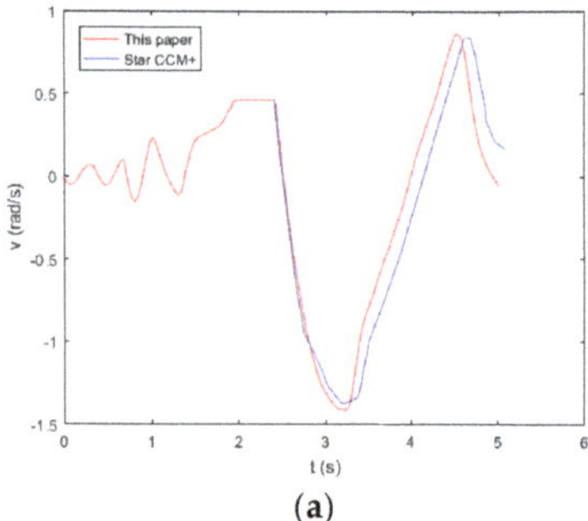
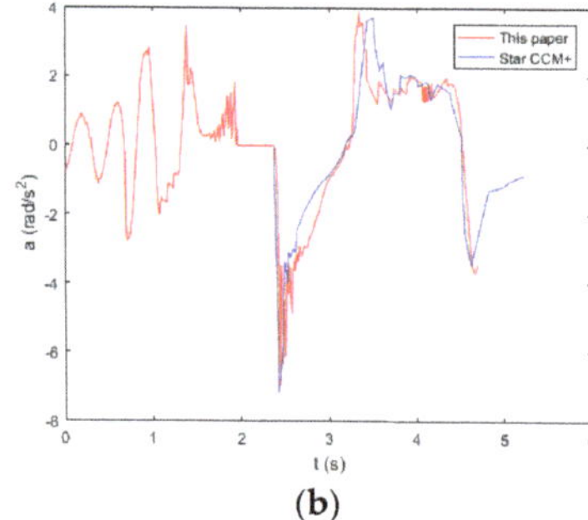

(a)   (b)

Figure 10. Velocity (**a**) and acceleration (**b**) of pitch of the lifeboat.

In summary, the position and velocity of the center of gravity of the boat, the angle, and velocity and acceleration of pitch calculated by the two methods are in good agreement. There is a little difference between the values of translation acceleration calculated by the two methods. The difference is acceptable. This shows that the numerical algorithm in this paper has good accuracy.

The boat experiences three water impacts (bow impact, stern impact, bow impact) in the previous analysis. In order to analyze the local acceleration of different positions in the boat, the local acceleration of the bow, midship, and stern is calculated by Equation (1). Figure 11 shows the acceleration of the bow (a,b), midship (c,d), and stern (e,f) in the boat coordinate system. The acceleration values in all figures are the ratio of actual values to the acceleration of gravity g.

During the whole water entry process, the acceleration curves have three peaks because of three water impacts (Figure 11a,c) The first two peaks are close; there is a positive maximum acceleration in the direction of b_3 at the bow at the first impact as the first peak of the curve (Figure 11a). There is a positive maximum acceleration in the direction of b_3 at the midship at the second impact as the second peak of the curve (Figure 11c). There are two contrary extreme values of the acceleration at the stern in the direction of b_3 (Figure 11e): a negative extremum due to the first impact, and a positive extreme value due to the second impact. The negative maximum acceleration in the direction of b_1 at the bow, midship, and stern is because of the first impact (Figure 11b,d,f). The maximum acceleration is about 3 g in the direction of b_3, derived from the bow's first impact.

4.3. Qualitative Analysis

According to the numerical calculation method used in this paper, the local acceleration extremum is calculated under different initial conditions. The wave heights are 1, 2, 3, 4, and 5 m. By adjusting the initial phase of the wave, the points of water entry of the boat are shown in Figure 12. Figure 12a shows the ship heading the wave and Figure 12b shows the ship following the wave. When the ship is heading or following the wave, only the surge, heave, and pitch of the ship are considered. Other initial conditions (period of wave, wind velocity, ship's initial position, ship's initial velocity) are the same as in condition A.

Figures 13–17 show the trajectory and pitch angle of the lifeboat when the ship is heading the wave. Figure 18 shows the maximum value of the acceleration in the directions of b_1 and b_3 at the bow (Figure 18a,b), midship (Figure 18c,d), and stern (Figure 18e,f) of the boat at eight different points of water entry when the ship is heading waves. Figures 19–23 show the trajectory and pitch angle of the lifeboat when the ship is following waves. Figure 24 shows the maximum value of the acceleration in the directions of b_1 and b_3 at the bow (Figure 24a,b), midship (Figure 24c,d), and stern (Figure 24e,f) of the boat at eight different points of water entry when the ship is following waves. As there are no experimental data for comparison, the numerical calculation results may not be very accurate, but can be used in a qualitative analysis.

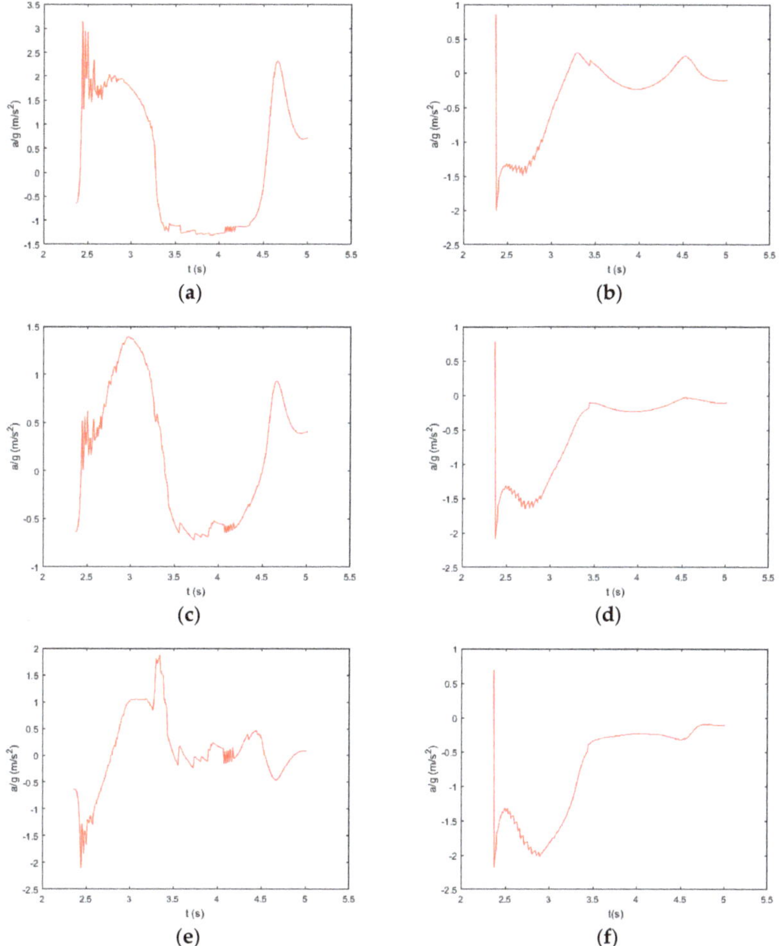

Figure 11. The acceleration of the bow, midship, and stern in the coordinate system of the lifeboat in the direction of b_3 (**a,c,e**) and in the direction of b_1 (**b,d,f**).

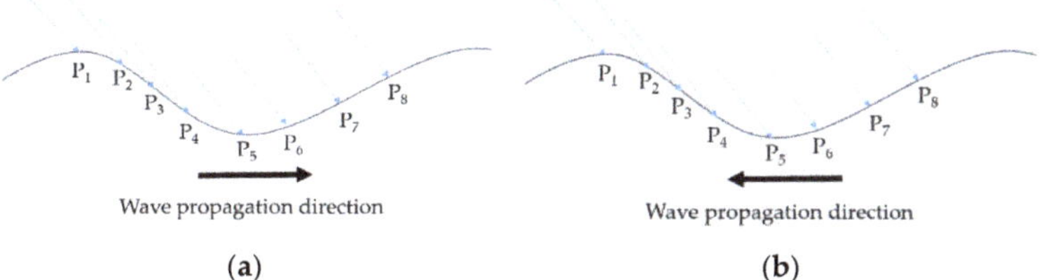

Figure 12. Eight different points of water entry of the boat when the ship is heading the wave (**a**) and following the wave (**b**).

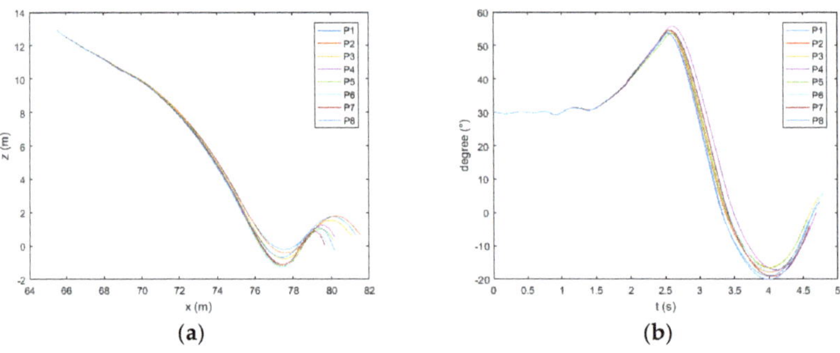

Figure 13. Trajectory (**a**) and pitch (**b**) of the lifeboat freefall when the wave height is 1 m.

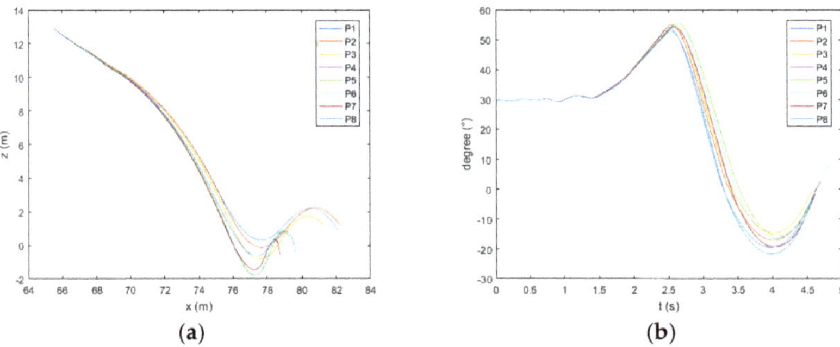

Figure 14. Trajectory (**a**) and pitch (**b**) of the lifeboat freefall when the wave height is 2 m.

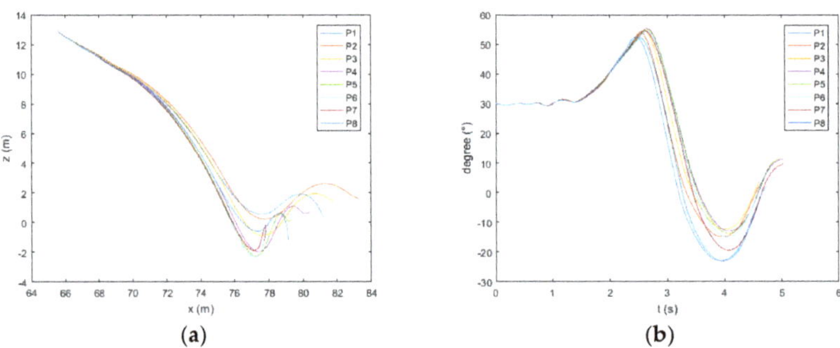

Figure 15. Trajectory (**a**) and pitch (**b**) of the lifeboat freefall when the wave height is 3 m.

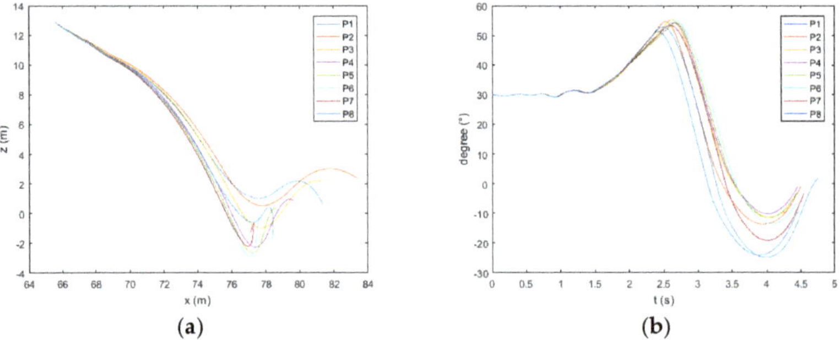

Figure 16. Trajectory (**a**) and pitch (**b**) of the lifeboat freefall when the wave height is 4 m.

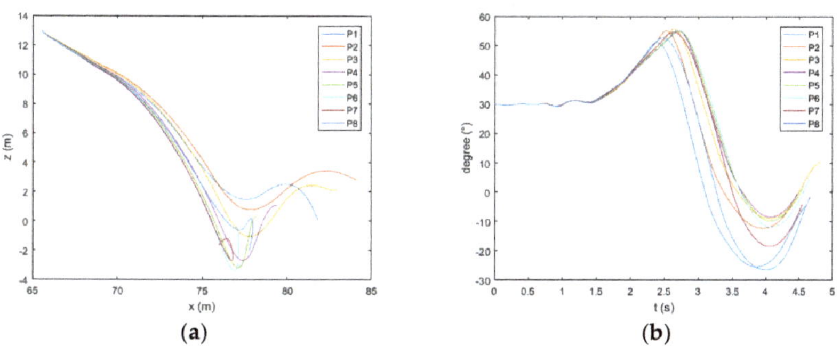

Figure 17. Trajectory (**a**) and pitch (**b**) of the lifeboat freefall when the wave height is 5 m.

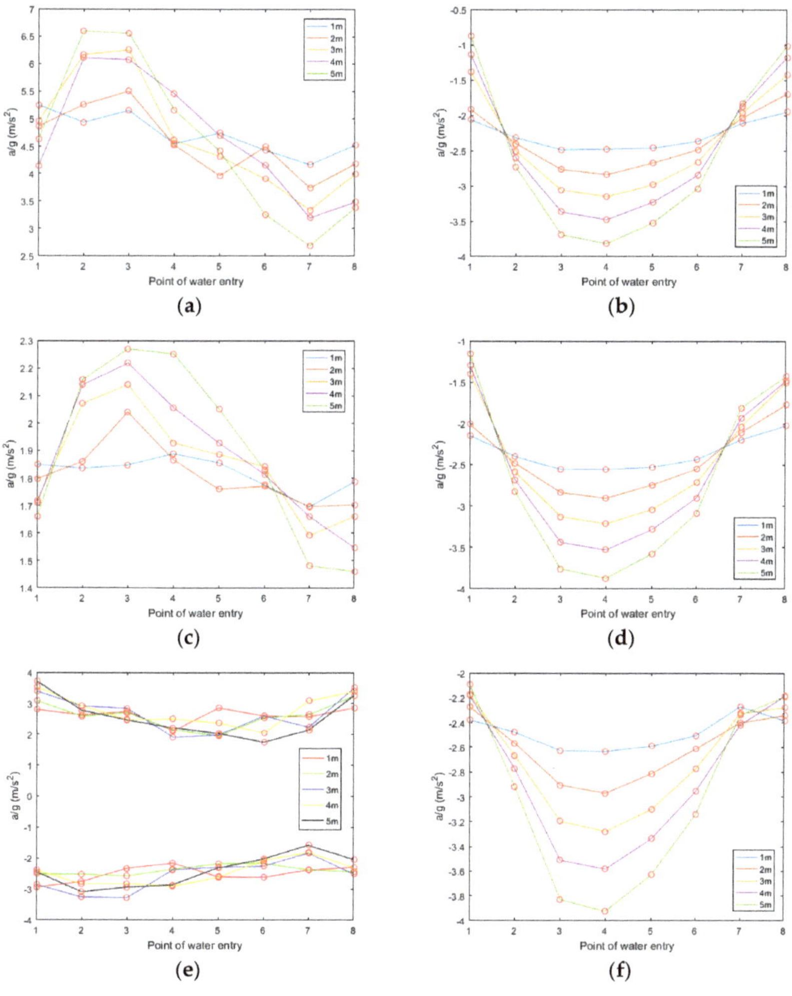

Figure 18. The maximum acceleration of the bow, midship, and stern in the coordinate system of the lifeboat in the direction of b_3 (**a,c,e**), and in the direction of b_1 (**b,d,f**), at different wave heights.

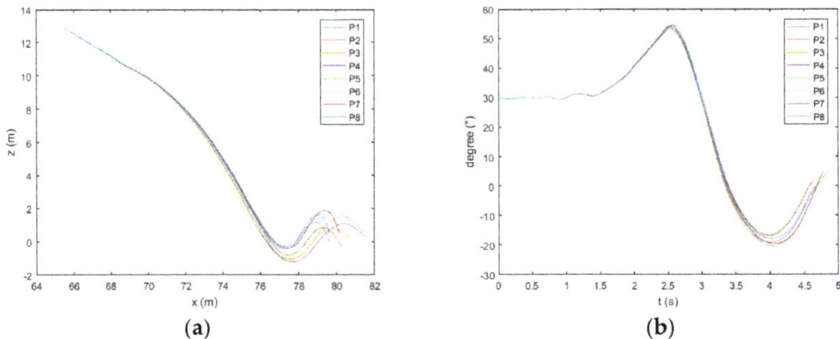

Figure 19. Trajectory (**a**) and pitch (**b**) of the lifeboat freefall when the wave height is 1 m.

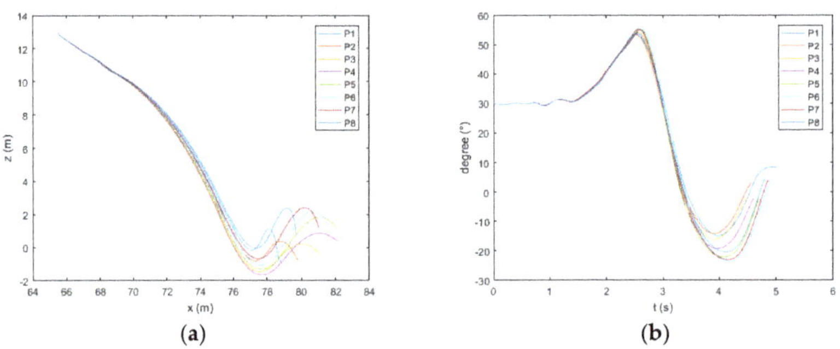

Figure 20. Trajectory (**a**) and pitch (**b**) of the lifeboat freefall when the wave height is 2 m.

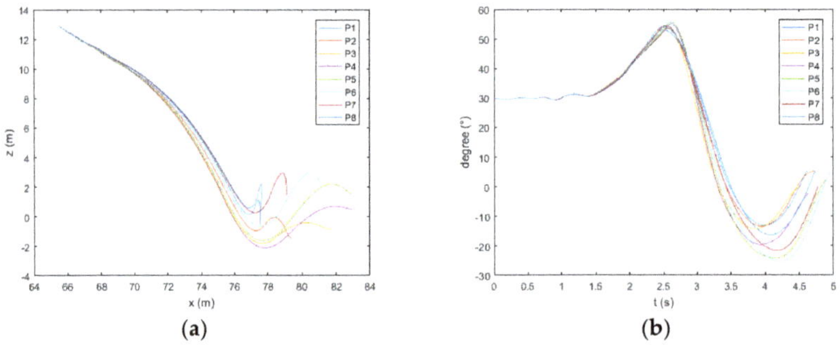

Figure 21. Trajectory (**a**) and pitch (**b**) of the lifeboat freefall when the wave height is 3 m.

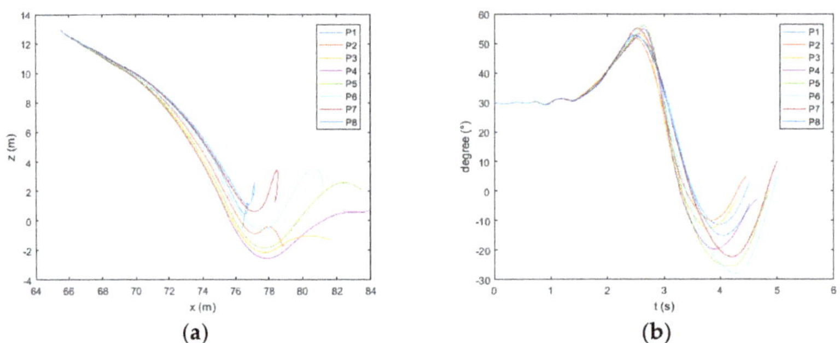

Figure 22. Trajectory (**a**) and pitch (**b**) of the lifeboat freefall when the wave height is 4 m.

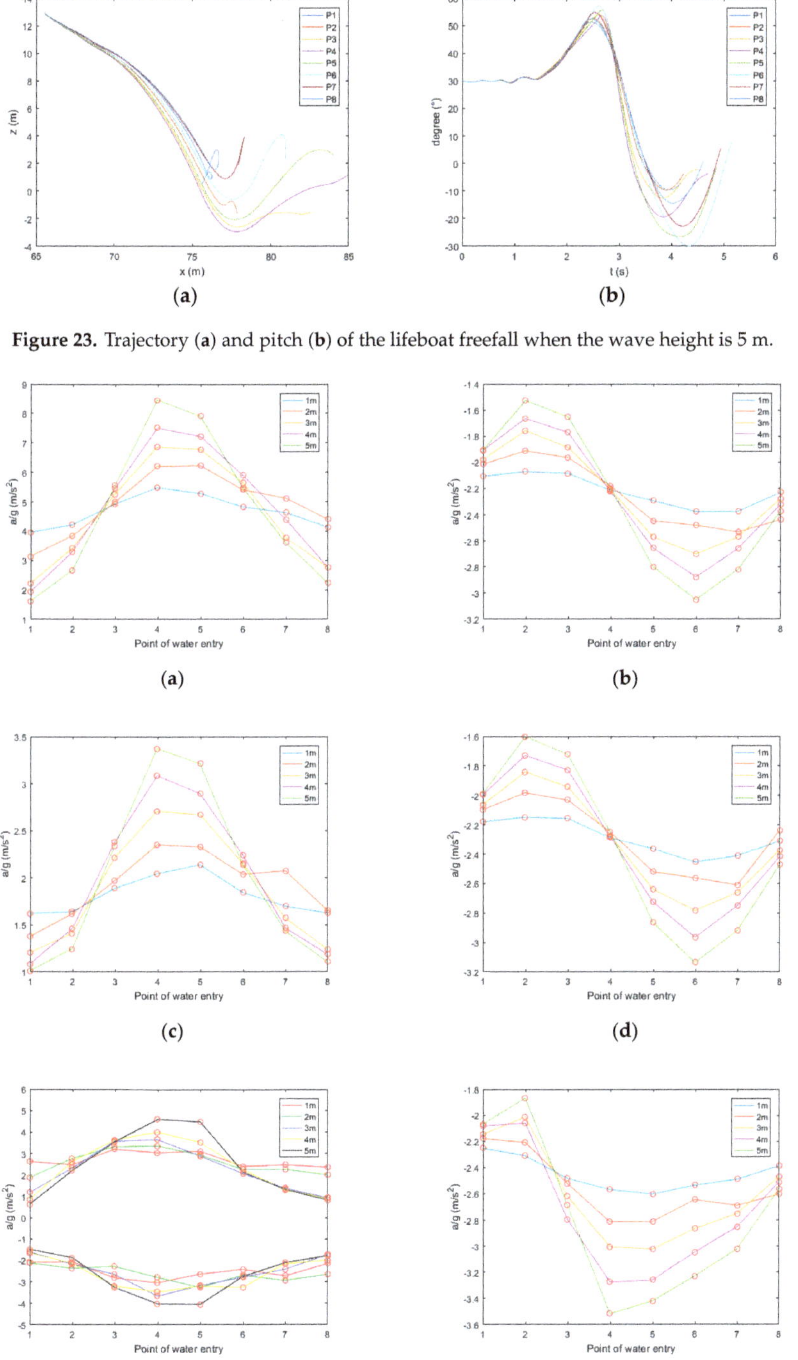

Figure 23. Trajectory (**a**) and pitch (**b**) of the lifeboat freefall when the wave height is 5 m.

Figure 24. The maximum acceleration of the bow, midship, and stern in the coordinate system of the lifeboat in the direction of b_3 (**a**,**c**,**e**), and in the direction of b_1 (**b**,**d**,**f**), at different wave heights.

The ship heads the wave. As shown in Figures 15–17a, when the wave height exceeds 3 m, the boat will move to the side of the ship after water entry at P7 and P8. As shown in Figures 16a and 17a, when the wave height exceeds 4 m, the same phenomenon appears at point P6. As shown in Figures 13–17b, at points P1 and P8, the pitch angle of the boat has a large range of change. As shown in Figure 18, in the direction of b_1, the maximum values of acceleration at the bow, midship, and stern of the boat appear at point P4; the minimum value generally appears at P1 and P8. The maximum values at the bow and midship are almost the same, slightly lower than the stern. In the direction of b_3, the maximum absolute values of acceleration at the bow and midship of the boat appear at points P3 and P2; the minimum value generally appears at points P7 and P8. There is no obvious rule about the values of acceleration at the stern; the acceleration is the largest at the bow and the smallest at the midship.

The ship follows the wave. As shown in Figures 21–23a, when the height of the wave exceeds 3 m, the boat will move to the side of the ship after water entry at points P1 and P8. As shown in Figures 22 and 23a, when the height of the wave exceeds 4 m, the same phenomenon appears at point P7. As shown in Figures 19–23b, the pitch angle of the boat has a large range of change at points P5 and P6. As shown in Figure 24, in the direction of b_1, the maximum values of acceleration at the bow and midship appear at points P6 and P7, and the minimum values generally appear at points P2 and P3. The maximum values of acceleration at the stern appear at points P4 and P5, and the minimum values generally appear at points P1 and P2. The maximum values at the bow and midship of the boat are almost the same, and slightly lower than for the stern. In the direction of b_3, the maximum values of acceleration at the bow and midship of the boat appear at points P4 and P5; the minimum value generally appears at points P1 and P8. The maximum absolute values of the two extremal values at the stern appear at P4 and P5. The minimum values generally appear at P1 and P8; the acceleration is the largest at the bow and the smallest at the midship.

Therefore, when the ship is heading the wave, P7 and P8 can be selected to enter the water at a low wave. In order to prevent the boat from moving to the side, one can choose point P1 at a high wave, which will withstand a large change in pitch. When the ship is following the wave, P1 and P8 can be selected at a low wave. In order to prevent the boat moving to the side, one can choose P2 and P3 at a high wave. In summary, it is safer to select points at the crest and behind the crest when entering the water.

5. Application

This paper applies an established mathematical model to a ship's lifesaving training system, improving the immersion and reality of the system. The system takes the Panama bulk carrier as a physical prototype, and uses 3D Studio Max (3ds Max) to build 3D models of the ship, lifeboat, and skid. The system constructs a virtual scene of a ship life-saving drill, and releases the lifeboat through a three-dimensional virtual operation, as shown in Figure 25.

 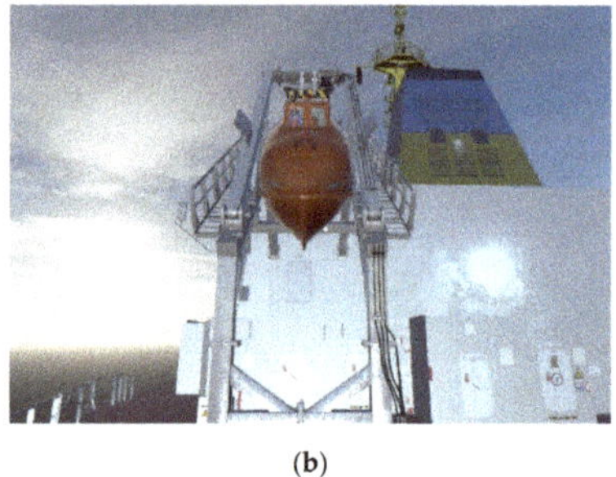

(a)   (b)

Figure 25. Scene of virtual human operating lifeboat (**a**) and lifeboat moving on the skid (**b**) in a ship's lifesaving training system.

6. Conclusions

This paper presents a computational model for the simulation of lifeboat freefall during its launching from a ship in rough seas. The model is applied to the ship's lifesaving training system. We can draw the following conclusions:

(1) The mathematical model in this paper can simulate the entire process of the water entry of the ship's lifeboat and can acquire the parameters of the boat's trajectory, pitch angle, velocity, local acceleration, etc.

(2) The results of a numerical simulation experiment are compared with the calculation results of the hydrodynamics software Star CCM+ at water entry under initial condition A. It shows that our numerical algorithm has good accuracy. The model can be applied to other ships by adjusting the parameters.

(3) Under different wave heights and two situations of the ship encountering waves, a qualitative analysis is performed to determine the safe point of water entry. It is safer to select points at the crest and behind the crest when entering the water.

(4) Since the motion of the boat on the skid is only three-degrees-of-freedom, the effects of the ship's roll, sway, and yaw are not considered; only the simulation experiments of the ship heading and following the wave are analyzed. Future research will consider the effect of the ship's roll, sway, and yaw.

Author Contributions: S.Q. led the writing of the manuscript, contributed to data analysis and research design, and serves as the corresponding author. H.R. supervised this study, contributed to the funding acquisition, and serves as the corresponding author. H.L. contributed to the investigation and corrected the format of the paper. All authors have read and agreed to the published version of the manuscript.

Funding: The authors are thankful for the financial support provided by National Natural Science Foundation of China (Grant No. 51679024), the Fundamental Research Funds for the Central Universities (Grant No. 3132020372), and Natural Science Foundation Guidance Project of Liaoning Province (Grant No. 2019-ZD-0152).

Conflicts of Interest: The authors declare no conflict of interest.

References

1. Nelson, J.K., Jr.; Waugh, P.J.; Schweickhardt, A.J. Injury criteria of the IMO and the Hybrid III dummy as indicators of injury potential in free-fall lifeboats. *Ocean Eng.* **1996**, *23*, 385–401. [CrossRef]

2. Billard, R.; Musharraf, M.; Veitch, B.; Smith, J. Using Bayesian methods and simulator data to model lifeboat coxswain performance. *WMU J. Mar. Aff.* **2020**, 1–18. [CrossRef]

3. Qiu, S.; Ren, H. Ship Life-Saving Training System Based on Virtual Reality Technology. In Proceedings of the 2018 IEEE 4th International Conference on Control Science and Systems Engineering, Wuhan, China, 21–23 August 2018; pp. 559–563.

4. Billard, R.; Smith, J.; Veitch, B. Assessing Lifeboat Coxswain Training Alternatives Using a Simulator. *J. Navig.* **2019**, 1–16. [CrossRef]

5. Re, A.S.; MacKinnon, S.; Veitch, B. Free-Fall Lifeboats: Experimental Investigation of the Impact of Environmental Conditions on Technical and Human Performance. In Proceedings of the 27th ASME International Conference on Offshore Mechanics and Arctic Engineering, Estoril, Portugal, 15–20 June 2008; pp. 81–88. [CrossRef]

6. Re, A.S.; Veitch, B. A comparison of three types of evacuation system. *Trans. Soc. Nav. Archit. Mar. Eng.* **2007**, *115*, 119–139.

7. Re, A.S.; Veitch, B. Experimental investigation of free-fall lifeboat performance. In Proceedings of the 17th 2007 International Offshore and Polar Engineering Conference, Lisbon, Portugal, 1–6 July 2007; pp. 3608–3615.

8. Hollyhead, C.J.; Townsend, N.C.; Blake, J.I.R. Experimental investigations into the current-induced motion of a lifeboat at a single point mooring. *Ocean Eng.* **2017**, *146*, 192–201. [CrossRef]

9. Hwang, J.K.; Chung, D.U.; Ha, S.; Lee, K.Y. Study on the safety investigation of the free-fall lifeboat during the skid-launching test. In Proceedings of the IEEE OCEANS 2012—Yeosu, Yeosu, Korea, 21–24 May 2012. [CrossRef]

10. Mørch, S.E.H.; Peric, M.; Schreck, E. Simulation of Lifeboat Launching under Storm Conditions. In Proceedings of the 6th International Conference on CFD in Oil and Gas, Trondheim, Norway, 10–12 June 2008.

11. Tregde, V.; Nestegrd, A. Prediction of Irregular Motions of Free-Fall Lifeboats during Drops from Damaged FPSO. In Proceedings of the ASME 2014 33rd International Conference on Ocean, Offshore and Arctic Engineering, San Francisco, CA, USA, 8–13 June 2014.

12. Ringsberg, J.; Heggelund, S.; Lara, P.; Jang, B.S.; Hirdaris, S.E. Structural response analysis of slamming impact on free fall lifeboats. *Mar. Struct.* **2017**, *54*, 112–126. [CrossRef]

13. Abrate, S. Hull Slamming. *Appl. Mech. Rev.* **2011**, *64*, 060803. [CrossRef]

14. Zakki, A.; Fauzan, A.; Bae, D.M.; Myung, D. The development of new type free-fall lifeboat using fluid structure interaction analysis. *J. Mar. Sci. Technol.* **2016**, *24*, 575–580. [CrossRef]

15. Nelson, J.K.; Hirsch, T.J.; Phillips, N.S. Evaluation of Occupant Accelerations in Lifeboats. *ASME J. Offshore Mech. Arct. Eng.* **1989**, *111*, 344–349. [CrossRef]

16. Karman, V. *The Impact on Seaplane Floats during Landing*; National Advisory Committee for Aeronatics, NO.321; NACA: Seattle, WA, USA, 1929; pp. 1–8.

17. Boef, W.J.C. Launch and impact of free-fall lifeboats. Part I. Impact theory. *Ocean Eng.* **1992**, *19*, 119–138. [CrossRef]

18. Boef, W.J.C. Launch and impact of free-fall lifeboats. Part II. Implementation and applications. *Ocean Eng.* **1992**, *19*, 139–159. [CrossRef]

19. Arai, M.; Khondoker, M.R.H.; Inoue, Y. Water Entry Simulation of Free-fall Lifeboat (First Report: Analysis of Motion and Acceleration). *Nihon Zosen Gakkai Ronbunshu* **1995**, *178*, 193–201. [CrossRef]

20. Arai, M.; Khondoker, M.R.H.; Inoue, Y. Water Entry Simulation of Free-fall Lifeboat (2nd Report: Effects of Acceleration on the Occupants). *Nihon Zosen Gakkai Ronbunshu* **1995**, *179*, 205–211.

21. Khondoker, M.R.H.; Khalil, G. Effect of guide rail on the motion and acceleration of a free-fall lifeboat. *Indian J. Eng. Mater. S* **1998**, *5*, 49–54.

22. Khondoker, M.R.H. Effects of Launching Parameters on the Performance of a Free-Fall Lifeboat. *Nav. Eng. J.* **2009**, *110*, 67–73. [CrossRef]

23. Khondoker, M.R.H.; Arai, M. A comparative study on the behaviour of free-fall lifeboat launching from a skid and from a hook. *Proc. Inst. Mech. Eng. Part C J. Mech. Eng.* **2000**, *214*, 359–370. [CrossRef]

24. Karim, M.M.; Iqbal, K.S.; Khondoker, M.R.H. Numerical investigation into the effect of launch skid angle on the behaviour of free-fall lifeboat in regular waves. *Trans. RINA Part B1 Intl. J. Small Craft Tech.* **2009**, *151*, 37–48.

25. Karim, M.M.; Iqbal, K.S.; Khondoker, M.R.H.; Rahman, S.M.H. Influence of falling height on the behavior of skid- launching free-fall lifeboat in regular waves. *J. Appl. Fluid Mech.* **2012**, *4*, 77–88.

26. Raman-Nair, W.; White, M. A model for deployment of a freefall lifeboat from a moving ramp into waves. *Multibody Syst. Dyn.* **2013**, *29*, 327–342. [CrossRef]

27. Faltinsen, O.M. *Sea Loads on Ships and Offshore Structures*; Cambridge University Press: Cambridge, UK, 1990; pp. 300–306.

28. Sumer, B.M.; Fredsoe, J. *Hydrodynamics around Cylindrical Structures*; World Scientific: Singapore, 1997; pp. 123–201.

29. Dymarski, C.; Dymarski, P. Computational simulation of motion of a rescue module during its launching from ship at rough sea. *Pol. Marit. Res.* **2014**, *21*, 54–60. [CrossRef]

30. Kane, T.R.; Likins, P.W.; Levinson, D.A. *Spacecraft Dynamics*; McGraw Hill Inc.: New York, NY, USA, 1983; p. 30.

31. Uğur, G.; Bülent, Ö.; Yilmaz, T. A spring force formulation for elastically deformable models. *Comput. Graph.* **1997**, *21*, 335–346. [CrossRef]

32. McMillan, A.J. A non-linear friction model for self-excited vibrations. *J. Sound Vib.* **1997**, *205*, 323–335. [CrossRef]

33. Jia, X.; Yan, G.Y. *Ship Motion Mathematical Model: The Mechanism Modeling and Identification Modeling*; Dalian Maritime University Press: Dalian, China, 1999; pp. 49–138.

34. Mo, J. Numerical Simulation of Ship Maneuvering Motion with Six Degree of Freedom in Waves. Master's Thesis, Harbin Engineering University, Harbin, China, 2009.

35. Wu, G. Hydrodynamic force on a rigid body during impact with liquid. *J. Fluid Struct.* **1998**, *12*, 549–559. [CrossRef]

36. Cointe, R.; Fontaine, E.; Molin, B.; Scolan, Y.M. On energy arguments applied to the hydrodynamic impact force. *J. Eng. Math.* **2004**, *48*, 305–319. [CrossRef]
37. Casetta, L.; Pesce, C.P.; Flávia, M.D.S. On the hydrodynamic vertical impact problem: An analytical mechanics approach. *Mar. Syst. Ocean Technol.* **2011**, *6*, 47–57. [CrossRef]
38. Scolan, Y.; Korobkin, A.A. Energy distribution from vertical impact of a three dimensional solid body onto the flat free surface of an ideal fluid. *J. Fluids Struct.* **2003**, *17*, 275–286. [CrossRef]

Article

Numerical Simulation of a Polar Ship Moving in Level Ice Based on a One-Way Coupling Method

Bao-Yu Ni *, Zi-Wang Chen, Kai Zhong, Xin-Ang Li and Yan-Zhuo Xue

College of Shipbuilding Engineering, Harbin Engineering University, Harbin 150001, China;
chenzw@hrbeu.edu.cn (Z.-W.C.); zhongkai.kai@foxmail.com (K.Z.); Lixinang@hrbeu.edu.cn (X.-A.L.);
xueyanzhuo@hrbeu.edu.cn (Y.-Z.X.)
* Correspondence: nibaoyu@hrbeu.edu.cn

Received: 31 July 2020; Accepted: 4 September 2020; Published: 7 September 2020

Abstract: In most previous ice–ship interaction studies involving fluid effects, ice was taken as unbreakable. Building breakable level ice on water domain is still a big challenge in numerical simulation. This paper overcomes this difficulty and presents a numerical modeling of a ship moving in level ice on the water by using a one-way CFD-DEM (computational fluid dynamics-discrete element method) coupling method. The detailed numerical processes and techniques are introduced. The ice crack propagation process including radial and circular cracks have been observed. Numerical results are compared with previous experimental data and good agreement has been achieved. The results show that water resistance is an order of magnitude smaller than ice resistance during the ice-breaking process. Ice resistance shows strong oscillation along with ice failure process, which are affected by ship speed and ice thickness significantly.

Keywords: one-way coupling; CFD-DEM; ice resistance; ice crack

1. Introduction

With global warming, the sea ice has melted gradually [1], making the Arctic shipping routes more navigable [2]. More and more ships are beginning to sail in the polar ice regions (NSRA) [3]. Different from open water areas, ships would encounter not only water loads but also ice loads in the polar ice region. In particular, the effect of ice on ships presents different character from that of water, which affects the motion response and structural safety of ships. Therefore, it is of significance to understand ice–water–ship interaction and predict the ice and water loads of a polar ship more accurately, which has become one of the core problems for the design and operation of polar ships [4,5].

The problem of ice–water–ship interactions is very complex [5]. To solve this problem, various experimental, analytical, and numerical methods have been developed. Experimental studies, including field tests [6–8] and model tests in ice tanks [9–14], are generally viewed as reliable; however, they are strict regarding the experimental facilities and methods used and are quite expensive so cannot be done easily. Analytical methods have a relatively long history from simple models to complex models [15–20]. However, they are usually based on many simplifications of body shape and ice model, which made them hard to extend to complex ship hulls and actual ice conditions. Numerical methods [21–25] have been developing quickly in recent years with the rapid development of computer capacity. In contrast to experimental and analytical methods, numerical methods are easy to extend to various body configurations.

From the perspective of ice mechanics, the finite element method (FEM) and discrete element method (DEM) are two mainstream numerical methods used to simulate ice–ship interactions. FEM is a relatively mature method to solve the continuum mechanics problem and has been used in ice–ship interaction recently, especially adopted by various commercial software. Valanto [26] adopted FEM to calculate the ice resistance of a ship moving in a level ice channel. Liu [27] developed a material model for icebergs and inserted it into the FEM software LS-DYNA and simulated the collision between an iceberg and a ship structure. Kim et al. [11,28] simulated the resistance performance of a cargo ship sailing in a broken ice channel using LS-DYNA software. The numerical results were compared with the results of the nonfrozen model ice test in a water tank and the cut ice test in an ice tank, and agreements were both achieved. On the basis of FEM, Zhou et al. [29] adopted the cohesive element method to simulate collision between ice and propeller.

There have been several attempts to consider water effects (mainly buoyancy effects) in ice–ship interaction by using FEM. Guo et al. [30] adopted the arbitrary Lagrange–Euler (ALE) approach in LS-DYNA software to consider the fluid–structure coupling and calculated the resistance of an ice-going container ship in a broken ice channel. Ni et al. [31] adopted a similar method to calculate the total resistance of a ship turning in a level ice region, with and without water effects. It was found that the existence of water increased the ice resistance of the ship on all directions whether in straight or rotational motion. More work on this topic could refer to the review from Xue et al. [32].

On the other hand, DEM has become a popular method in simulating ice dynamics, by virtue of its granular characteristics, and has been extensively used in simulating ice–structure interaction [33]. Hansen and Løset [34] applied a two-dimensional disk discrete element to simulate broken ice, by using a linear viscoelastic force model between ice elements and studied broken ice–ship collision firstly. Zhan et al. [35] and Lau et al. [36] adopted DEM to study the ice force and moments of a ship maneuvering in level ice. Morgan [37] calculated the interaction between level ice and structures by using an open source DEM software LIGGGHTS. Cai and Ji [38] used DEM to simulate the navigation process of ships in level ice and discussed the influence of ship speed and ice thickness on ship resistance. Further work extended the interaction between level ice and conical structures [39–41]. Gong et al. [42] modeled an ice ridge by using DEM and calculated the ice force of a ship colliding on an unconsolidated and deformable ice ridge.

Although DEM is used extensively during ice–ship interaction, it is difficult for DEM itself to include water effect. As an alternative of neglecting water effect directly, a common treatment is to add buoyancy force and/or drag force of water on the discrete ice elements based on empirical equations [33]. However, this simplified treatment cannot account for water effects fully, such as ship-generated wave effects on ice movement [43]. To solve this problem, a combined CFD and DEM method has developed rapidly recently, which solves the fluid flow by using the Euler method and ice particle movement by using Lagrangian method.

Currently, STAR-CCM+ software has developed a DEM module, which provides a combined CFD-DEM method to simulate ice–ship interaction. Vroegrijk [44] adopted STAR-CCM+ to simulate the movement of a ship in a broken ice channel by using a combined CFD-DEM method. By comparing numerical results with measured data, the combined CFD-DEM model was validated. Huang et al. [43] simulated a ship advancing in floating ice floe regions by using the combined CFD-DEM approach based on STAR-CCM+. They developed two algorithms for generating probability-distributed ice floe fields. The influences of ship speed and ice concentration on ice resistance were investigated. Luo et al. [45] calculated the resistance of an ice-strengthened bulk carrier in a brash ice channel using the CFD-DEM coupling method. Unbreakable brash ice blocks were constructed by bonding ice particles in various shapes. Numerical results were compared with HSVA (Hamburgische Schiffbau Versuchsanstalt) ice tank results and good agreements were achieved.

All the previous studies have proved that the CFD-DEM coupling method is feasible in simulating ice–water-ship interaction. However, to our best knowledge, all the previous CFD-DEM studies just simulated unbreakable ice, not considering broken ice channel, floating ice floes, or brash ice regions. It seems little work has simulated a ship moving in a breakable level ice region by using the CFD-DEM method. On the other hand, a ship moving in breakable level ice with water effects has its own distinction. It needs to consider not only the ice breakup and crack propagation under ship impact but also the drift and overturn of ice fragments under water effects. Therefore, it is of significance to explore this process by using the CFD-DEM approach. This also forms the prime motivation and distinct innovation of this paper. Based on STAR-CCM+, this paper develops a method to bond a breakable level ice using DEM. A one-way CFD-DEM coupling framework is also built and solved. An ice breaker advancing in the level ice and the resistance of the ship under various parameters are studied. Numerical results are further compared with experimental data.

2. Computational Modeling

The finite volume method (one of the CFD methods) is used to describe fluid flow, and DEM is used to describe ice particle movement. In this part, the modeling of fluid phase and solid phase is described, respectively, and the one-way coupling scheme is illustrated subsequently. Considering that the establishment of breakable level ice is innovative, its modeling process is stressed.

2.1. Fluid Model (CFD)

The fluid domain is governed by the equation of continuity and Navier–Stokes equation for an incompressible fluid, as expressed in Equations (1) and (2) [46],

$$\frac{\partial \eta \rho_f}{\partial t} + \nabla \cdot (\eta \rho_f \boldsymbol{u}_f) = 0 \tag{1}$$

$$\frac{\partial (\eta \rho_f \boldsymbol{u}_f)}{\partial t} + \nabla \cdot (\eta \rho_f \boldsymbol{u}_f \boldsymbol{u}_f) = \eta \rho_f g - \eta \nabla p + v_f \nabla^2 (\eta \boldsymbol{u}_f) - \boldsymbol{R}_{pf} \tag{2}$$

where η is the volume fraction of the fluid term in the control volume and it satisfies $\eta = 1 - \varepsilon = 1 - (\sum_{i=1}^{n} \varphi_{pi} V_{pi})/V_{cell}$, in which ε is the volume fraction of the particle term in the control volume, or the commonly called 'void ratio', V_{cell} is the total volume of the calculated cell, n is the number of the particles in the cell, φ_{pi} is the weighting coefficient of the particle i according to the ratio of the particle volume in the cell to its total particle volume V_{pi}. $\boldsymbol{u}_f$ is the fluid velocity, p is the fluid pressure, ρ_f is the fluid density, v_f is the kinematic viscous coefficient of the fluid, and $\boldsymbol{R}_{pf}$ represents the momentum exchange between the fluid phase and the particle phase. Standard $k - \varepsilon$ turbulence model is selected in this paper. For more details on turbulent models, refer to reference [46].

There are many methods to calculate the momentum exchange term $\boldsymbol{R}_{pf}$ and a common method is introduced here [47], which is relatively accurate and easy to implement.

$$\boldsymbol{R}_{pf} = \frac{\left|\boldsymbol{F}_{pf}\right|}{\left|\boldsymbol{u}_p - \boldsymbol{u}_f\right|} (\boldsymbol{u}_p - \boldsymbol{u}_f), \tag{3}$$

where $\boldsymbol{F}_{pf}$ is the interaction force between the fluid and particle, which is obtained by integrating the pressure of the fluid on the particle, and $\boldsymbol{u}_p$ is the particle velocity.

The volume of fluid (VOF) method is used to deal with the interface between water and air. The VOF model defines α_w and α_a as the water volume fraction and the air volume fraction, respectively. Considering volume fraction of the fluid term $\eta = (V_w + V_a)/V_{cell}$, where V_w and V_a are water volume and air volume in the cell, α_i is defined as $\alpha_i = V_i / \sum_{i=1}^{2} V_i$, where phase $i = 1$ and 2 denote water and air, respectively. Thus, in a control volume, it satisfies

$$\sum_{i=1}^{2} \alpha_i = 1 \tag{4}$$

If the cell is full of water, one has $\alpha_w = 1$. If there is no water in the cell, one has $\alpha_w = 0$. Otherwise, one has $0 < \alpha_w < 1$. The interface is tracked by solving the volume fraction transport equation [48], as below:

$$\frac{\partial}{\partial t}\int_V \alpha_i \eta dV + \oint_A \alpha_i \eta v \cdot nds = \int_V S_{\alpha_i} dV - \int_V \frac{1}{\rho_i} \nabla \cdot (\alpha_i \eta \rho_i v_{dr,i}) dV \tag{5}$$

where A is the surface of the control volume, n is the unit normal vector of the surface, S_{α_i} is user-defined source term for phase i, $v_{dr,i}$ is the diffusion velocity for phase i.

Ship is viewed as a rigid body upright floating in the water at a given draught in this simulation, which can move at a given speed in just one direction, and the hull form comes from an icebreaker. The principal dimensions of the hull are shown in Table 1.

Table 1. The main geometric parameters of the hull of an icebreaker.

Parameter	Value	Unit
Length overall	124	m
Breadth	22	m
Draught	7.8	m
Stem angle	20	Deg.
Waterline angle	34	Deg.
Flare angle	53	Deg.

In numerical simulations of a ship moving in a brash ice region, to make the numerical modeling easier, part of previous work [43] kept the ship stationary, set water moving in a constant velocity against the hull, and released the ice particles into the water with the same initial velocity. In this way, water carried ice particles towards the ship, which was sometimes used as an alternative in simulating a ship moving in a brash ice region. However, this does not work for level ice, given that the push of water on the level ice is not large enough to support the collision between ship and level ice, inducing results wildly inconsistent with reality. In order to solve this problem, it is necessary to set the ship to move in and break the level ice floating on the water.

The overset grid technique is used in this paper to achieve this process. The overset grid is composed of two sets of grids, the background grid, and the moving grid. Information is exchanged between these two sets of grids using interpolation methods. In this paper, the linear interpolation method is used. The grid accepting information is named as 'acceptor', while the grid providing information is named as 'provider'. In linear interpolation formulation, one has

$$\Phi_{\text{acceptor}} = \sum \chi_i \Phi_{\text{provider}\_i}, \tag{6}$$

where Φ is the physical quantity, χ is weighting factor, and the subscript i denotes the number of the provider. As shown in Figure 1, the physical quantity of a grid C comes from those of the neighboring grids N_1, N_2, and N_3 of the same set of grids and the neighboring grids N_4, N_5, and N_6 of the other set of grids. The equation is solved iteratively until the residual error is small enough. During the construction, it should be noted that the overlapping area between two sets of grids should contain at least 4–5 grid cell layers, and the boundary mesh sizes should be similar, which reduces the error of interpolation as much as possible. More information about overset grid technique can refer to Hadzic [49].

The flow field is truncated as far as possible in order to reduce the influence of the outer boundary of the flow field. In this paper, the flow field is taken as 600 m × 600 m × 450 m. The fluid domain is meshed with a trimmed meshing model and boundary layer grids are divided around the ship surface, as shown in Figure 2. According to the ship speed, the Reynolds number in this paper is $1.94 \times 10^8 \leq Re \leq 4.54 \times 10^8$, which is a high-Reynolds-number problem. According to the user's manual of STAR-CCM+ [48], y+ for a high-Reynolds-number problem is recommended to be larger than 30 in standard $k - \varepsilon$ turbulence model. Thus, in this paper, the thickness of the first layer grid of the boundary layer is 8×10^{-4} m, and the y+ value is chosen around 60 based on our numerical experience. The meshes on the ship surface are refined near waterline, stem, and stern area, as shown in Figure 2a. In the flow domain, grids are refined near the water surface and in Kelvin wave area, as shown in Figure 2b,c.

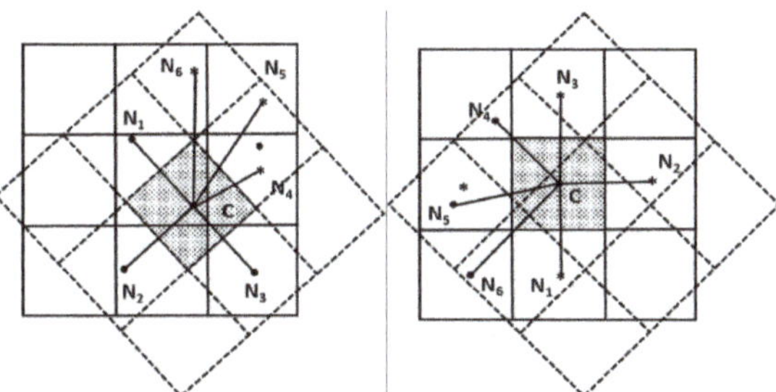

Figure 1. Sketch map of the information exchange between two sets of grids.

(**a**) The surface on the hull

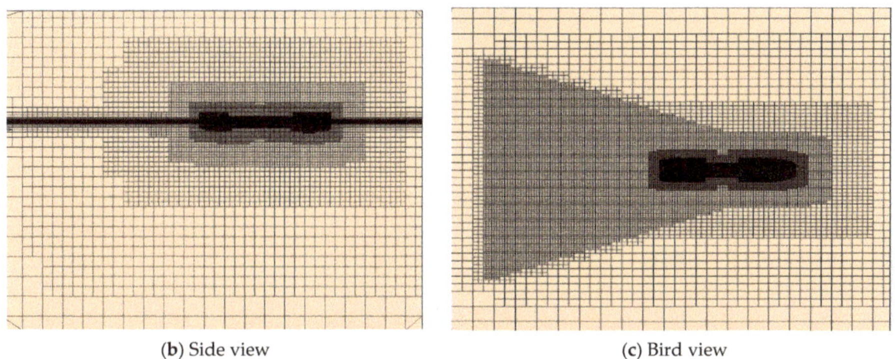

(**b**) Side view   (**c**) Bird view

Figure 2. Mesh distribution of the numerical model with meshes on (**a**) the surface on the hull, and meshes of the flow domain in (**b**) side view and (**c**) bird view, respectively.

The truncated surfaces of the flow domain are defined as front, rear, left, right, top, and bottom surfaces based on the ship, where the front surface is the surface that ship bow points. The boundary conditions of the flow domain include a velocity inlet for the rear, left, right, top, and bottom surfaces, and a pressure outlet for the front surface. Rigid wall boundary condition is exerted on the ship surface, which satisfies both impermeable condition and nonslip condition, namely,

$$u_f \cdot n = u_s \cdot n \tag{7}$$

$$u_f \cdot \tau = u_s \cdot \tau \tag{8}$$

where u_f and u_s are the fluid velocity and ship velocity, respectively, and n and τ are unit normal and tangential vectors of the ship surface. On the ice surfaces, no boundary conditions are satisfied because the coupling method is based on force and moment exchange balance and not the interface track or capture method [50]. More details about the coupling method will be discussed in Section 2.3. The initial conditions of the flow field are stationary.

An appropriate number of grids should be chosen to ensure the calculation accuracy and save calculation time. After grid independence verification based on the water resistance of the ship in 5 kn, as shown in Table 2, the overall grid is taken about 1.63 million, namely mesh3 in Table 2.

Table 2. Water resistance of the ship in 5 kn under different mesh parameters.

	Basic Size/m	Total Mesh Number	Resistance/N	Resistance Deviation (with Finest Mesh)
mesh1	7	6.42×10^5	1.21×10^5	9.93%
mesh2	4.95	9.82×10^5	1.10×10^5	5.33%
mesh3	3.5	1.63×10^6	1.05×10^5	1.01%
mesh4	2.48	2.80×10^6	1.04×10^5	-

2.2. Ice Model (DEM)

Ice is modeled by using DEM, and the governing equation is Newton's second law. Taking a particle element i as an example, one has

$$m_i \frac{du_{p,i}}{dt} = \sum_j F_{c,ij} + \sum_k F_{lr,ik} + F_{pf,i} + F_{g,i},$$

(9)

$$I_i \frac{d\omega_i}{dt} = \sum_j (M_{t,ij} + M_{r,ij}),$$

(10)

where subscript i, j, and k denote the particle number, m is the particle mass, F_c and F_{lr} are the contact and non-contact forces between particles, respectively, F_{pf} is the interaction force between the fluid and particle, same to which in Equation (3), and F_g is the gravity force of the particle. I is the moment of inertia of the particle, ω is angular velocity of the particle, M_t and M_r are the moments of the sliding friction and the rolling friction, respectively. Considering F_{lr}, such as electromagnetic force or van der Waals forces, is just involved in special cases, F_{lr} is not included in this paper. Here the method to calculate the contact force F_c is introduced briefly.

There are many methods to calculate the contact force F_c, and a common model 'Hertz–Mindlin' collision model is adopted here [47]. In the Hertz–Mindlin collision model, the contact force F_c between two spheres, A and B, are described by the following set of equations:

$$F_c = F_n \boldsymbol{n} + F_t \boldsymbol{\tau},$$

(11)

where subscript n and t denote normal and tangential components, respectively, and $\boldsymbol{n}$ and $\boldsymbol{\tau}$ are unit vector in normal and tangential directions, respectively. Normal and tangential forces F_n and F_t are expressed as:

$$F_n = -K_n d_n - N_n v_n,$$

(12)

$$F_t = \begin{cases} -K_t d_t - N_t v_t, & |K_t d_t| < |K_n d_n| C_{fs} \\ |K_n d_n| C_{fs} d_t / |d_t|, & |K_t d_t| \geq |K_n d_n| C_{fs} \end{cases},$$

(13)

where d and v are relative displacement and velocity, respectively. Figure 3 provides the normal and tangential relative displacements d_n and d_t, and they are obtained by the integral of normal and tangential relative velocities v_n and v_t, which are obtained by using Equation (9). K and N are spring stiffness and damping, respectively, and C_{fs} is a static friction coefficient, which is taken as 0.15 according to the recommended value for sea ice [51]. K and N are not constant in Hertz-Mindlin collision model and they are calculated each step by using $K_n = \frac{4}{3} E_{eq} \sqrt{d_n R_{eq}}$, $K_t = 8 G_{eq} \sqrt{d_t R_{eq}}$, $N_n = \sqrt{5 K_n M_{eq}} N_{n\ damp}$,

$N_t = \sqrt{5K_t M_{eq}} N_{t\ damp}$, in which equivalent Young modulus $E_{eq} = \frac{1}{\frac{1-v_A^2}{E_A}+\frac{1-v_B^2}{E_B}}$ with v as Poisson's ratio and subscript A and B denoting particles number, equivalent radius $R_{eq} = \frac{1}{\frac{1}{R_A}+\frac{1}{R_B}}$, equivalent shear modulus $G_{eq} = \frac{1}{\frac{2(2-v_A)(1+v_A)}{E_A}+\frac{2(2-v_B)(1+v_B)}{E_B}}$, equivalent mass $M_{eq} = \frac{1}{\frac{1}{M_A}+\frac{1}{M_B}}$, normal damping coefficient $N_{n\ damp} = \frac{-\ln(C_{n\ rest})}{\sqrt{\pi^2+\ln(C_{n\ rest})^2}}$ and tangential damping coefficient $N_{t\ damp} = \frac{-\ln(C_{t\ rest})}{\sqrt{\pi^2+\ln(C_{t\ rest})^2}}$, in which $C_{n\ rest}$ and $C_{t\ rest}$ are the normal and tangential coefficients of restitution, which are taken as 0.5 [45].

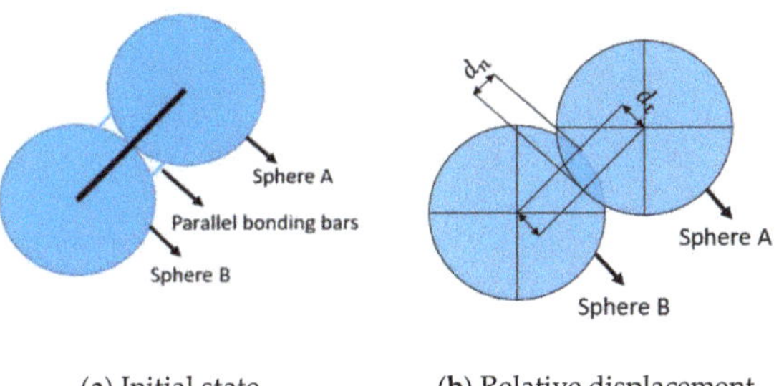

(**a**) Initial state (**b**) Relative displacement

Figure 3. Sketch map of parallel bonding model, where (**a**) is the initial state and (**b**) is the relative displacement of two particles in movement.

The generation of level ice is one of the main difficulties in the process of ship-level ice–water interaction. Care must be taken to build breakable level ice in DEM. In STAR-CCM+, there are a variety of particle models in the DEM module, and the composite particle model is widely used because it can be freely combined into the shape of the target object. However, the composite particle model cannot be broken, which is inconsistent with the properties of the level ice in this paper. On the other hand, the particle clumps model can realize the fracture between particles, but it requires initial input of the position and size of each particle to define the initial shape of the object. It would be a huge amount of work to generate such a model of level ice with tens of thousands of particles. As an alternative, in this paper, a simple spherical particle model is adopted, and the bonding bond and fracture criterion between the particles are applied to establish the model of breakable level ice, which will be proved to be successful and relatively easy. The model is chosen as viscoelastic with ice density of 900 kg/m^3 and Young's modulus of 1 Gpa.

The detailed processes are described. It includes two main steps: 'particle generation' and 'particle bonding'. For the first step, particles are generated based on seed points. For the success of bonding a level ice, it is necessary to ensure that the particles are relatively compact and no large gaps exist. Generating particles with a conventional 'part ejector' does not guarantee needed particle density. To solve this problem, a 'random ejector' with the maximum filling form is used. This 'random ejector' injects a specified number of particle seeds with small starting volume into the space in random positions, in which the number is obtained by the area of the space and the prescribed radius of the particle. Then, spherical particles start to grow from the seeds. As the maximum filling form is adopted, the spherical particles grow until there is no room left. For the 'random ejector', the space to fill must be real. A direct idea is to build a real space where the level ice locates, before filling it. However, building such a real space in the fluid domain will affect the computation of the fluid. To avoid this disturbance, a technique was developed to create a new space in a spare fluid domain. This space and this fluid domain are not used for calculation, but just for generating ice particles. After the particles are generated by 'random ejector' with the maximum filling form there, the positions and sizes of these particles can be derived to make a table, which consists of all the information of the particles. Then, the original fluid domain is returned, and another ejector, 'table ejector', is used to generate ice particles at the above locations and sizes in the original fluid domain. There is another problem with this method. The 'table ejector' does not generate ice particles simultaneously, which makes the early ejected ice particles deviate from initial positions when all the ice particles have been ejected, inducing the bonding step to fail. Therefore, another technique was put forward, which is to use a custom function to generate these particles simultaneously. In this way, all the particles are kept at their initial positions before bonding, laying a good foundation for a successful bonding step subsequently. These two techniques above are of significance to this step.

After all the ice particles have been released to their prescribed positions in the first step 'particle generation', they need to be bonded together as a level ice. Thus, the second step, 'parallel bonding', is implemented. As shown in Figure 3a, the parallel bonding model uses the concept of massless bars connecting a pair of bonded particles. The bars can transmit force and torque between particles. Alongside force Equation (11), torque equation is expressed as:

$$M_c = M_n \boldsymbol{n} + M_t \boldsymbol{\tau} \tag{14}$$

where M_n and M_t denote normal and tangential torque components, respectively. One has

$$\Delta M_n = -K_t J \Delta \Omega_n \tag{15}$$

$$\Delta M_t = -K_n L \Delta \Omega_t \tag{16}$$

where Δ is increment, $J = \frac{1}{2}\pi R_{eq}^4$ and $L = \frac{1}{4}\pi R_{eq}^4$, K is still spring stiffness and the value is the same to that in Equations (12) and (13), Ω is relative angular displacement.

Based on 'parallel bonding' model, a simple rupture model is used. It means that once the tensile or shear stresses between particles exceeds the maximum limits, the bar breaks and corresponding force and torque disappear. Then the particles separate, and cracks generate. In this way, a breakable ice model becomes available. It should be stated that the interparticle tensile and shear bonding strength of ice in the DEM model are affected by many factors, including particle shape, particle radius, number of layers, and arrangement modes (regular or random arrangement) [52]. According to previous studies [53], the interparticle tensile and shear bonding strength of ice in random arrangement mode should be larger than macroscopic measured tensile and shear strength of sea ice [54], so that the simulated ice behaviors including compression and bending agree with those of real ice. Based on the suggested values of reference [53] and our numerical experience, interparticle tensile and shear bonding strength were both taken as 3.0 Mpa. The final successful bonded level ice model with two layers of ice particles is shown in Figure 4, which is also adopted in this paper.

Figure 4. Numerical model of bonded level ice of two layers, with the upper part in bird's eye view and the lower part in side view.

2.3. Coupling Scheme

As mentioned in Luo et al. [45] and Ni et al. [5], there are usually two kinds of coupling schemes in the coupling framework of CFD and DEM: two-way coupling (TWC) and one-way coupling (OWC).

TWC considers both the full interaction between fluid (solved by CFD) and particles (solved by DEM), while OWC just considers the force of fluid on the particle but ignores the force of ice on the fluid. In other words, fluid exerts forces on the particles, but the particles do not affect the movement of the fluids. Luo et al. [45] adopted both TWC and OWC methods to calculate a ship moving in brash ice channel and compared the influences of these two schemes. It was found that although the minimum error of TWC was slightly smaller than that of OWC relative to experimental resistance, the average error of them was very close. However, TWC occupied much more computing resources than OWC. The total solving CPU time of TWC rose more sharply and became much longer, up to five times or longer, than that of OWC along with the calculation [55]. As a result, Luo et al. [45] recommended OWC for the case of a ship moving in ice at a low speed. Considering the speed of a ship moving in level ice is much lower than that in brash ice, the OWC method is chosen in the simulation in this paper.

Here the framework of OWC is introduced briefly, as shown in Figure 5, and that of TWC can refer to [34]. To start with, all components of simulation, including DEM, CFD and coupling parts, are initialized. All the initial conditions and initial value, including stationary fluid with hydrostatic pressure, stationary level ice floating on the water surface and rigid ship moving forward in a given speed, are assigned. The OWC starts with calculating the fluid porosity, namely the ratio of the particle volume in each fluid cell, based on the particle information (position and size) and fluid mesh information (mesh size and node location). As mentioned in Section 2.1, the interface between water and ice particle is not tracked or captured, and the fluid-particle interaction force and momentum exchange are calculated based on fluid porosity, so it is important to calculate fluid porosity. Thereafter, velocity of particles and fluid as well as the pressure and stress tensor of fluid at the current time step are used to calculate the fluid-particle interaction force $F_{pf,i}$ in Equation (9). The next step is the iteration loop of the DEM. After the DEM loop involving Equations (9) and (10) is completed, the new position and translational and rotational velocities of all particles in the next fluid time step are obtained. On this basis, the fluid phase Equations (1) and (2) are solved. In OWC, R_{pf} in Equation (2) is taken as zero. In this way, the solving process becomes much easier and enormous computation time and storage are saved. Then the obtained particle and fluid information will be used in next loop, until the whole simulation terminates.

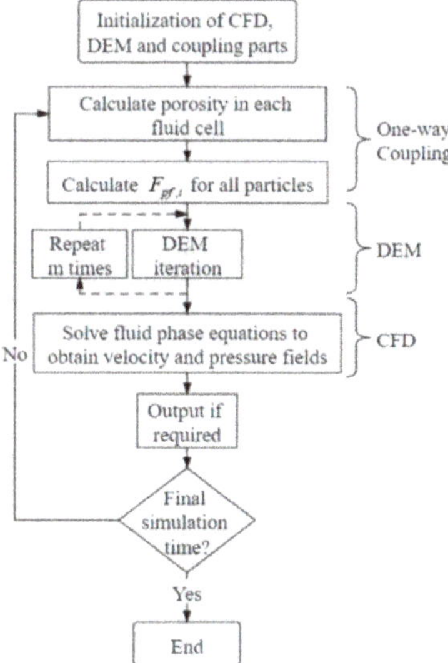

Figure 5. Framework of one-way coupling in the discrete element method-computational fluid dynamics (CFD-DEM) model, revised based on [50].

3. Validation

Before studying the motion of a ship in the level ice, the coupling method is firstly validated by comparing numerical results with experimental data. There have been previous studies of model ships moving in level ice in an ice tank as mentioned in Section 1. Considering the hull form information, the experimental study from ice tank of Tianjin University [14] was chosen. The corresponding prototype used by Huang et al. [14] is that same as that used in this paper, as shown in Table 1 and Figure 2, and the scale ratio is 37.5 for their model. Considering that they have predicted the resistance of the prototype from model resistance under the similarity laws, we calculated the prototype and compared our numerical results with their predicted resistance of prototype. The experimental picture and the state of the ship before moving into level ice region in numerical simulation are shown in Figure 6a,b, respectively.

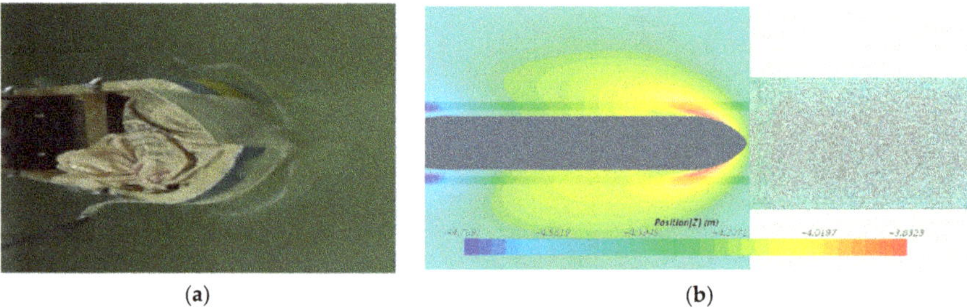

(a)   (b)

Figure 6. Bird's eye view of experimental picture (**a**) (reproduced from [14], with permission from ELSEVIER, 4 September 2020), and the state of the ship before moving into level ice region in numerical simulation (**b**).

Two main physical quantities are concerned. One is the damage of the level ice under impact of the ship, including the crack development and broken ice movement. The other is the total resistance. The former can just be compared qualitatively while the later can be compared quantitatively. In order to validate the former, the case with ice thickness 1.5 m and ship speed 5 kn in prototype is chosen as an example. The damage of ice obtained by numerical simulation in this paper is compared with that in the model test, as shown in Figure 7. Comparison of ice damage between model tests (a) [14] and (b) [56] and numerical simulation (c) is examined, where the general outline of the ice crack is marked by lines.

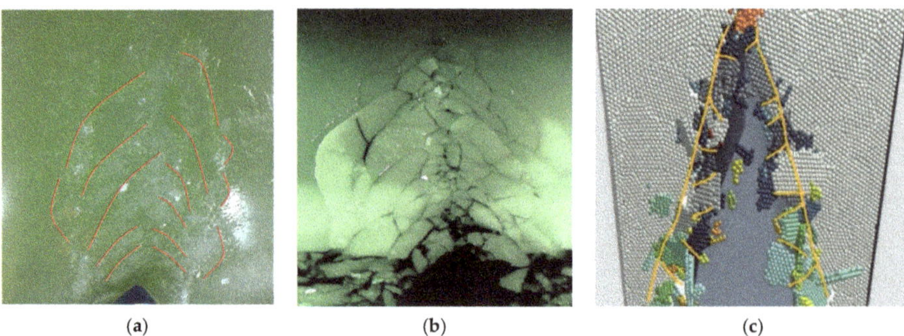

(a)   (b)   (c)

Figure 7. Comparison of ice damage between model tests (**a**) (reproduced from [14], with permission from ELSEVIER, 4 September 2020), and (**b**) (reproduced from [56], with permission from ELSEVIER, 4 September 2020), and numerical simulation (**c**), where the general outline of the ice crack is marked by lines.

In Figure 7, two cases of experimental data are chosen for comparison as shown in Figure 7a,b, respectively. Figure 7a provides the experimental data from [14], where the red lines are the outlines of the circular cracks. Experimental scene was taken by camera, while the model vessel was driven back to prepare the next test run. Figure 7c is the numerical result with the same parameters in Figure 7a. In Figure 7c, ice particles in the same fragment are shown in the same color, so different colors can be seen to distinguish the shapes of the broken ice. Considering the cracks in Figure 7a from [14] are not clear enough, although we tried to highlight them by red lines, the picture from another reference [56] is adopted as supplementary in Figure 7b. It is worth mentioning that there is no direct link between numerical simulation and model test in Figure 7b, as the ship forms are different. As a result, the characteristics of the ice breakup in numerical simulation Figure 7c are compared qualitatively with those in Figure 7a,b. It can be found that the simulated ice destruction area and the shape of the broken ice are similar to those of the model test. There are mainly two points of qualitative similarities. One is the ice-breaking areas extending outwards in a V-shape, as shown by the general outlines of the crack in Figure 7a,c. Circular cracks and radial cracks cross each other, forming crushed ice of various sizes. The other similarity is that crushed ice has a smaller size when it is closer to the center line of the ship, especially shown by the fragments in Figure 7b,c.

Secondly, the ice resistance values of numerical simulation are compared with those transformed from model test, as plotted in Figure 8. In Figure 8, the squares denote the average resistance at ship velocity 2 kn, 3 kn, 4 kn, and 5 kn in prototype transformed from model test, while the dots denote the average resistance at ship velocity 3 kn, 5 kn, and 7 kn in numerical simulation. Considering the ship with lower speed taking much longer calculation time and resources, we did not calculate cases 2 kn and 4 kn and took cases 3 kn and 5 kn as examples for comparisons. As can be seen from Figure 8, when the speed is 5 kn, the mean value of numerical simulation resistance is about 1.9 MN, which is very close to the measured resistance 2 MN in ice tank, with the deviation just around 5%. The total ice resistance deviation is only 2% when the speed is 3 knots. The good agreement validates the accuracy of the coupling method to some extent.

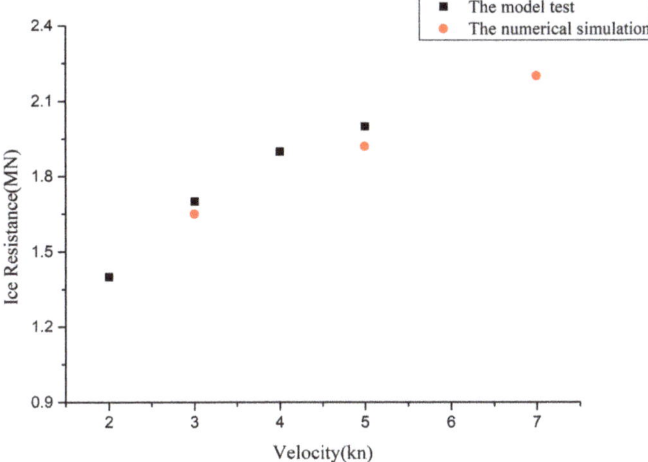

Figure 8. Comparison of ice resistance between model test (the squares), transformed into prototype by Huang et al. [14] and numerical simulation (the dots) at various ship velocity.

Based on the above analysis, it can be seen agreement has achieved between numerical results and experimental data both qualitatively and quantitatively. In this way, it is considered that the coupling numerical model can simulate ship navigation in level ice effectively.

4. Results and Discussion

The one-way CFD-DEM coupling model established in this paper is applied to the numerical simulation of a ship moving in the level ice region. The responses of ice in this process, including crack propagation, are obtained. In addition, the characteristics of the resistance of the ship in this process are also one focus of this paper.

4.1. Ice-Breaking Process

The interaction between a ship and level ice is a very complicated problem. The process of ice breaking and crack propagation is accompanied by various failure modes. To study this problem, the case with ship speed of 5 kn and the ice thickness of 1.5 m is selected as a typical case study in this section. Figure 9 shows the ice evolution during the ice-breaking process. As mentioned before, different colors denote different ice fragments.

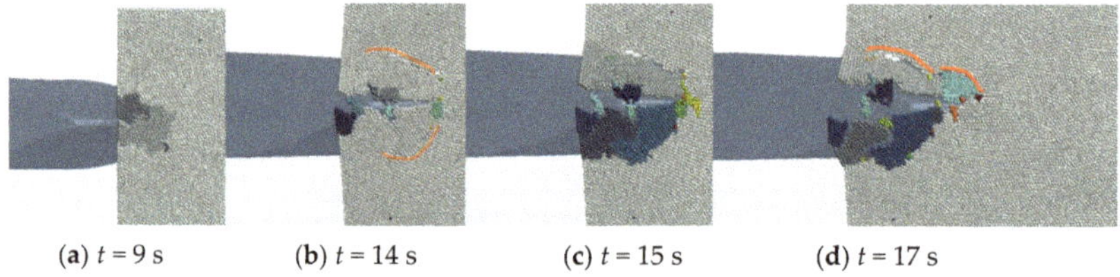

(a) t = 9 s   (b) t = 14 s   (c) t = 15 s   (d) t = 17 s

Figure 9. Ice evolution during the ice-breaking process at (**a**) t = 9 s, (**b**) t = 14 s, (**c**) t = 15 s and (**d**) t = 17 s, respectively, where some circular cracks are highlighted by curves in (**b**,**d**), respectively.

As mentioned in Section 2.2, ice is taken as viscoelastic model. When the ship comes into contact with the ice, the elastic deformation of the ice occurs first, and no cracks appear at this time. As the ship continues forward, the contact force between ship and ice increases further. When the stress inside the ice exceeds its ultimate stress, the particle bond of the ice in DEM model breaks. With the increasing of the broken bonds, the crack of the ice starts to expand. Radial cracks are first observed propagating from the contact points between ship and the ice sheet, as Figure 9a t = 9 s shows. Along with the forward movement of the ship, radial cracks extend outwards gradually. A first-order circular crack appears at the ship's shoulder, as Figure 9b t = 14 s shows, and a small amount of ice fragments appear at the bottom of the bow, forming the so-called local crushing zone [13]. As the ship continues to move forward, the interaction between the ship's shoulder and the ice sheet gets severer. The ice bends under the action of ships. The first-order circular crack continues to extend forward, inducing large pieces of ice separating from the ice sheet, as shown at Figure 9c t = 15s. These ice fragments overturn under the bow and may collide with ship bottom or other ice fragments and get broken again, resulting in several smaller pieces of ice. Then a second-order circular crack appears on the basis of the first-order circular crack around ship bow, as shown at t = 17 s in Figure 9d. Many pieces of crushed ice with various sizes are formed between the radial cracks and the circular cracks. As the ship continues to move further, more radial and circular cracks generate in a similar and repeated ways as described above, which have also been observed in experiments [13]. Finally, a V-shaped crushing region will be formed as shown in Figure 9b.

4.2. Total Resistance and Ice Resistance

The total resistance is one of the most significant problems for ship performance in ice regions. It is a common way to divide the total resistance into ice resistance and water resistance [57]. Although water resistance is usually small compared with ice resistance, especially for a ship moving in level ice, water effects cannot be easily ignored. On the one hand, the ice sheet is not fixed in the direction of gravity and needs the support of water to keep it afloat and stable. On the other hand, the broken ice fragments usually drift and overturn under the action of water, which affects the change of ice resistance, especially the so-called submersion component [58]. That is also the reason why we choose the coupling model, which considers the interaction between ship, ice and water.

In the numerical modeling, on the ship surface, there are two forces, one is the fluid force and the other is the ice force. Therefore, one can obtain two resistances from the integral of fluid and ice forces, respectively. They are defined as 'water resistance' and 'ice resistance' in this paper. Because OWC method is adopted in this paper, fluid motion and pressure are not affected by the ice, so the 'water resistance' is not affected by ice either. 'Ice resistance' is affected by the fluid motion under OWC method, so it can be seen as the ice resistance considering the influence of fluid. This division method is easy to compare the contribution of ice and water in a rough way. Therefore, water and ice resistances are checked separately.

Figure 10a,b provides curves of water resistance and ice resistance, respectively. As shown in Figure 10a, one can see that water resistance when stable is about 0.1–0.2 MN, which is at least an order of magnitude smaller than the ice resistance shown in Figure 10b, which is in MN. This verifies again that the water resistance component is very small in the total resistance for a ship moving in level ice. Considering that the one-way coupling method is adopted in this paper, the water resistance is not affected by ice and remains the same under different ice conditions. Therefore, the following discussion will focus on the characteristics of ice resistance instead of water resistance.

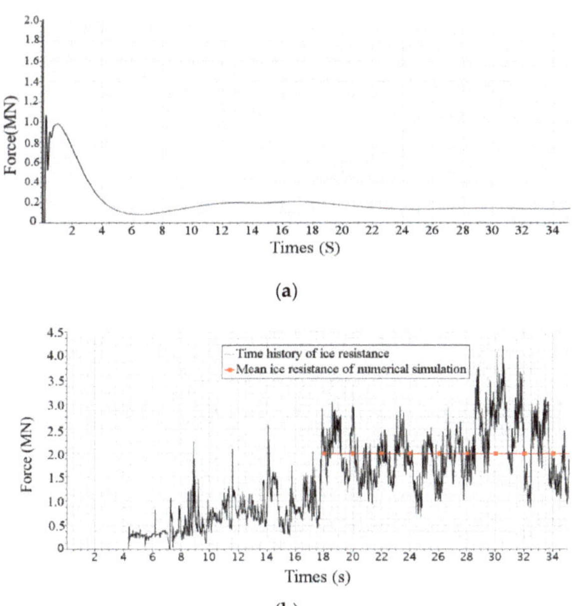

Figure 10. Time history of water resistance (**a**) and ice resistance (**b**), in which the red dotted line is the mean value of ice resistance after the whole ship bow enters the level ice region.

Figure 10b shows the time-history curve of ice resistance. Corresponding to Figure 9, the influence of various ice failure patterns on the ice resistance can be vividly seen in the curve. Firstly, it can be seen from the figure that at the first 9 s, the ice resistance is rising gradually in a small extent. Before 9 s, the ice is bent under the moment of ship bow. Local large elastic deformations occur but not the whole cracks. At about 9 s, the first crack starts to form and propagate, as shown in Figure 9a, so the ice resistance reaches its first peak before it decreases sharply along with the unloading of the ice force. As the ship moves further and contacts with more ice, the ice resistance rises slowly with fluctuations due to cracks. Around 14 s, the ice resistance reaches another peak, when the first-order circular crack just generates, as shown in Figure 9b. Then the ice is completely broken and the circular crack has been completely formed around 15 s, as shown in Figure 9c, so the ice resistance reaches bottom sharply due to unloading of the ice force. The generation of secondary circular crack propagation in Figure 9d also induces a rise in the resistance curve but with a smaller magnitude than that induced by the first order circular crack. Then there exists a distinct rise of ice resistance around 18 s, when the whole ship bow enters the ice region. After that moment, although the ice resistance still fluctuates strongly, its mean value tends to be stable. This indicates that the parallel middle body of the ship has a smaller contribution to the ice resistance than bow, because the force resulted from ice collision in this area is perpendicular to the motion direction. Although the frictional resistance between broken force and hull surface also has a contribution, it is small compared with ice–ship collision force. To compare with the mean ice resistance in model test mentioned above, the mean value after the whole ship bow enters the level ice region is defined as the average ice resistance here, as denoted by the red dotted line in Figure 10.

4.3. Effect of Ship Speed

Ship speed is a key factor that affects ice breaking and ice resistance. Other parameters are kept the same as those in Section 4.1, and ship speed is changed from 3, 5, to 7 kn to determine its effect.

Figure 11 shows the ice conditions under different ship speed, in which the distances of the ship into the level ice are the same with different moments marked below each figure. When the ship speed is 3 kn, the damage range of the ice sheet in the lateral direction is largest. The ice far away from the ship side is also affected and damaged, with larger pieces of broken ice and longer cracks propagated. As the ship speed increases to 7 kn, it can be found that the ice-breaking channel is narrowest. The ice beyond ship breadth has been little affected by the ship motion. Furthermore, the difference in size between the pieces of broken ice becomes smaller and the very large ice fragments get fewer.

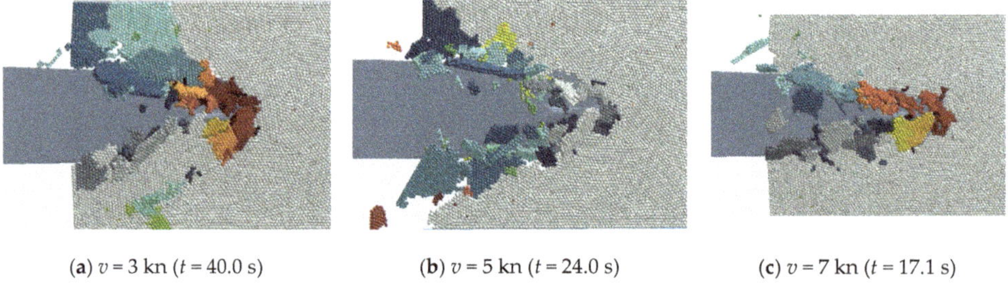

(a) v = 3 kn (t = 40.0 s)   (b) v = 5 kn (t = 24.0 s)   (c) v = 7 kn (t = 17.1 s)

Figure 11. Ice-breaking conditions at different ship speed (**a**) v = 3 kn, (**b**) v = 5 kn and (**c**) v = 7 kn, respectively.

The change of ship speed can be considered as the change of the loading rate on the ice. As one may know, the responses of the ice are affected by the loading rate significantly [54]. When ship speed increases, the loading rate of the ice increases and the ice sheet presents a greater brittle response [54]. As a result, the size of ice fragments gets small, the crack propagation becomes short, and the lateral damage area reduces, and the ice-breaking channel gets relatively narrow.

Figure 12 shows the time histories and typical values of ice resistance at different ship speed. From time history curves, one can still see two stages for each velocity. Although the time when they enter the second stage differs, the distance they take is the same, which is just equal to the length of the ship bow. This verifies the analysis in Section 4.2 again. The variation of maximum and average ice resistance in the second stage of each case are further checked in Figure 12d. It can be seen that ship speed has a significant effect on the ice resistance. Both maximum and mean values of ice resistance increase with the increase of speed, almost in a linear relationship. Similar trends are also observed in the previous experimental studies [14,56]. Furthermore, the rising slopes of maximum and mean values are close.

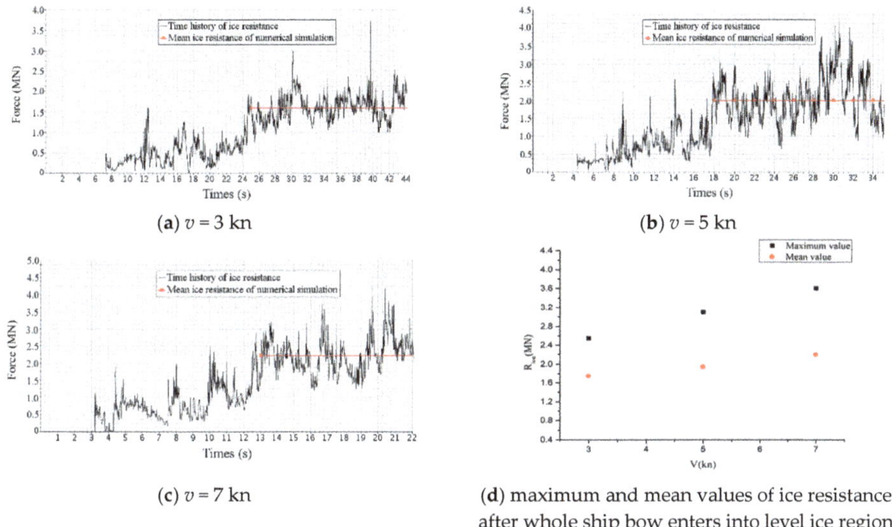

(**a**) v = 3 kn

(**b**) v = 5 kn

(**c**) v = 7 kn

(**d**) maximum and mean values of ice resistance after whole ship bow enters into level ice region

Figure 12. Ice resistance at different ship speeds (**a**) v = 3 kn, (**b**) v = 5 kn and (**c**) v = 7 kn, respectively, and (**d**) provides maximum and mean values of ice resistance after whole ship bow enters into level ice region.

4.4. Effect of Ice Thickness

Ice thickness is also an important factor affecting ice breaking. The ice sheet in this paper is modeled by using DEM particles. The mechanical strength of the ice sheet is determined by the bonding bars between the particles. The values of the bonding parameters are closely related to the particle size. In order to eliminate this error, the particle size is kept unchanged, and the ice thickness is changed by increasing or decreasing the number of particle layers. Considering the diameter of the particle is 0.75 m, single-layer, double-layer, and three-layer ice can be modeled, so the ice thickness is 0.75 m, 1.5 m, and 2.25 m, respectively. Other parameters are kept the same with those in Section 4.1.

Figure 13 shows the ice conditions under different ice thicknesses at the same time t = 20 s. It can be found that when the ice thickness is 0.75 m, there are more long radial cracks formed on the ice surface. The ice fragments are relatively larger in size and the total amount decreases instead. As the ice grows to 1.5 m in thickness, the size of the crushed ice varies. Both large and small ice fragments exist, and the amount of crushed ice increases compared with 0.75 m case. When the ice thickness rises to 2.25 m, the damage of ice sheet reduces distinctively. Many small-sized ice fragments are generated just around the ship–ice contact region.

The ice resistance under different ice thickness is shown in Figure 14. It can be found that both maximum and mean ice resistances increase with ice thickness sharply. As the ice thickness rises, the difference between the maximum and average ice resistance increases also. It denotes that maximum ice resistance is more sensitive to ice thickness.

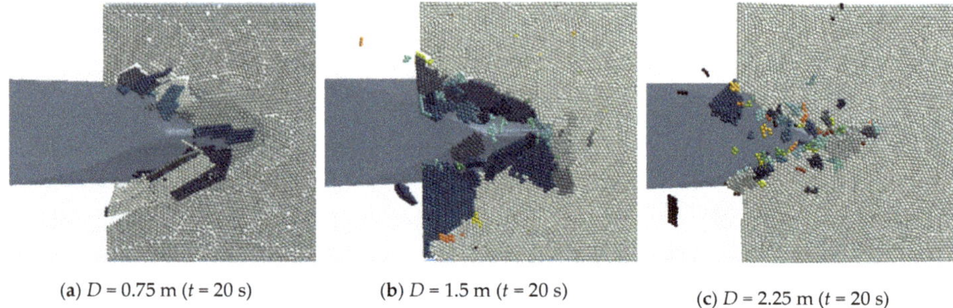

(**a**) $D = 0.75$ m ($t = 20$ s)   (**b**) $D = 1.5$ m ($t = 20$ s)   (**c**) $D = 2.25$ m ($t = 20$ s)

Figure 13. Ice conditions at different ice thickness (**a**) $D = 0.75$ m, (**b**) $D = 1.5$ m and (**c**) $D = 2.25$ m, respectively.

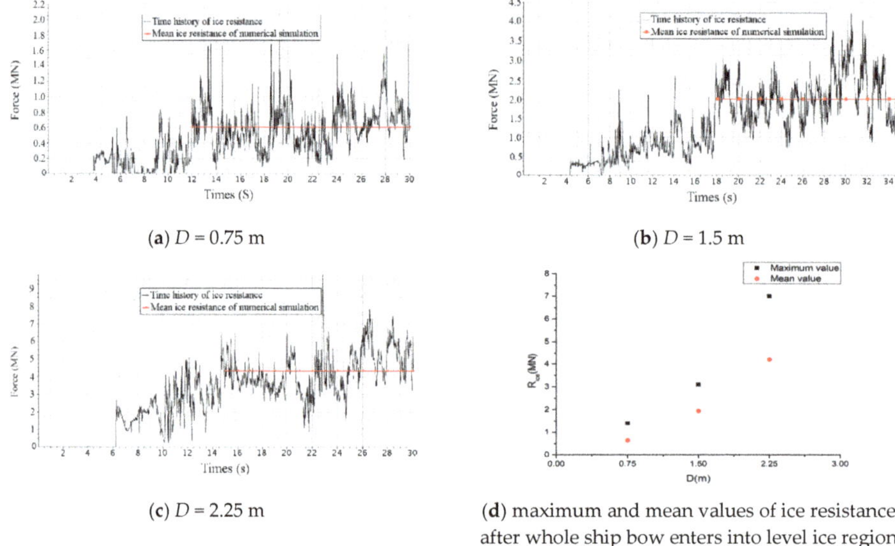

(**a**) $D = 0.75$ m

(**b**) $D = 1.5$ m

(**c**) $D = 2.25$ m

(**d**) maximum and mean values of ice resistance after whole ship bow enters into level ice region

Figure 14. Ice resistance at different ice thickness(**a**) $D = 0.75$ m, (**b**) $D = 1.5$ m and (**c**) $D = 2.25$ m, respectively, and (**d**) provides maximum and mean values of ice resistance after whole ship bow enters into level ice region.

5. Conclusions

In this paper, an icebreaker moving in level ice has been simulated by using the one-way CFD-DEM coupling method. Numerical results are compared with experimental data and good agreement has been achieved. On this basis, influences of various parameters on ice resistance and ice breaking conditions are further investigated. The following conclusions are drawn preliminarily:

(1)   One-way CFD-DEM coupling method is firstly used to simulate a ship moving in level ice. To solve the challenge of building level ice on the water domain, this paper puts forward two numerical techniques, which are crucial for the success of the modeling.

(2)   For a ship moving in level ice, water resistance is an order of magnitude smaller than ice resistance. However, this does not mean water can be easily ignored. Actually, ice resistance is affected by the fluid motion in the OWC method. In other words, the ice resistance in OWC method has included the influence of fluid already.

(3)   Ice resistance shows strong oscillation in the process of ice breaking, presenting cycling of 'gradual rising—reaching maximum—sudden drop' generally. This is closely related to the ice failure process of 'elastic deformation-reaching stress, limit-crack generation, and unloading'. After the ship bow fully enters into the level ice region, the average ice resistance reaches stable broadly. Both ice resistance and ice destruction are affected by ship speed and/or ice thickness significantly.

In future work, TWC method will be studied further and used in the ship moving in level ice. Furthermore, the variation speed of the ship will be studied, as well as the motion of the ship in six degree of freedom.

Author Contributions: Conceptualization & Methodology, B.-Y.N.; Writing & Validation, Z.-W.C. and X.-A.L.; Software, K.Z. and X.-A.L.; Supervision & Funding acquisition, B.-Y.N. and Y.-Z.X. All authors have read and agreed to the published version of the manuscript.

Funding: This research was funded by National Key R&D Program of China, grant number 2017YFE0111400, and National Natural Science Foundation of China, grant number 51979051, 51979056 and 51639004.

Acknowledgments: The authors would like to acknowledgment College of Shipbuilding Engineering, Harbin Engineering University in purchasing and support in using STAR-CCM+ software.

Conflicts of Interest: The authors declare no conflict of interest.

References

1.   Melia, N.; Haines, K.; Hawkins, E. Sea ice decline and 21st century trans-Arctic shipping routes. *Geophys. Res. Lett.* **2016**, *43*, 9720–9728. [CrossRef]

2.   Smith, L.C.; Stephenson, S.R. New trans-arctic shipping routes navigable by midcentury. In Proceedings of the National Academy of Sciencesin, Washington, DC, USA, 27 April 2014; pp. E1191–E1195.

3.   Available online: www.nsra.ru (accessed on 7 September 2020).

4.   Xue, Y.Z.; Ni, B.Y. Review of mechanical issues for polar region ships and floating structures. *J. Harbin Eng. Univ.* **2016**, *37*, 36–40. (In Chinese)

5.   Ni, B.Y.; Han, D.F.; Di, S.C.; Xue, Y.Z. On the development of ice-water-structure interaction. *J. Hydrodyn.* **2020**, *32*, 629–652. [CrossRef]

6.   Chernov, A.V. Measuring total ship bending with a help of tensometry during the full-scale in situ ice impact study of ice-breaker 'kapitan nikolaev'. In Proceedings of the 20th International Conference on Port and Ocean Engineering under Arctic Conditions, Luleå, Sweden, 9–12 June 2009. POAC09-027.

7.   Kujala, P.; Arughadhoss, S. Statistical analysis of ice crushing pressures on a ship's hull during hull-ice interaction. *Cold Reg. Sci. Technol.* **2012**, *70*, 1–11. [CrossRef]

8.   Kellner, L.; Herrnring, H.; Ring, M. Review of ice load standards and comparison with measurements. In Proceedings of the 36th International Conference on Offshore Mechanics and Arctic Engineering (OMAE), Trondheim, Norway, 25–30 June 2017.

9.   *Brash Ice Tests for a Panmax Bulker with Ice Class 1B*; HSVA Report: Hamburg, Germany, 2012.

10.   Tian, Y.F.; Huang, Y. The dynamic ice loads on conical structures. *Ocean Eng.* **2013**, *59*, 37–46. [CrossRef]

11.   Kim, M.C.; Lee, S.K.; Lee, W.J. Numerical and experimental investigation of the resistance performance of an icebreaking cargo vessel in pack ice conditions. *Int. J. Nav. Archit. Ocean Eng.* **2014**, *5*, 116–131. [CrossRef]

12.   Huang, Y.; Sun, J.Q.; Ji, S.P. Experimental study on the resistance of a transport ship navigating in level Icejournal. *Mar. Sci. Appl.* **2016**, *15*, 105–111. [CrossRef]

13.   Huang, Y.; Li, W.; Wang, Y.; Wu, B. Experiments on the resistance of a large transport vessel navigating in the arctic region in pack ice conditions. *J. Mar. Sci. Appl.* **2016**, *15*, 269–274. [CrossRef]

14.   Huang, Y.; Huang, S.Y.; Sun, J.Q. Experiments on navigating resistance of an icebreaker in snow covered level ice. *Cold Reg. Sci. Technol.* **2018**, *152*, 1–14. [CrossRef]

15.   Sturova, I.V. Generation of long waves in ice-covered lakes by moving disturbances of atmospheric pressure. *J. Hydrodyn.* **2010**, *22*, 34–39. [CrossRef]

16.   Sturova, I.V. Radiation of waves by a cylinder submerged in water with ice floe or polynya. *J. Fluid Mech.* **2015**, *784*, 373–395. [CrossRef]

17.   Korobkin, A.A.; Malenica, S.; Khabakhpasheva, T. The vertical mode method in the problems of flexural-gravity waves diffracted by a vertical cylinder. *Appl. Ocean Res.* **2019**, *84*, 111–121. [CrossRef]

18. Pogorelova, A.V.; Zemlyak, V.L.; Kozin, V.M. Moving of a submarine under an ice cover in fluid of finite depth. *J. Hydrodyn.* **2019**, *31*, 562–569. [CrossRef]

19. Li, Z.F.; Shi, Y.Y.; Wu, G.X. A hybrid method for linearized wave radiation and diffraction problem by a three dimensional floating structure in a polynya. *J. Comput. Phys.* **2020**, *412*, 109445. [CrossRef]

20. Li, Z.F.; Wu, G.X.; Ren, K. Wave diffraction by multiple arbitrary shaped cracks in an infinitely extended ice sheet of finite water depth. *J. Fluid Mech.* **2020**, *893*, A14. [CrossRef]

21. Su, B.; Riska, K.; Moan, T. A numerical method for the prediction of ship performance in level ice. *Cold Reg. Sci. Technol.* **2010**, *60*, 177–188. [CrossRef]

22. Lubbad, R.; Løset, S. A numerical model for real-time simulation of ship-ice interaction. *Cold Reg. Sci. Technol.* **2011**, *65*, 111–127. [CrossRef]

23. Zhou, L.; Riska, K.; Ji, C. Simulating transverse icebreaking process considering both crushing and bending failures. *Mar. Struct.* **2017**, *54*, 167–187. [CrossRef]

24. Zhou, L.; Diao, F.; Song, M.; Han, Y.; Ding, S. Calculation methods of icebreaking capability for a double-acting polar ship. *J. Mar. Sci. Eng.* **2020**, *8*, 179. [CrossRef]

25. Song, M.; Shi, W.; Ren, Z.; Zhou, L. Numerical study of the interaction between level ice and wind turbine tower for estimation of ice crushing loads on structure. *J. Mar. Sci. Eng.* **2019**, *7*, 439. [CrossRef]

26. Valanto, P. Spatial distribution of numerically predicted ice loads on ship hulls in level ice. In *Report for Deliverable D6-3 of SAFEICE Project*; Espoo, Finland, 2007. Available online: https://trimis.ec.europa.eu/sites/default/files/project/documents/20120320_100305_87982_D11-2.1%20Final%20scientific%20report-vol2.pdf (accessed on 7 September 2020).

27. Liu, Z.H. *Analytical and Numerical Analysis of Iceberg Collisions with Ship Structures*; Norwegian University of Science and Technology: Trondheim, Norway, 2011.

28. Kim, M.C.; Lee, W.-J.; Shin, Y.J. Comparative study on the resistance performance of an icebreaking cargo vessel according to the variation of waterline angles in pack ice conditions. *Int. J. Nav. Archit. Ocean Eng.* **2014**, *6*, 876–893. [CrossRef]

29. Zhou, L.; Wang, F.; Diao, F.; Ding, S.; Yu, H.; Zhou, Y. Simulation of ice-propeller collision with cohesive element method. *J. Mar. Sci. Eng.* **2019**, *7*, 349. [CrossRef]

30. Guo, C.Y.; Zhang, Z.T.; Tian, T.P.; Li, X.Y.; Zhao, D.G. Numerical simulation on the resistance performance of ice-going container ship under brash ice conditions. *China Ocean Eng.* **2018**, *32*, 546–556. [CrossRef]

31. Ni, B.Y.; Huang, Q.; Chen, W.S.; Xue, Y.Z. Numerical simulation of ice load of a ship turning in level ice considering fluid effects. *Chin. J. Ship Res.* **2020**, *15*, 1–7.

32. Xue, Y.; Liu, R.; Li, Z.; Han, D. A review for numerical simulation methods of ship–ice interaction. *Ocean Eng.* **2020**, *215*, 107853. [CrossRef]

33. Tuhkuri, J.; Polojärvi, A. A review of discrete element simulation of ice–structure interaction. *Philos. Trans. R. Soc. A Math. Phys. Eng. Sci.* **2018**, *376*, 20170335. [CrossRef]

34. Hansen, E.H.; Løset, S. Modelling floating offshore units moored in broken ice: Model description. *Cold Reg. Sci. Technol.* **1999**, *29*, 97–106. [CrossRef]

35. Zhan, D.; Agar, D.; He, M.; Spencer, D.; Molyneux, D. Numerical simulation of ship maneuvering in pack ice. In Proceedings of the ASME 2010 29th International Conference on Ocean, Offshore and Arctic Engineering, Shanghai, China, 6–11 June 2010. OMAE2010-21109.

36. Lau, M.; Lawrence, K.P.; Rothenburg, L. Discrete element analysis of ice loads on ships and structures. *Ships Offshore Struct.* **2011**, *6*, 211–221. [CrossRef]

37. Morgan, D. An improved three-dimensional discrete element model for ice-structure interaction. In Proceedings of the 23rd IAHR International Symposium on Ice, Ann Arbor, MI, USA, 31 May–3 June 2016.

38. Cai, K.; Ji, S.Y. Analysis of interaction between level ice and ship hull based on discrete element method. *Nav. Archit. Ocean Eng.* **2016**, *32*, 5–14.

39. Di, S.C.; Xue, Y.Z.; Wang, Q.; Bai, X.L. Discrete element simulation of ice loads on narrow conical structures. *Ocean Eng.* **2017**, *146*, 282–297. [CrossRef]

40. Liu, L.; Ji, S.Y. Ice load on floating structure simulated with dilated polyhedral discrete element method in broken ice field. *Appl. Ocean Res.* **2018**, *75*, 53–65. [CrossRef]

41. Long, X.; Liu, S.; Ji, S. Discrete element modelling of relationship between ice breaking length and ice load on conical structure. *Ocean Eng.* **2020**, *201*, 107152. [CrossRef]

42. Gong, H.; Polojärvi, A.; Tuhkuri, J. Discrete element simulation of the resistance of a ship in unconsolidated ridges. *Cold Reg. Sci. Technol.* **2019**, *167*, 102855. [CrossRef]

43. Huang, L.; Tuhkuri, J.; Igrec, B.; Li, M.; Stagonas, D.; Toffoli, A.; Cardiff, P.; Thomas, G. Ship resistance when operating in floating ice floes: A combined CFD&DEM approach. *Mar. Struct.* **2020**, *74*, 102817. [CrossRef]

44. Vroegrijk, E. *Validation of CFD+ DEM against Measured Data*; International Conference on Offshore Mechanics and Arctic Engineering; American Society of Mechanical Engineers: New York, NY, USA, 2015; Volume 56567, p. V008T07A021.

45. Luo, W.; Jiang, D.; Wu, T.; Guo, C.; Wang, C.; Deng, R.; Dai, S. Numerical simulation of an ice-strengthened bulk carrier in brash ice channel. *Ocean Eng.* **2020**, *196*, 106830. [CrossRef]

46. Stephen, B.P. *Turbulent Flowsi*; Cambridge University Press: Cambridge, UK, 2010.

47. Kloss, C.; Goniva, C.; Hager, A. Models, algorithms and validation for open source DEM and CFD–DEM. *Prog. Comput. Fluid Dyn.* **2012**, *12*, 140–152. [CrossRef]

48. Siemens PLM Software. *Simcenter STAR-CCM+®Documentation Version 2019*; Siemens PLM Software: Munich, Germany, 2019; Volume 1, pp. 7482–7486.

49. Hadzic, H. *Development and Application of Finite Volume Method for the Computation of Flows around Moving Bodies on Unstructured, Overlapping Grids*; Technische Universität: Harburg, Germany, 2006.

50. Norouzi, H.R.; Zarghami, R. *Coupled CFD-DEM Modeling*; John Wiley & Sons, Ltd: Chichester, UK, 2016.

51. *Regulations for Offshore Ice Condition Application in China Sea*; China National Offshore Oil Corporation: Beijing, China, 2002. (In Chinese)

52. Di, S.; Xue, Y.; Bai, X.; Wang, Q. Effects of model size and particle size on the response of sea-ice samples created with a hexagonal-close-packing pattern in discrete-element method simulations. *Particuology* **2018**, *36*, 106–113. [CrossRef]

53. Yu, J.Y. *Study on Ice Load of Ship in Ice Zone Based on Discrete Element Method*; Harbin Engineering University: Harbin, China, 2020. (In Chinese)

54. Ding, D.W. *Introduction to Engineering Sea Ice*; Ocean Press: Beijing, China, 1999. (In Chinese)

55. Xu, P.; Guo, C.Y.; Wang, C. Numerical simulation of propeller-crushing ice-water interaction based on CFD-DEM coupling. *China Shipbuild.* **2019**, *60*, 120–140. (In Chinese)

56. Tan, X.; Riska, K.; Moan, T. Effect of dynamic bending of level ice on ship's continuous-mode icebreaking. *Cold Reg. Sci. Technol.* **2014**, *106*, 82–95. [CrossRef]

57. Spencer, D. A standard method for the conduct and analysis of ice resistance model tests. In *Laboratory Memorandum*; NRCC, Institute for Marine Dynamics: Ottawa, ON, Canada, 1992.

58. Lindqvist, G. A straight forward method for calculation of ice resistance of ships. In Proceedings of the Port and Ocean Engineering under Arctic Conditions, Luleaa, Sweden, 12–16 June 1989; p. 722.

Journal of
Marine Science and Engineering

Article

Three-Dimensional Fluid–Structure Interaction Case Study on Elastic Beam

Mahdi Tabatabaei Malazi [1,*], Emir Taha Eren [2], Jing Luo [1], Shuo Mi [1] and Galip Temir [2]

[1] State Key Laboratory of Hydraulics and Mountain River Engineering, Sichuan University, Chengdu 610500, China; luojing@scu.edu.cn (J.L.); medievals@163.com (S.M.)
[2] Faculty of Mechanical Engineering, Yildiz Technical University, Besiktas, Istanbul 34349, Turkey; emirtahaeren@gmail.com (E.T.E.); galip@yildiz.edu.tr (G.T.)
* Correspondence: m.tabatabaei.malazi@scu.edu.cn; Tel.: +86-131-8381-1574

Received: 10 August 2020; Accepted: 7 September 2020; Published: 15 September 2020

Abstract: A three-dimensional T-shaped flexible beam deformation was investigated using model experiments and numerical simulations. In the experiment, a beam was placed in a recirculating water channel with a steady uniform flow in the inlet. A high-speed camera system (HSC) was utilized to record the T-shaped flexible beam deformation in the cross-flow direction. In addition, a two-way fluid-structure interaction (FSI) numerical method was employed to simulate the deformation of the T-shaped flexible beam. A system coupling was used for conjoining the fluid and solid domain. The dynamic mesh method was used for recreating the mesh. After the validation of the three-dimensional numerical T-shaped flexible solid beam with the HSC results, deformation and stress were calculated for different Reynolds numbers. This study exhibited that the deformation of the T-shaped flexible beam increases by nearly 90% when the velocity is changed from 0.25 to 0.35 m/s, whereas deformation of the T-shaped flexible beam decreases by nearly 63% when the velocity is varied from 0.25 to 0.15 m/s.

Keywords: fluid–structure interaction; flexible beam; high speed imaging; system coupling

---

1. Introduction

Deformation of structures has been an area of interest for engineering. Because the coupling process between fluid and structure plays an important role in many engineering fields. Interaction of fluid and structures in underwater flexible beams is a complex problem in industrial applications. Experimental and numerical research has been conducted to examine fluid–structure interaction (FSI). A number of previous studies are summarized in this section.

Important FSI problems were simulated numerically by Chimakurthi et al. [1]. The authors applied Ansys workbench system coupling and two-way fluid–structure interaction to various multiphysics coupled problems, such as FSI in an oscillating solid structure and sub-sea pipeline vibrations. This study is important, because it shows that Ansys workbench system coupling is highly suited to multiphysics coupled problems. Furthermore, the study results were also validated with other experimental and numerical studies. Gluck et al. [2] studied fluid–structure interaction numerically in different plate forms, such as vertical and L-shaped plates. The authors used two code samples for simulation flow and structure, one using the finite volume method for the flow side and the other using the finite element method for the solid side. The one-way fluid–structure interaction method was employed for the simulation effect of waves on a ship's hull by Dhavalikar et al. [3]. It was assumed that the ship's hull was a rigid body and wave loads were then simulated on a rigid body. Narayanan et al. [4] numerically investigated flow behavior past a cylinder with a flexible filament. The commercial software STAR-CCM+ was used for solving the governing equations. Large deformation of a flexible rod in fluid flow was studied numerically and experimentally by

Hassani et al. [5]. The authors applied a wind tunnel for testing a flexible rod and a mathematical model was developed by coupling with the Kirchhoff rod theory. The deformation of plants under different combinations of wave periods was examined experimentally by Juan et al. [6]. An oscillatory tunnel and volumetric particle image velocimetry system were used in this study. Results showed the velocity distribution around plants. Mantecon et al. [7] numerically investigated the fluid–structure interaction of nuclear fuel plates under axial flow conditions. The authors employed the commercial software Ansys CFX for modeling the fluid flow and Ansys Mechanical for modeling solid plates. Results showed the maximum deflection of the plates happened at the leading edge. The fluid–structure interaction of a square sail was investigated experimentally and numerically by Ghelardi et al. [8]. Initially, the square sail was tested in a wind tunnel at various velocities and then the ADINA commercial program was applied for numerical simulation under the same conditions. The authors obtained a good overall agreement between experimental and numerical results. Liu et al. [9] studied the FSI of a single flexible cylinder in an axial flow at different inlet velocities. Ansys Fluent commercial software was applied to the flow field. They used a user-defined function (UDF) code sample for the deformation of the cylinder. Results showed that increasing or decreasing vibration depends on flow velocity. Xu et al. [10] applied a new method for solving fluid–structure interaction. The lattice Boltzmann method (LBM) and immersed boundary method (IBM) were combined in this method. The authors used a large-eddy turbulence model (LEM) for high Reynolds numbers. This new method was tested with different benchmarks. The results proved that it has good accuracy. Wang et al. [11] studied three different risers (one steel riser and two composite risers) for their Vortex Induced Vibration (VIV) characteristics using the fluid–structure interaction method. They simulated 2D and 3D models using the Ansys Fluent commercial software. Deformations of the models were obtained. The results show that the displacements of the Fiber Reinforced Polymer (FRP) composite risers are significantly larger than those of other models. The fluid–structure interaction problems of two side-by-side flexible plates were also numerically investigated by Dong et al. [12]. The authors presented results of drag force and energy capture performance in a three-dimensional model. A two-dimensional immersed boundary method was used for simulation fluid–structure interaction of a large structure by Wang et al. [13]. A finite difference method was applied to solve compressible Navier–Stokes equations. The authors validated the results using a flexible plate in a hypersonic flow. Good agreement was found between experimental and numerical data. Turek et al. [14] defined a new benchmark for comparing various methods in fluid–structure interaction problems. They located an elastic object in a laminar incompressible channel flow. A good comparison was obtained in this new benchmark study. Wang et al. [15] investigated the flow past a circular cylinder with a flexible splitter plate. A two-dimensional model was used for simulation of the fluid–structure interaction using two different Reynolds numbers. The authors also studied a circular cylinder with a plate, with a gap between the cylinder and the plate. Zheng et al. [16] used a new fluid–structure interaction method that coupled a finite-element method and immersed boundary method (IBM) to study the flow-induced vibrations of the vocal folds during phonation. Wang et al. [17] presented a fluid–structure interaction (FSI) methodology to simulate flexible submerged vegetation stems and kinetic turbine blades. The structural dynamic solver was based on the combined finite element method–discrete element method (FEM-DEM), and solid and fluid solvers were coupled using an immersed boundary method (IBM) iterative algorithm. The flow solver was a ghost-cell-based sharp-interface immersed boundary method (IBM) described by Mittal et al. [18]. The unsteady, incompressible Navier–Stokes equations were solved and a second-order, central-difference scheme was used for all spatial derivatives. The fluid–structure interaction of a three-dimensional flexible membrane was studied by Nestola et al. [19]. The authors used the immersed boundary method for solving complex structures immersed in laminar, transitional, and turbulent flows. Peskin [20] applied the immersed boundary method to computer simulation of fluid–structure interaction. Eulerian and Lagrangian variables were coupled in this method. A fixed Cartesian mesh for the Eulerian variables and a moving curvilinear mesh for the Lagrangian variables have been used in the immersed boundary (IB) method, particularly in biological fluid dynamics. A monolithic FEM/Multigrid method for the fluid–structure interaction of an incompressible elastic object in laminar incompressible viscous flow was presented by Hron and Turek [21]. Griffith and Luo [22] studied a coupling scheme for the immersed boundary method to link the Lagrangian and Eulerian variables. They used this method for solving FSI problems. The left ventricle of the heart was stimulated by this method. The arbitrary Lagrangian-Eulerian finite element method was applied by Nassiri [23]. They used this method for numerical simulation and experimental investigation of wavy interfacial morphology during high velocity impact welding. Tabatabaei et al. [24] studied the hydrodynamics behavior of an axisymmetric squid model numerically. They applied SST k-w turbulence model for simulation. Various fineness ratios, jet propulsion, and drag force were investigated for different swimming velocities. Squid's flow characteristics were studied numerically and experimentally by Olcay et al. [25]. Digital particle image velocimetry (DPIV) was used for obtaining velocity contours in the experimental region. Ansys Fluent commercial software was applied for solving governing equations in the flow field. They showed that numerical results were so close to the experimental data. Hydrodynamic forces of a moving cylinder and fixed cylinder were investigated numerically by Eren et al. [26]. They used the dynamic mesh method for recreating mesh and moving a cylinder in the incompressible flow.

To improve our understanding of fluid–structure interaction methodology for flow past a flexible beam, we studied the fluid–structure interaction experimentally and numerically for a single flexible beam. We use a high-speed camera system (HSC) for obtaining the deformation of the beam. The three-dimensional flexible beam model was then investigated numerically at three different velocities. Two-way FSI method was applied for numerical simulation, and also, a validation test was carried out. All in all, this study helps to improve our understanding of flexible beams deformations. The paper was organized as follows: Section 2 defines experimental setup, computational domain governing equations, and numerical methods. Section 3 presents validation of deformation T-shaped flexible beam, numerical results, and discussions. Section 4 contains the conclusions.

2. Materials and Methods

2.1. Experimental Arrangement and Analysis Methodology

Details of the experimental setup are explained in Sections 2.2 and 2.3. Section 2.2 presents the water channel, the high-speed camera, and the technique of high-speed imaging method. Section 2.3 describes the data analysis of the experiment results.

2.2. Experimental Setup

Experiments were carried out in a recirculating water channel of State Key Laboratory of Hydraulics and Mountain River Engineering (SKLHMRE) at Sichuan University (SCU), China. The water channel has a 12 m (length) × 0.5 m (width) × 0.6 m (height) test section with the mean flow velocity up to 0.25 m/s. The water level was maintained at 0.5 m during the experiments and a propeller velocity meter was used to measure the inflow velocity. Froude number is 0.11 (Fr = 0.11), so flow is subcritical in this study. Multiple measurement points in the same vertical plane of the channel at different depths are chosen to measure the velocity in a long period, and the mean flow velocity is calculated by time averaging and space averaging. A T-shaped flexible beam made of polyurethane was fixed in a vertical plane facing to the approaching flow. The dimensions of the T-shaped flexible beam are shown in Figure 1. Water was used for the fluid part and a polyurethane beam was applied for the solid part. The experiment was conducted indoors; physically essential properties of selected materials at room temperature are provided in Table 1.

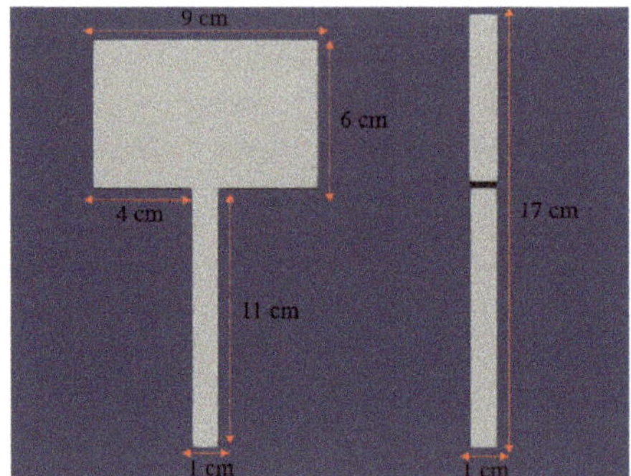

Figure 1. Schematics showing the dimensions of the beam.

Table 1. Characteristics of fluid and solid in numerical and experiment methods.

Water	
Density (ρ)	1000 kg/m^3
Dynamic viscosity (μ)	0.001 kg/m^{-s}
Flexible Beam	
Density (ρ_s)	1.17 g/cm^3
Young's modulus (E)	18 MPa
Poisson's ratio (ν)	0.3

For the beam motion and image recognition process, illumination was provided by an LED light sheet, and the beam was marked black. A non-intrusive technique of high-speed imaging method was employed to record the deformation of the flexible beam in this study. Images were captured at 1000 frames per second (fps) using a high-speed camera (Fastcam Mini UX100, Photron Inc., Chiyoda-Ku, Tokyo, Japan, maximum acquisition rate: 4000 fps with the resolution of 1280 × 1024 pixels). The resolution of the images is cut down to 616 pixels × 1024 pixels for saving camera memory space to only cover the area where the beam exists, as shown in Figure 2.

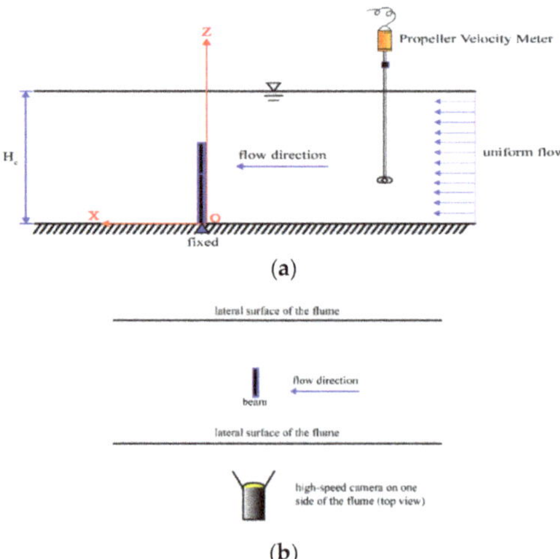

Figure 2. *Cont.*

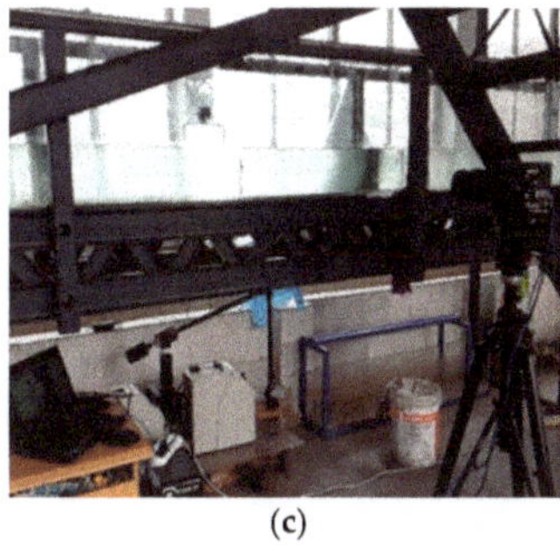
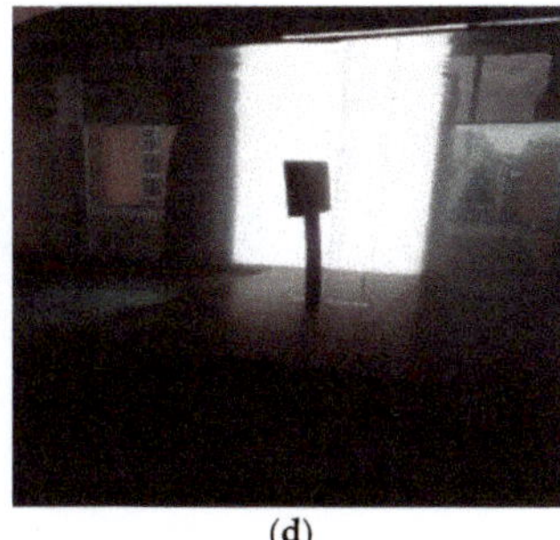

(c)   (d)

Figure 2. Schematics of the experiment setup: (**a**) experimental arrangement in the recirculating water flume; (**b**) high-speed camera on one side of the flume; (**c**) and (**d**) side views.

The standardization tests were carried out in still water in order to determine the actual distance per pixel (Figure 3). The field of view was approximately 12.07 cm × 20.07 cm, leading to a spatial resolution of 0.0196 cm pixel^{-1}.

Figure 3. Standardization test.

2.3. Data Analysis

To track the deformation of the T-shaped flexible beam, an image recognition Python code was developed to obtain quantitative data from images captured in the experiment (Figure 4). A bilateral filter is a basic theory of image noise reduction, and it is better in edge-preserving than other filters, so we used a bilateral filter to reduce the noise of experimental images. Then, the Canny edge detector, which is an edge detection operator that uses a multi-stage algorithm, was used to detect edges of the beam in experimental images. Image binarization was set the grayscale value of the pixel on the image to 0 or 255, which is the process of presenting the whole image with an obvious black and white effect. The erosion, closing operation, and dilation are the morphological operations to enhance image features. We used erosion, closing operation, and dilation to make the edges detected by the Canny edge detector clearer, then the pixel coordinates of the edges were recorded. The pixel size of a picture can be detected. Thus, the scale S of actual distance and Pixel distance can be calculated through Figure 3. S is defined by S = actual distance/pixel distance. Therefore, the actual displacement D of the tracking point is calculated by D = S*L, where L is the pixel distance between the current and initial position of the tracking point. Details about image processing are illustrated in Howes [27].

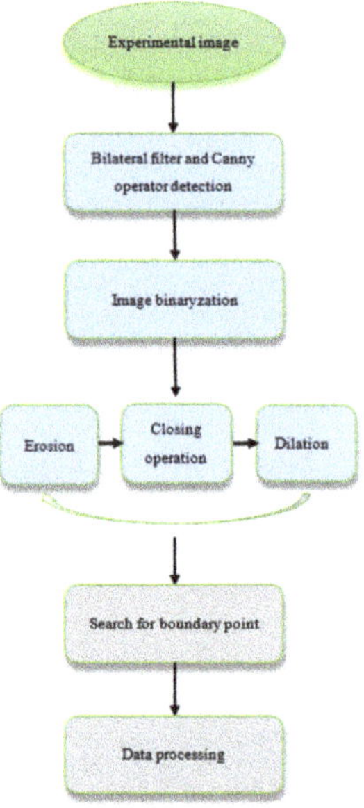

Figure 4. Flow chart of image processing program.

2.4. Numerical Methods

The two-way fluid-structure interaction (FSI) numerical method is explained in Sections 2.5–2.8. The solution of the three-dimensional fluid domain is described in Section 2.5. The structural dynamics of the T-shaped flexible beam is explained in Section 2.6. Section 2.7 presents system coupling between fluid and solid domain. Details of the computational domain and boundary conditions are investigated in Section 2.8.

2.5. Computational Fluid Dynamics (CFD)

The realizable k-ε turbulence model was applied for the turbulent flow simulation in the three-dimensional fluid domain (Olcay et al. [28] and ANSYS Fluent Theory Guide [29]). The governing equations representing the continuity and momentum formulas as given below:

$$\frac{\partial \rho}{\partial t} + \frac{\partial (\rho u_i)}{\partial x_i} = 0 \tag{1}$$

$$\frac{\partial (\rho u_i)}{\partial t} + \frac{\partial \left(\rho u_i u_j\right)}{\partial x_j} = -\frac{\partial P}{\partial x_i} + \rho g_i + \frac{\partial}{\partial x_j}(\mu + \mu_t)\left(\frac{\partial u_i}{\partial x_j} + \frac{\partial u_j}{\partial x_i}\right) + S_i \tag{2}$$

where ρ is the density and u_i and u_j are the average velocity component of the fluid. P is pressure, S_i is the source term for the momentum equation μ is the dynamic viscosity, μ_t is the eddy viscosity, and it is defined as $\mu_t = \rho C_\mu \frac{k^2}{\varepsilon}$. The transport equation for k and ε are for the realizable $k - \varepsilon$ model given as,

$$\frac{\partial}{\partial t}(\rho k) + \frac{\partial}{\partial x_i}\left(\rho k u_i\right) = \frac{\partial}{\partial x_j}\left[\left(\mu + \frac{\mu_t}{\sigma_k}\right)\frac{\partial k}{\partial x_j}\right] + G_k + G_b - \rho \varepsilon - Y_M + S_k \tag{3}$$

$$\frac{\partial}{\partial t}(\rho \varepsilon) + \frac{\partial}{\partial x_i}(\rho \varepsilon u_i) = \frac{\partial}{\partial x_j}\left[\left(\mu + \frac{\mu_t}{\sigma_\varepsilon}\right)\frac{\partial \varepsilon}{\partial x_j}\right] + C_{1\varepsilon}\frac{\varepsilon}{k}(G_k + C_{3\varepsilon}G_b) - C_{2\varepsilon}\rho\frac{\varepsilon^2}{k} + S_\varepsilon \tag{4}$$

where k is the turbulent kinetic energy and ε is rate of dissipation. G_k is turbulent kinetic energy generation because of the mean velocity gradients, G_b is turbulent kinetic energy generation because of buoyancy, and Y_M is fluctuating dilatation contribution to the overall dissipation rate. The model constants for realizable k-ε turbulence model are $C_{1\varepsilon} = 1.44$, $C_{2\varepsilon} = 1.92$, $\sigma_k = 1.0$, $C_\mu = 0.09$, and $\sigma_\varepsilon = 1.3$.

2.6. Computational Structural Dynamics (CSD)

Deformation of a three-dimensional flexible solid structure is described by the equation of motion, which can be expressed as follows:

$$[M]\{\ddot{u}\} + [C]\{\dot{u}\} + [K]\{u\} = \{F\} \tag{5}$$

where $[M]$ is the structural mass matrix, $[C]$ is the structural damping matrix, $[K]$ is the structural stiffness matrix, and $\{F\}$ is the applied load vector acting on the structure caused by fluid. $\{\ddot{u}\}$ is the nodal acceleration vector, $\{\dot{u}\}$ is the nodal velocity vector, and $\{u\}$ is the nodal displacement vector. Newmark time integration method with an improved algorithm (HHT) was used for the solution of Equation (5). The Newmark method and HHT method were applied for implicit transient analyses. The Newmark method applies finite-difference expansions in the time interval Δt. It is presumed that (Bathe [30]):

$$\{\dot{u}_{n+1}\} = \{\dot{u}_n\} + \left[(1-\delta)\{\ddot{u}_n\} + \delta\{\ddot{u}_{n+1}\}\right]\Delta t \tag{6}$$

$$\{u_{n+1}\} = \{u_n\} + \{u_n\}\Delta t + \left[\left(\frac{1}{2} - \alpha\right)\{\ddot{u}_n\} + \alpha\{\ddot{u}_{n+1}\}\right]\Delta t^2 \tag{7}$$

where α and δ are Newmark integration parameters, Δt is $t_{n+1} - t_n$, $\{u_n\}$ is the nodal displacement vector at time t_n, $\{\dot{u}_n\}$ is the nodal velocity vector at time t_n, $\{\ddot{u}_n\}$ is the nodal acceleration vector at time t_n, $\{\ddot{u}_{n+1}\}$ is the nodal acceleration vector at time t_{n+1}, $\{\dot{u}_{n+1}\}$ is the nodal velocity vector at time t_{n+1}, and $\{u_{n+1}\}$ is the nodal displacement vector at time t_{n+1}.

Since the primary aim is the calculation of displacement $\{u_{n+1}\}$, the governing Equation (5) can be computed at time t_{n+1} as:

$$[M]\{\ddot{u}_{n+1}\} + [C]\{\dot{u}_{n+1}\} + [K]\{u_{n+1}\} = \{F\} \tag{8}$$

The solution of displacement at time t_{n+1} can be obtained by first rearranging Equations (6) and (7), such that:

$$\{\ddot{u}_{n+1}\} = a_0(\{u_{n+1}\} - \{u_n\}) - a_2\{\dot{u}_n\} - a_3\{\ddot{u}_n\} \tag{9}$$

$$\{\dot{u}_{n+1}\} = \{\dot{u}_n\} + a_6\{\ddot{u}_n\} + a_7\{\ddot{u}_{n+1}\} \tag{10}$$

where $a_0 = \frac{1}{\alpha\Delta t^2}$, $a_2 = \frac{1}{\alpha\Delta t}$, $a_3 = \frac{1}{2\alpha} - 1$, $a_6 = \Delta t(1-\delta)$, $a_7 = \delta\Delta t$.

Once a solution is obtained for $\{u_{n+1}\}$, velocities and accelerations are computed as defined Equations (9) and (10). For the nodes where the velocity or the acceleration is obtained, a displacement constraint is computed from Equation (7). The HHT time integration method can help to have the desired property for the numerical damping in the full transient analysis (Chung and Hulbert [31]).

The basic form of the HHT method is defined as Equation (11)

$$[M]\{\ddot{u}_{n+1-\alpha_m}\} + [C]\{\dot{u}_{n+1-\alpha_f}\} + [K]\{u_{n+1-\alpha_f}\} = \{F_{n+1-\alpha_f}\} \tag{11}$$

where α_m and α_f are two extra integration parameters for the interpolation of the acceleration and the displacement, velocity, and loads. It was also realized that the transient dynamic equilibrium equation considered in the HHT method is a linear combination of two successive time steps of n and n + 1 after comparing Equations (5) and (11).

2.7. CFD-CSD Coupling

We used Ansys Workbench-system coupling for simulation two-way fluid–structure interactions (Chimakurthi et al. [1]). Fluid Flow (Ansys Fluent) and the Transient Structural systems (Ansys Mechanical) are connected in system coupling. Reynolds-averaged Navier-Stokes (RANS) equations with the realizable k-ε turbulence model are solved in the computational domain by using the CFD solver (Fluent), and also, transient structural analysis is used to solve the T-shaped flexible beam deformation under the action of loads. Forces or stresses on the fluid side of the interface are transformed on to the solid side, and also, displacements or velocities on the solid side of the interface are transformed on to the fluid side in the system coupling method. Transferring in this system coupling includes the computation of weights and their subsequent use in the interpolation of data. It can happen between topologically similar and/or dissimilar element types, distributions, and dimensions such as surface to surface, volume to volume, point to volume, surface to volume, and vice-versa. Figure 5 shows the calculation procedure and detailed overview of the partitioned system coupling. The induced force on the beam was obtained after the flow field was calculated by using CFD solver from Ansys Fluent. Then, the displacement of the beam was solved by using a structure transient from Ansys Mechanical. This process gets continued until convergence is obtained in system coupling.

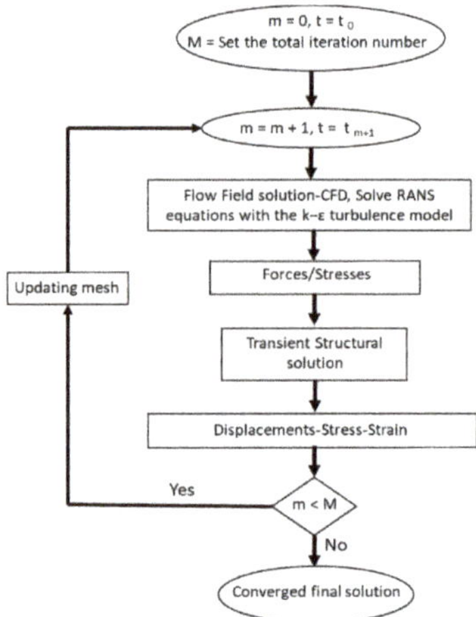

Figure 5. Numerical steps of system coupling.

2.8. Computational Model Geometry, Boundary Conditions, and Meshing

The T-shaped flexible beam was fixed on the channel bottom wall. Figure 6 shows the computational domain and boundary conditions schematically. The height (H) of the T-shaped flexible beam is 0.17 m. The beam was put on a cuboid-shape computational domain. The height, length, and width of the domain were defined as 5 H, 20 H, and 10 H. The velocity inlet boundary condition was located at 5H upstream of the beam, and the pressure outlet boundary condition was located at 15 H downstream of the T-shaped flexible beam. No slip boundary condition was selected for walls (bottom wall and T-shaped flexible beam). Free slip boundary condition was applied for top and to the sides of the computational. Dimensions and material properties of T-shaped flexible beam and flow properties that were used in all simulations were the same with experiment case Table 1.

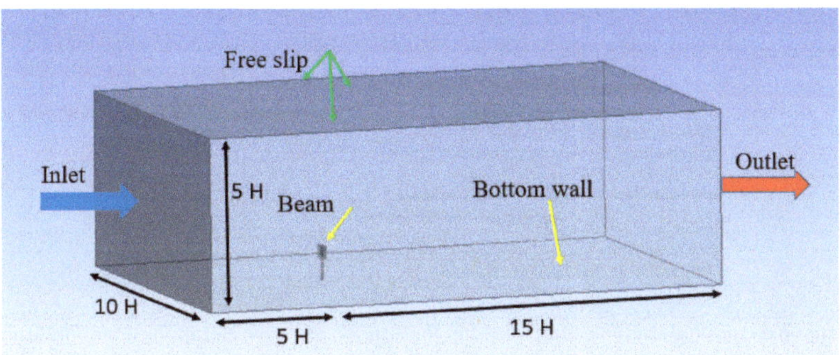

Figure 6. Details of the computational domain and boundary conditions.

The non-dimensional Reynolds number was applied for defining flow characteristics in this study. The Reynolds number was defined as Equation (12)

$$Re = \frac{\rho U L}{\mu} \tag{12}$$

where ρ is density, μ dynamic viscosity of the fluid, U free stream velocity, and L is the characteristic length (i.e., height of beam). The highest water velocity was 0.25 m/s in the experimental setup, because the close-circuit water channel could supply 0.25 m/s as the highest water velocity throughout the channel. In the numerical study, three different velocities were used for the understanding of beam behaviors in various velocities. U = 0.15, 0.25, and 0.35 m/s were chosen for numerical simulations after validation. The Reynolds numbers were defined as 25,500, 42,500, and 59,500 for U = 0.15, 0.25, and 0.35 m/s Table 2. Commercial computational fluid dynamic (CFD) code Ansys Fluent and Ansys Mechanical programs were employed to solve the flow domain and solid part. A system coupling method was used to connect between flow domain and solid part. The coupled scheme was selected among five pressure–velocity coupling algorithms. The second-order upwind scheme was used for discretization of advective terms of the transport equations. Criteria of convergence were set to 10^{-6} for the continuity and momentum equations. The solution of continuity and momentum equations were continued until criteria of convergence were achieved.

Table 2. Shows the relation between velocities and applied models.

Velocity (m/s)	Reynolds Number	Applied
0.15	25,500	Numerical model
0.25	42,500	Numerical and experimental model
0.35	59,500	Numerical model

Tetrahedron and prism with triangle base elements were set for meshing the fluid solution domain with high-density mesh near walls, and Tetrahedron mesh was used for the solid domain. The dynamic mesh method was applied to simulate the deformation of the T-shaped flexible beam. Totally 800,000–1,200,000 elements were employed to solve the fluid domain, and 21,210–73,000 elements were used to solve the solid domain, as illustrated in Figure 7. A mesh sensitivity study was also carried out for all models in the fluid domain. Table 3 shows the variety of total deformation along with the various number of elements at 0.25 m/s inlet velocity. It was identified that 1,200,000 elements for the fluid solution domain, and 72,425 elements for the solid domain were needed for obtaining good results at maximum velocity in our study. Three different coupling time steps were inspected at 0.01, 0.001, and 0.0001. All simulations were run for a total time of 10 s in this study. A time step of 0.001 was selected for all simulations after mesh refinement studies.

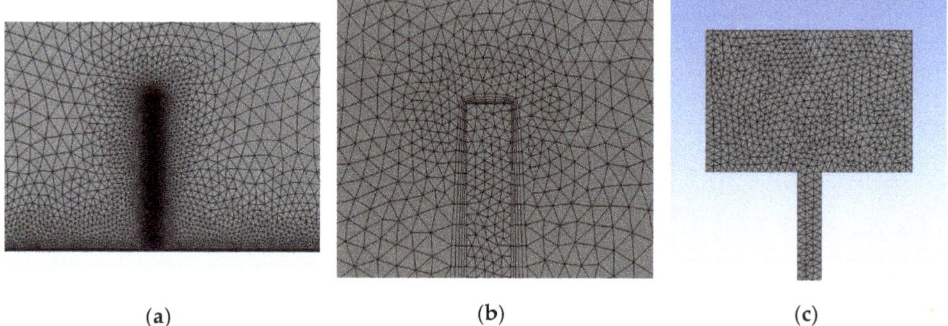

(a)   (b)   (c)

Figure 7. (a) Computational mesh of fluid domain, (b) enlarged view around beam surface, and (c) computational mesh of beam.

Table 3. Mesh convergence study for the computational domain.

Mesh Resolution	Deformation at t = 10 s
810,000	0.022400
890,000	0.01989
920,000	0.01934
1,020,000	0.019147
1,040,000	0.019144

3. Results and Discussion

3.1. Comparison between Experimental and Numerical Results

High-speed camera (HSC) measurements were carried out for the three-dimensional flexible beam model at one Reynolds number (42,500). The close-circuit water channel could supply 0.25 m/s as the highest velocity, so the highest Reynolds number for flexible beam was 42,500 in this experiment work. The numerical simulation was done at the same conditions, as used for the experiment. The results of numerical and experimental data were compared regarding the total deformation, Figure 8. The total deformation shape in the numerical model agrees well with the deformation obtained from HSC measurements for a Reynolds number of 42,500. A point (red point) was selected at the top of the T-shaped flexible beam for the tracking maximum displacement of the T-shaped flexible beam. When the T-shaped flexible beam has a vertical position, the red point is at position 1. After deformation, the red point changes from position 1 to position 2. The maximum displacement is the distance between position 1 and position 2. Table 4 shows the maximum deformation of the T-shaped flexible beam at t = 6 s and t = 10 s for numerical and experimental models.

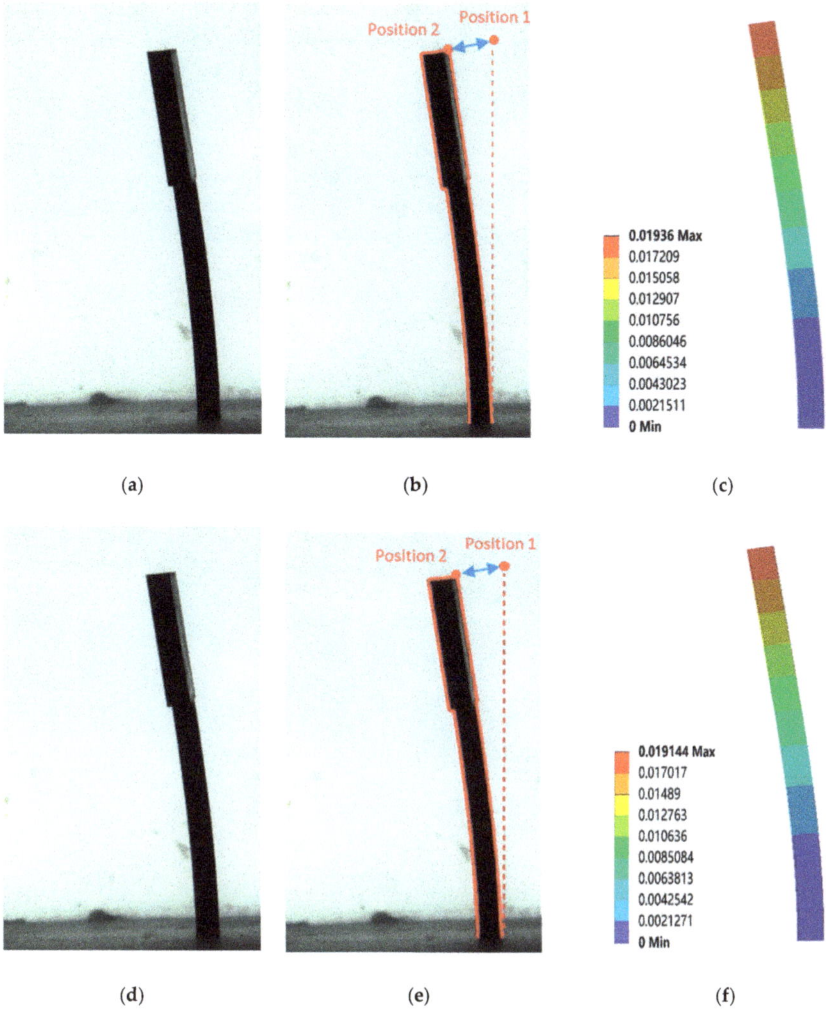

Figure 8. Comparison of total deformation of T-shaped flexible beam for Re = 42,500 and t = 6 s. (**a**) Experiment (high-speed camera system (HSC)), (**b**) experiment (HSC) with image track program, and (**c**) numerical (CFD). Comparison of total deformation of T-shaped flexible beam for Re = 42,500 and t = 10 s. (**d**) Experiment (HSC), (**e**) experiment (HSC) with image track program, and (**f**) numerical (CFD).

Table 4. Comparison of maximum deformation of T-shaped flexible beam for Re = 42,500, t = 6 s, and t = 10 s.

Re = 42,500	t = 6 s (Figure 8a–c)	t = 10 s (Figure 8d–f)
Experimental model deformation (m)	0.0205 ± 0.001	0.0202 ± 0.001
Numerical model deformation (m)	0.0193	0.0191

3.2. Deformation and Stress Study

The T-shaped flexible beam was validated at a Reynolds number of 42,500 (U = 0.25 m/s), and then, it was investigated at two different Reynolds numbers of 25,500 (U = 0.15 m/s) and 59,500

($U = 0.35$ m/s). Totally, The T-shaped flexible beam was studied in three various Reynolds numbers numerically. Total deformation was calculated numerically in this study. Total deformation can be computed by using Equation (13)

$$U = \sqrt{U_x^2 + U_y^2 + U_z^2} \tag{13}$$

where U_x is component deformation in the x direction, U_y is component deformation in the y direction, U_z is component deformation in the z direction. Figure 9 shows total deformation record from t = 0 s to t = 10 s at three different velocities of 0.15, 0.25, and 0.35 m/s. In all velocities, deformation of the T-shaped flexible beam increased in time, then it started to decrease and, finally, it had a constant value. The deformation of the T-shaped flexible beam increased with increasing velocity. The outer load also changed with velocity. The T-shaped flexible beam bent more and more when velocity increases. It seems that the maximum value of deformation happens early for minimum velocity. We can observe that the maximum value of deformation occurs at t = 0.47 s when velocity is $U = 0.15$ m/s, and also, it occurs at t = 0.511 s when velocity is $U = 0.35$ m/s. Maximum stress (Von Mises stress) of the T-shaped flexible beam happened at maximum inlet velocity between three different inlet velocities.

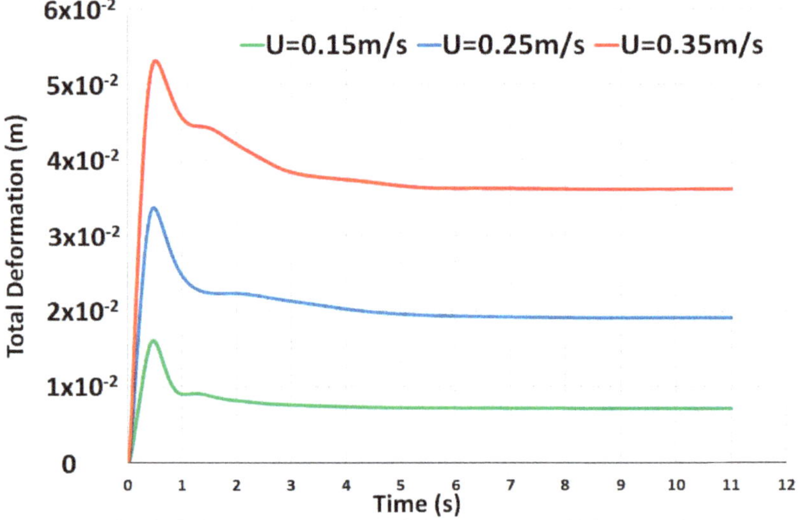

Figure 9. Variation of total deformation of T-shaped flexible beam between t = 0 s and 10 s.

Von Mises stress was calculated numerically in this study. Von Mises stress can be calculated by using Equation (14)

$$\sigma_e = \left[\frac{(\sigma_1 - \sigma_2)^2 + (\sigma_2 - \sigma_3)^2 + (\sigma_3 - \sigma_1)^2}{2} \right]^{1/2} \tag{14}$$

where σ_1 stress statestress in the x direction, σ_2 is stress statestress in the y direction, and σ_3 is stress statestress in the z direction. The maximum stress and maximum deformation of the beam have similar behavior, so we can figure out that stress changes like deformation in all velocities. Figure 10 shows maximum stress of beam that changes by time, and it occurs obviously on the bottom of the beam. Figure 11a–c show maximum principal stress, middle principal stress, and minimum principal stress of beam that changes by time.

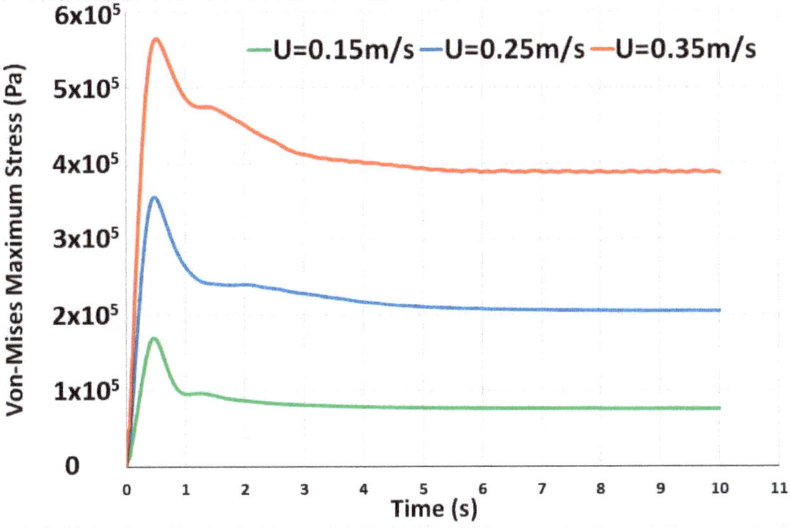

Figure 10. Variation of Von Mises maximum stress of T-shaped flexible beam between t = 0 s and 10 s.

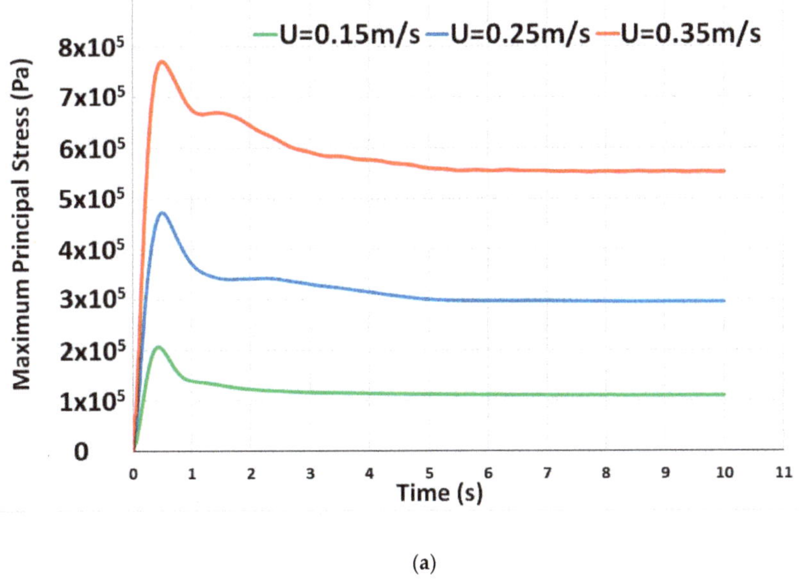

(a)

Figure 11. *Cont.*

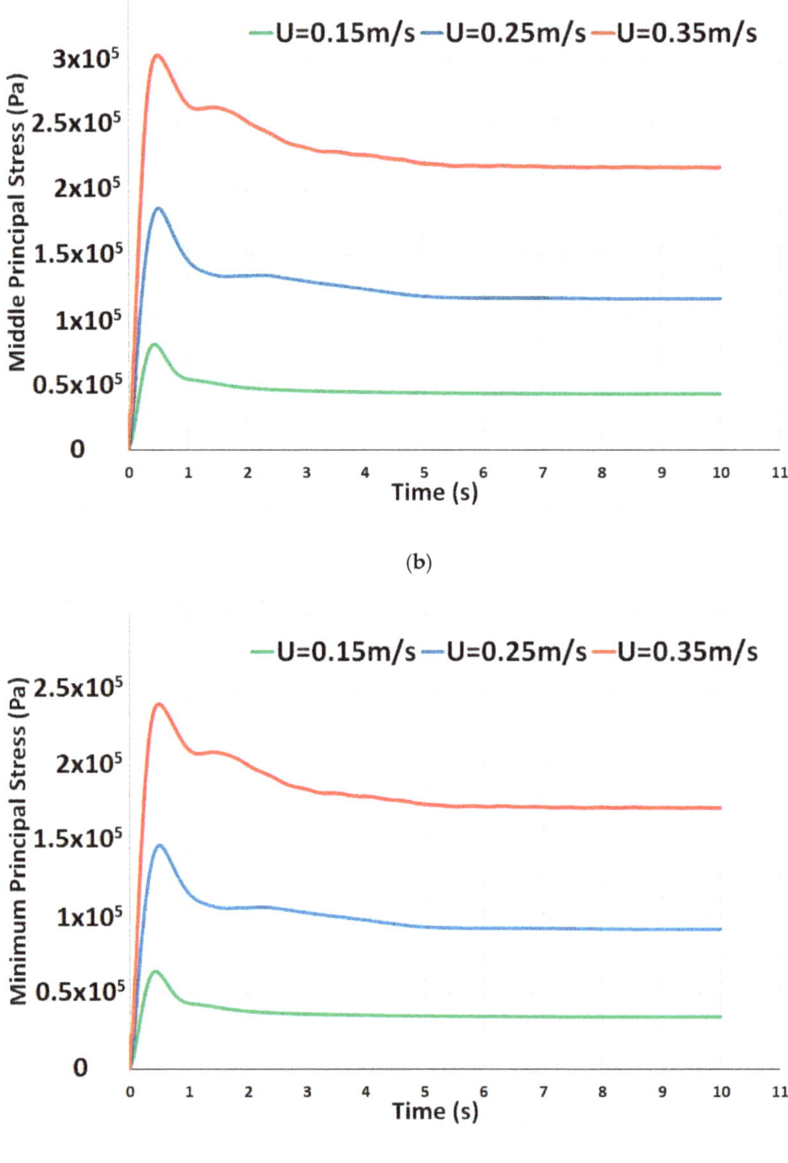

Figure 11. (a) Variation of maximum principal stress of T-shaped flexible beam between t = 0 s and 10 s. (b) Variation of middle principal stress of T-shaped flexible beam between t = 0 s and 10 s. (c) Variation of minimum principal stress of T-shaped flexible beam between t = 0 s and 10 s.

Equivalent strain was computed numerically in this study. Equivalent strain can be calculated by using Equation (15)

$$\varepsilon_e = \frac{1}{1+v'}\left(\frac{1}{2}\left[(\varepsilon_1 - \varepsilon_2)^2 + (\varepsilon_2 - \varepsilon_3)^2 + (\varepsilon_3 - \varepsilon_1)^2\right]\right)^{\frac{1}{2}} \tag{15}$$

where ε_1 is principal strain in the x direction, ε_2 is principal strain in the y direction, ε_3 is principal strain in the z direction, and v' is effective Poisson's ratio. Figure 12 shows equivalent strain of beam that changes by time.

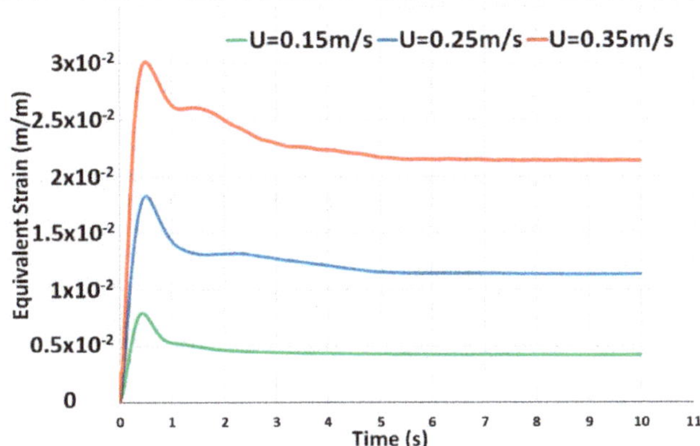

Figure 12. Variation of equivalent strain of T-shaped flexible beam between t = 0 s and 10 s.

3.3. Contours Plots of Numerical Study

Deformation of the T-shaped flexible beam was shown for three different Reynolds numbers of 25,500 (U = 0.15 m/s), 42,500 (U = 0.25 m/s), and 59,500 (U = 0.35 m/s) at t = 10 s, Figure 13. The stress of the T-shaped flexible beam was illustrated for three different Reynolds numbers of 25,500, 42,500, and 59,500 at t = 10 s, Figure 14. It was realized that maximum deformation and stress occurred at the maximum Reynolds number, because the T-shaped flexible beam has a large pressure in the front surface.

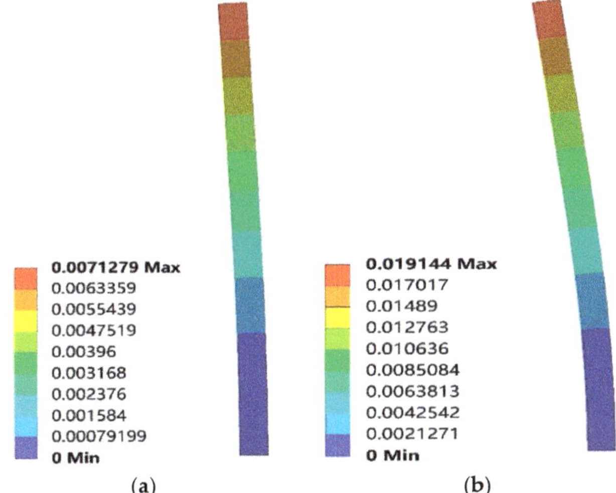

Figure 13. *Cont.*

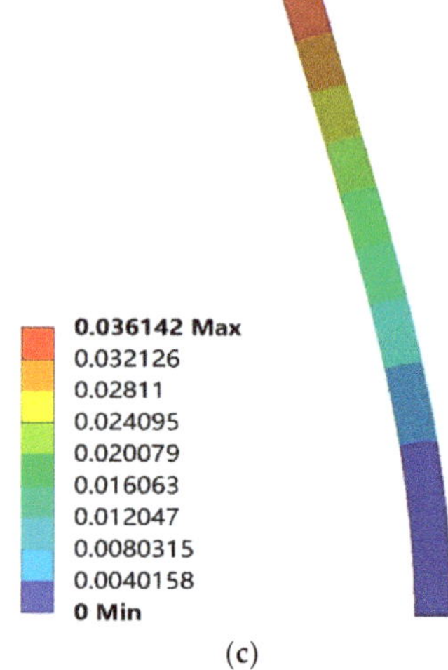

(c)

Figure 13. Deformation (m) of T-shaped flexible beam at t = 10 s (**a**) Re = 25,500, (**b**) Re = 42,500, and (**c**) Re = 59,500.

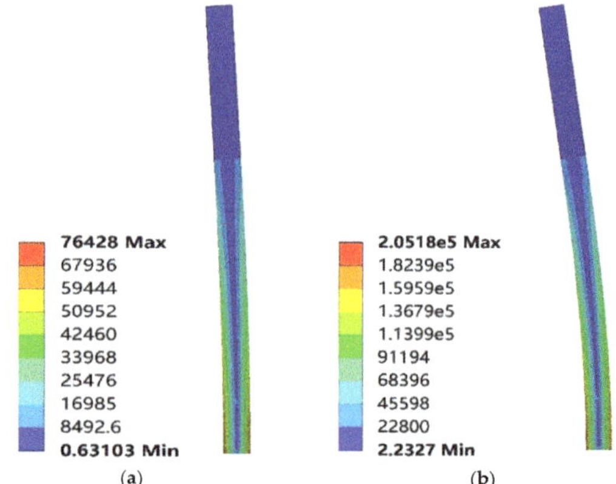

(a)

(b)

Figure 14. *Cont.*

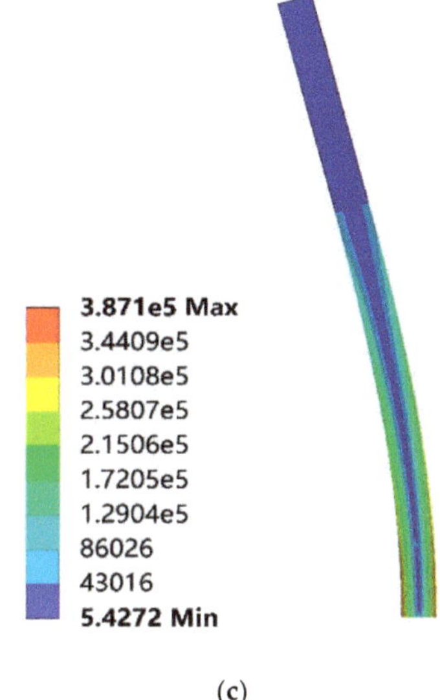

(**c**)

Figure 14. Stress (Pa) of T-shaped flexible beam at t = 10 s (**a**) Re = 25,500, (**b**) Re = 42,500, and (**c**) Re = 59,500.

Pressure contours at the solution domain were plotted in Figure 15 for t = 10 s. When flow around a T-shaped flexible beam was studied, there was a large pressure on the front surface of T-shaped flexible beams. Figure 16 also showed lower pressure regions at the top region of the T-shaped flexible beam implying flow separation. It was noted that the pressure difference between the front and back surface of T-shaped flexible beams was large. In addition to pressure contours, streamline contours were plotted in Figure 16 for t = 10 s. It was revealed that recirculation regions were formed behind the T-shaped flexible beams. Recirculation regions were near the top region of the T-shaped flexible beam, because the separation happened near the head of the T-shaped flexible beam.

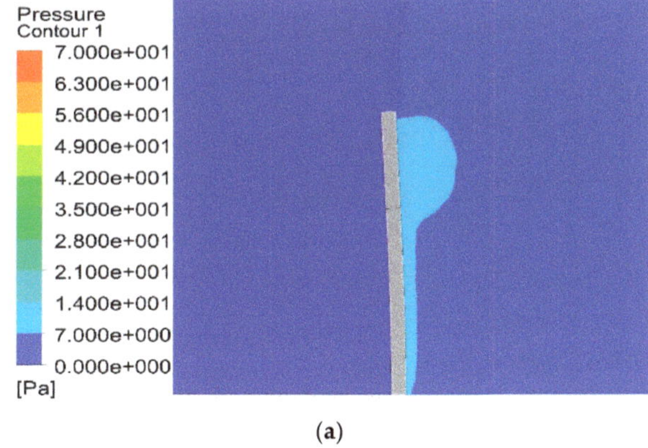

(**a**)

Figure 15. *Cont.*

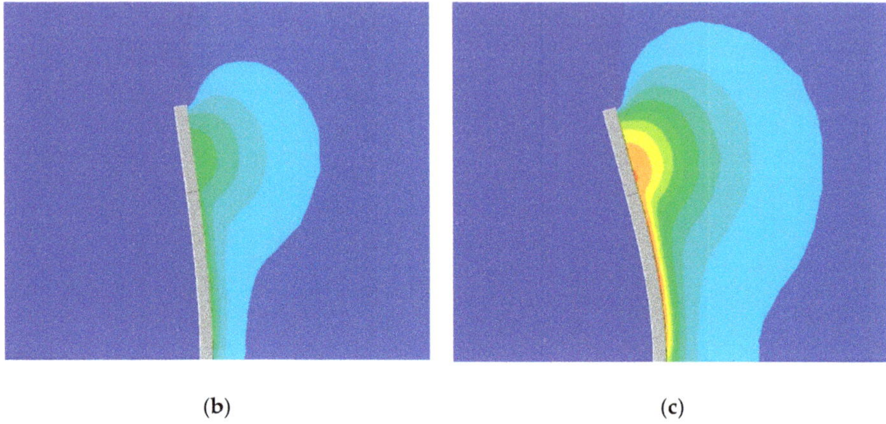

(b) **(c)**

Figure 15. Pressure distribution (Pa) of T-shaped flexible beam at t = 10 s (**a**) Re = 25,500, (**b**) Re = 42,500, and (**c**) Re = 59,500.

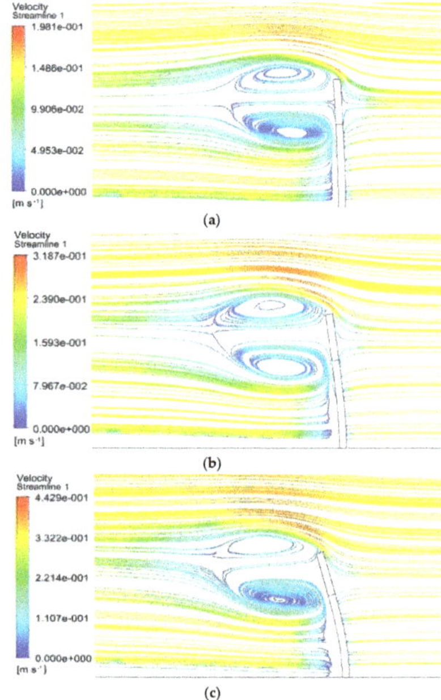

Figure 16. Streamline (m) of T-shaped flexible beam at t = 10 s (**a**) Re = 25,500, (**b**) Re = 42,500, and (**c**) Re = 59,500.

3.4. Drag Coefficients Study

When the body locates in fluid flow, it can experience a certain amount of drag force (Olcay et al. [28], Batchelor [32], and Vasudev et al. [33]). Drag force was given by

$$F_D = F_{D\_pressure} + F_{D\_viscous} = \oint P\hat{n} \cdot \hat{e}_d dS + \oint \tau_w \hat{t} \cdot \hat{e}_d dS \tag{16}$$

where $F_{D\_pressure}$ and $F_{D\_viscous}$ are drag forces in the x-direction due to the pressure and viscous effects. Here, p is the pressure on the T-shaped flexible beam, and τ_w is the wall shear stress on the surface of T-shaped flexible beam.

Drag force studies are shown in Table 5 for t = 6 s and t = 10 s. It was noted that the drag force increased with increased Reynolds numbers for T-shaped flexible beam.

Table 5. Change in drag force with the Reynolds numbers for t = 6 s and t = 10 s.

Reynolds Umbers	Drag Force, t = 6 s	Drag Force, t = 10 s
25,500 (U = 0.15 m/s)	0.09591	0.09563
42,500 (U = 0.25 m/s)	0.25932	0.25663
59,500 (U = 0.35 m/s)	0.48103	0.47961

Once the drag force was obtained, the drag coefficient was studied using Equation (9).

$$C_d = \frac{F_{Drag}}{\frac{1}{2}\rho U^2 A} \tag{17}$$

where C_d is drag coefficient, F_{Drag} is Drag force, ρ is density of the fluid, U is velocity of fluid, and A is the reference area (the frontal area of the body). Drag coefficient was plotted in Figure 17. It was also realized that the drag coefficient decreased with increased Reynolds numbers for T-shaped flexible beam.

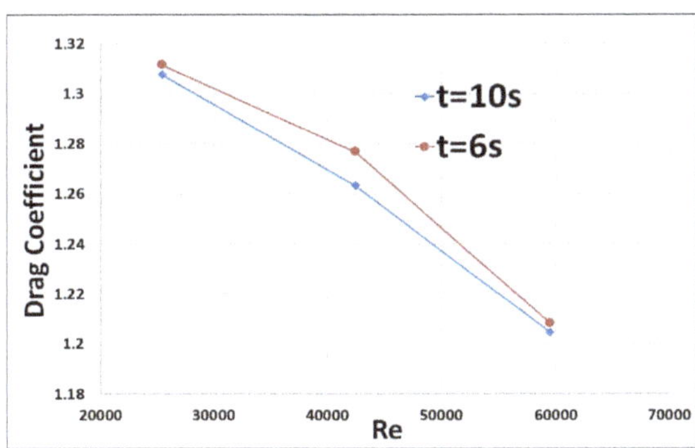

Figure 17. Change in drag coefficient with Reynolds numbers for t = 6 s and t = 10 s

4. Conclusions

In this study, the deformation of a T-shaped flexible beam was investigated at 0.25 m/s inlet velocity. A three-dimensional T-shaped flexible beam was placed into a close-circuit water channel for high-speed camera system (HSC) measurements. The results of a three-dimensional T-shaped flexible beam agreed well with the results of HSC measurements for 0.25 m/s inlet velocity. Then, two additional inlet velocities were noticed for a flexible beam, and those velocities were examined for the T-shaped flexible beam. A two-way FSI coupling method was employed for solving fluid and solid parts. The dynamic mesh method was used for grid, and mesh was updated in every time step in fluid and solid sides. Deformation, maximum stress, and minimum stress of the T-shaped flexible beam were calculated, and also, velocity distribution and pressure distribution of the flow around the T-shaped flexible beam were computed at various velocities in the numerical model. The results reveal that deformation and stress in the flexible beam has increased with increasing velocity. It was also found that a large pressure region was created on the front surface of the T-shaped flexible beam and flow separation happened in the head of the T-shaped flexible beam. It was concluded that high velocity caused the drag force to be larger when compared with low velocity, so high drag force caused a large deformation and high stress in the T-shaped flexible beam. We carried out a validation study. The results of experimental and numerical methods were compared in the present study. In this study, the percent error of maximum deformation between experimental and numerical methods is nearly 4%~5%. The study also revealed that the system coupling method can be used in fluid–structure interaction applications and a two-way FSI coupling method has high efficiency, so this method can be employed in various engineering fields such as mechanical, civil, and ocean engineering.

Author Contributions: Conceptualization, M.T.M., E.T.E., J.L., and S.M.; methodology, M.T.M., E.T.E., J.L., and S.M.; software, M.T.M. and E.T.E; validation, M.T.M., E.T.E., J.L., S.M., and G.T.; investigation, G.T.; data curation, M.T.M. and G.T.; writing—original draft preparation, M.T.M., E.T.E., J.L., S.M., and G.T. All authors have read and agreed to the published version of the manuscript.

Funding: This research received no external funding.

Conflicts of Interest: The authors declare no conflict of interest.

References

1. Chimakurthi, S.K.; Reuss, S.; Tooley, M.; Scampoli, S. ANSYS workbench system coupling: A state-of-the-art computational framework for analyzing multiphysics problems. *Eng. Comput.* **2017**, *34*, 385–411. [CrossRef]
2. Gluck, M.; Breuer, M.; Durst, F.; Halfmann, A.; Rank, E. Computation of fluid–structure interaction on lightweight structures. *J. Wind Eng. Ind. Aerodyn.* **2001**, *89*, 1351–1368. [CrossRef]
3. Dhavalikar, S.; Awasare, S.; Joga, R.; Kar, A.R. Whipping response analysis by one way fluid structure interaction—A case study. *Ocean Eng.* **2015**, *103*, 10–20. [CrossRef]
4. Narayanan, K.V.; Vengadesan, S.; Murali, K. Wall proximity effects on the flow past cylinder with flexible filaments. *Ocean Eng.* **2018**, *157*, 54–61. [CrossRef]
5. Hassani, M.; Mureithi, N.W.; Gosselin, F.P. Large coupled bending and torsional deformation of an elastic rod subjected to fluid flow. *J. Fluids Struct.* **2016**, *62*, 367–383. [CrossRef]
6. Juan, J.S.; Carrillo, G.V.; Tinoco, R.O. Experimental observations of 3D flow alterations by vegetation under oscillatory flows. *Environ. Fluid Mech.* **2019**, *19*, 1497–1525. [CrossRef]
7. Mantecon, J.G.; Neto, M.M. Numerical methodology for fluid-structure interaction analysis of nuclear fuel plates under axial flow conditions. *Nucl. Eng. Des.* **2018**, *333*, 76–86. [CrossRef]
8. Ghelardi, S.; Freda, A.; Rizzo, C.M.; Villa, D. A fluid structure interaction case study on a square sail in a wind tunnel. *Ocean Eng.* **2018**, *163*, 136–147. [CrossRef]
9. Liu, Z.G.; Liu, Y.; Lu, J. Fluid-structure interaction of single cylinder in axial flow. *Comput. Fluids* **2012**, *56*, 143–151. [CrossRef]
10. Xu, L.; Tian, F.-B.; Young, J.; Lai, J.C.S. A novel geometry-adaptive Cartesian grid based immersed boundary–lattice Boltzmann method for fluid–structure interactions at moderate and high Reynolds numbers. *J. Comput. Phys.* **2018**, *375*, 22–56. [CrossRef]
11. Wang, C.; Sun, M.; Shankar, S.; Xing, S.; Zhang, L. CFD Simulation of Vortex Induced Vibration for FRP composite riser with different modeling methods. *Appl. Sci.* **2018**, *8*, 684. [CrossRef]
12. Dong, D.; Chen, W.; Shi, S. Coupling motion and energy harvesting of two side-by-side flexible plates in a 3D uniform flow. *Appl. Sci.* **2016**, *6*, 141. [CrossRef]
13. Wang, L.; Currao, G.M.D.; Han, F.; Neely, A.J.; Young, J.; Tian, F.B. An immersed boundary method for fluid–structure interaction with compressible multiphase flows. *J. Comput. Phys.* **2017**, *346*, 131–151. [CrossRef]
14. Turek, S.; Hron, J. Proposal for numerical benchmarking of fluid-structure interaction between an elastic object and laminar incompressible flow. *Fluid Struct. Interact.* **2006**, *53*, 371–385.
15. Wang, H.; Zhai, Q.; Zhang, J. Numerical study of flow-induced vibration of a flexible plate behind a circular cylinder. *Ocean Eng.* **2018**, *163*, 419–430. [CrossRef]
16. Zheng, X.; Xue, Q.; Mittal, R.; Beilamowicz, S. A coupled sharp-interface immersed boundary-finite-element method for flow-structure interaction with application to human phonation. *J. Biomech. Eng.* **2010**, *132*. [CrossRef]

17. Wang, M.; Avital, E.J.; Bai, X.; Ji, C.; Xu, D.; Williams, J.J.; Munjiza, A. Fluid–structure interaction of flexible submerged vegetation stems and kinetic turbine blades. *Comput. Part. Mech.* **2019**. [CrossRef]

18. Mittal, R.; Dong, H.; Bozkurttas, M.; Najjar, F.M.; Vargas, A.; Von Loebbecke, A. A versatile sharp interface immersed boundary method for incompressible flows with complex boundaries. *J. Comput. Phys.* **2008**, *227*, 4825–4852. [CrossRef]

19. Nestola, M.G.C.; Becsek, B.; Zolfaghari, H.; Zulian, P.; Marinis, D.D.; Krause, R.; Obrist, D. An immersed boundary method for fluid-structure interaction based on variational transfer. *J. Comput. Phys.* **2019**, *398*, 108884. [CrossRef]

20. Peskin, C.S. The immersed boundary method. *Acta Numer.* **2002**, *11*, 479–517. [CrossRef]

21. Hron, J.; Turek, J. A monolithic FEM/multigrid solver for an ALE formulation of fluid-structure interaction with applications in biomechanics. *Fluid Struct. Interact.* **2006**, 146–170. [CrossRef]

22. Griffith, B.E.; Luo, X. Hybrid finite difference/finite element immersed boundary method. *Int. J. Numer. Methods Biomed. Eng.* **2017**, *33*. [CrossRef] [PubMed]

23. Nassiri, A.; Chini, G.; Vivek, A.; Daehn, G.; Kinsey, B. Arbitrary Lagrangian–Eulerian finite element simulation and experimental investigation of wavy interfacial morphology during high velocity impact welding. *Mater. Des.* **2015**, *88*, 345–358. [CrossRef]

24. Tabatabaei-Malazi, M.; Okbaz, A.; Olcay, A.B. Numerical investigation of a longfin inshore squid's flow characteristics. *Ocean Eng.* **2015**, *108*, 462–470. [CrossRef]

25. Olcay, A.B.; Tabatabaei-Malazi, M.; Okbaz, A.; Heperkan, H.A.; Firat, E.; Ozbolat, V.; Gokcen, M.G.; Sahin, B. Experimental and numerical investigation of a longfin inshore squid's flow characteristics. *J. Appl. Fluid Mech.* **2017**, *10*, 21–30. [CrossRef]

26. Eren, E.T.; Tabatabaei-Malazi, M.; Temir, G. Numerical investigation on the collision between a solitary wave and a moving cylinder. *Water* **2020**, *12*, 2167. [CrossRef]

27. Howse, J. *OpenCV Computer Vision with Python*; Packt Publishing Ltd.: Birmingham, UK, 2013.

28. Olcay, A.B.; Tabatabaei-Malazi, M. The effects of a longfin inshore squid's fins on propulsive efficiency during underwater swimming. *Ocean Eng.* **2016**, *128*, 173–182. [CrossRef]

29. *ANSYS Fluent Theory Guide*; ANSYS, Inc.: Canonsburg, PA, USA, 2016; pp. 39–136.

30. Bathe, K.J. *Finite Element Procedures*; Prentice-Hall: Englewood Cliffs, NJ, USA, 1996.

31. Chung, J.; Hulbert, G.M. A time integration algorithm for structure dynamic with improved numerical dissipation: The generalized-α method. *J. Appl. Mech.* **1993**, *60*, 371. [CrossRef]

32. Batchelor, G.K. *An Introduction to Fluid Dynamics*; Cambridge University Press: Cambridge, UK, 2000.

33. Vasudev, K.L.; Sharma, R.; Bhattacharyya, S.K. A multi-objective optimization design framework integrated with CFD for the design of AUVs. *Methods Oceanogr.* **2014**, *10*, 138–165. [CrossRef]

Journal of
Marine Science and Engineering

Article

Derivation of Engineering Design Criteria for Flow Field Around Intake Structure: A Numerical Simulation Study

Lee Hooi Chie [1,* and Ahmad Khairi Abd Wahab [1,2]

[1] School of Civil Engineering, Faculty of Engineering, Universiti Teknologi Malaysia, Johor Bahru 81310, Malaysia; akhairi@utm.my
[2] Centre for Coastal and Ocean Engineering, Universiti Teknologi Malaysia, Jalan Sultan Yahya Petra, Kuala Lumpur 54100, Malaysia
* Correspondence: hooichie@gmail.com

Received: 25 August 2020; Accepted: 16 October 2020; Published: 21 October 2020

Abstract: The primary environmental impact caused by seawater intake operation is marine life impingement resulting from the intake velocity. Environmental Protection Agency (EPA) of United State has regulated the use of velocity cap fitted at intake structures to reduce the marine life impingement. The engineering design parameters of velocity cap has not been well explored to date. This study has been set to determine the fundamental relationships between intake velocity and design parameters of velocity cap, using computational fluid dynamic (CFD) model. A set of engineering design criteria for velocity cap design are derived. The numerical evidence yielded in this study show that the velocity cap should be designed with vertical opening (H_{vc}) and horizontal shelf (ℓ_{vc}). The recommended intake opening ratio (O_r) shall be 0.36 $V_r^{-0.31}$, where $O_r = H_{vc}/\ell_{vc}$ and $V_r = V_0/V_{pipe}$. V_o is the velocity at the intake window and V_{pipe} is the suction velocity at the intake pipe. The volume ratio (ω_r) between the velocity cap (ω_{vc}) and intake tower (ω_{IT}) is recommended at 0.11 $V_r^{-1.23}$. The positive outlooks that yielded from this study can be served as a design reference for velocity cap to mitigate the detrimental impacts from the existing intake structure.

Keywords: coastal structure; fluid-structure interaction; engineering design parameters; environment protection; intake velocity; velocity cap

1. Introduction

Seawater intake structure are widely used by most coastal plants. A seawater intake structure usually made of reinforced concrete supporting the inlet of the withdrawal pipe. Figure 1 shows the photo examples of existing intake structures at Plant of Aguilas, Plant of Nungua, Plant of Skikda, and Plant of Campo De Cartagena [1]. Most of the intake towers are circular in shape with diameter ranging between 4.7 and 5.3 m. The intake tower height is between 5 and 7 m with intake rate between 1 and 7 m^3/s. However, depending on the intake rate, the diameter of the intake tower can be ranging between 2 and 20 m [2]. The shape of the intake structure can be rectangular if it is designed to be in current, with shorter sides angled against the current. The marine life impingement and entrainment resulting from the intake operation is a major environment concern [3]. An optimum intake structural design should be operated in a way that minimizes marine impacts, particularly considering the impingement and entrainment of marine life. There are substantial literatures that points to increase intake mortality with increase intake velocities [4–6]. Even though fish are excellent swimmers, they can often be drawn in by vertical currents generated by intake velocity. The Environment Protection Agency of United State (USEPA) has suggested the use of velocity caps fitted at the intake structures to convert the water flow from vertical to horizontal [6]. The velocity cap, which is usually made of steel or concrete, is a simple modification to the unscreened intake in the open sea to draw water in horizontally. The used of velocity cap can reduce marine entrainment because fish are adapted to respond to horizontal current rather than vertical current fluctuations [7]. A change in horizontal flow pattern created by velocity cap will triggers an avoidance response mechanism in fish, which aids to reduce the marine life impingement [8]. The use of velocity cap can reduce 80% to 90% of the fish impingement at the intake entrance [9]. Based on this record of performance, USEPA has promulgated final ruling under the Clean Water Act to restrict the intake velocity to a maximum of 0.15 m/s and specify that seawater intake structures should be equipped with a velocity cap [10]. Electric Power Research Institute [11] has performed comprehensive literature reviews on the swimming capabilities of marine, freshwater, and estuarine fishes to determine the appropriateness of regulating 0.15 m/s as the maximum allowable intake velocity to preclude impingement impacts. They have concluded that intake velocity is an appropriate regulatory parameter, and it should be measured, preferably, as a vector parallel to the main water flow at the intake window. They highlighted that the 0.15 m/s intake velocity criterion can be useful to delineate where significant impingement impacts are unlikely to happen under common environmental conditions. Several federal and state agencies in US have also developed the intake velocity criteria to protect local populations of fish from being impinged at the intake window, and they generally proposed maximum allowable intake velocity of 0.15 m/s as an acceptable indicator of likely low occurrence of impingement problems at the intake structure [12,13].

Plant of Aguilas (Murcia)   **Plant of Nungua (Ghana)**

Plant of Skikda (Algeria)

Plant of Campo De Cartagena

Figure 1. Photo examples of the existing seawater intake structure [1].

Previous literature and sources of information reviewed did mentioned that velocity cap is one of the best technology available (BTA) and it shall result in a design intake velocity that less than or equal to 0.15 m/s. Figure 2 shows the schematic illustration of the flow in a capped intake structure. It is reasonable to speculate that the local flow conditions induced by the intake structure are the important fundamental to protect marine life from impingement mortality. However, to the author's best knowledge, very little research work has been conducted in the area of engineering design criteria for velocity cap. Voutchkov (2018) [2] suggested the vertical opening of the velocity cap shall be between 1 and 3 m. Schuler and Larson (1975) [14] suggested that the horizontal distance of the velocity cap shall extend approximately 1.5 times the vertical opening. A key problem with many of these suggestions is that there are mostly based on "rules of thumb", which have not been scientifically tested. This raises many questions about whether those intake design suggestions are suitable to be used for various structural configuration and intake rate. A variety of engineering properties, i.e., vertical opening and horizontal shelf of velocity cap are important contributory factors to influence the intake velocity. The literature review revealed that no single source of information or document precisely contained velocity cap design standards. Intake velocity requirement is significant and that this paper is essential for the understanding of fundamental relationships between the design parameters and intake velocity to derive engineering design criteria for environment protection.

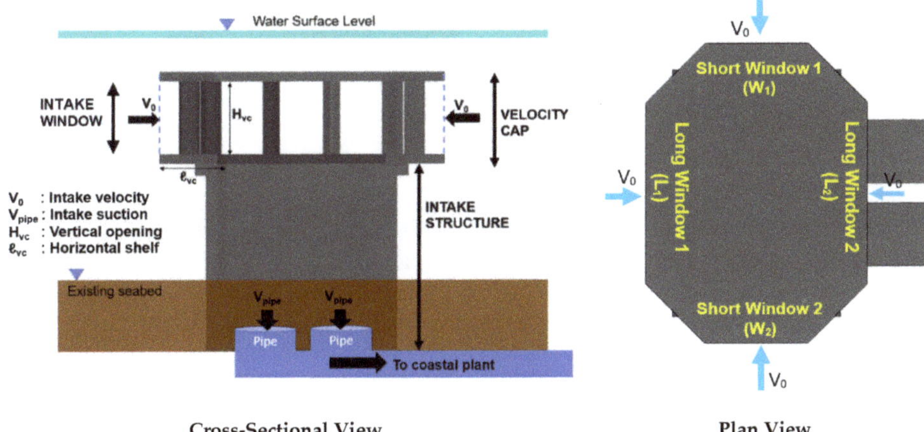

Figure 2. Schematic illustration of the flow in intake structure with velocity cap.

A vast research has been undertaken to study the flow field around the coastal structure by using CFD models [15–18]. They have concluded that, with the advancement in numerical modelling tools, it is possible to analyze the fluid–structure interactions with numerical models. This paper presents the application of CFD model to simulate the flow field around an existing capped intake structure that located in Penang Strait of Malaysia. The existing capped intake structure was used to draw ambient seawater for power plant usage. The feasibility of the present numerical model was validated by comparing with the measured field values that obtained from secondary data source [19]. With proper data checking and data screening, field data has the advantage of being able to observe the outcome in a natural setting rather than in a contrived laboratory environment. Subsequently, the relationships between influencing design parameters and intake velocity are systematically analyzed and presented in SI unit. Finally, the engineering design criteria for intake structure is recommended.

2. Method

2.1. Numerical Model

The numerical investigation was undertaken using FLOW3D based on Reynolds' Averaged Navier–Stokes (RANS) equations. The FLOW3D solver is structured upon the principle laws of mass, momentum and energy conservation, and the equations were solved using finite difference method. The central aim of any flow model is to provide projections of turbulent fluctuations on the flow quantities. This is commonly represented by including diffusion terms in the following mass and momentum transport equation:

$$V_F\frac{\partial \rho}{\partial t} + \frac{\partial}{\partial x}(\rho u A_x) + R\frac{\partial}{\partial y}(\rho v A_y) + \frac{\partial}{\partial z}(\rho w A_z) + \xi\frac{\rho u A_x}{x} = R_{DIF} + R_{SOR} \tag{1}$$

where u, v, and w are fluid velocities in the Cartesian coordinate directions (x, y, z), A_x, A_y, and A_z are the fractional areas open to flow in the x, y, and z axis. V_F is the fractional volume open to flow, ρ is the fluid density, then R and ξ are coefficients that depend upon the choice of coordinate system. R_{DIF} and R_{SOR} are respectively the turbulent diffusion and mass source terms and are defined in Equations (2) and (3) below:

$$R_{DIF} = \frac{\partial}{\partial x}\left(v_\rho A_x\frac{\partial \rho}{\partial x}\right) + R\frac{\partial}{\partial y}\left(v_\rho A_y R\frac{\partial \rho}{\partial y}\right) + \frac{\partial}{\partial z}\left(v_\rho A_z\frac{\partial \rho}{\partial z}\right) + \xi\frac{v_\rho A_x}{x}\frac{\partial \rho}{\partial x} \tag{2}$$

$$\frac{\partial}{\partial x}(u A_x) + R\frac{\partial}{\partial y}(v A_y) + \frac{\partial}{\partial z}(w A_z) + \xi\frac{u A_x}{x} = \frac{R_{SOR}}{\rho} \tag{3}$$

where $v_\rho = S_c\mu/\rho$, in which S_c is the turbulent Schmidt number and μ is the coefficient of momentum diffusion. The 3D equations of motion are solved with the following Navier–Stokes equations with some additional terms:

$$\begin{aligned}
\frac{\partial u}{\partial t} + \frac{1}{V_F}\left\{u A_x\frac{\partial u}{\partial x} + v A_y R\frac{\partial u}{\partial y} + w A_z\frac{\partial u}{\partial z}\right\} - \xi\frac{A_y v^2}{x V_F} &= -\frac{1}{\rho}\frac{\partial \rho}{\partial x} + G_x + f_x - b_x - \frac{R_{SOR}}{\rho V_F}(u - u_w - \partial u_s) \\
\frac{\partial v}{\partial t} + \frac{1}{V_F}\left\{u A_x\frac{\partial v}{\partial x} + v A_y R\frac{\partial v}{\partial y} + w A_z\frac{\partial v}{\partial z}\right\} + \xi\frac{A_y uv}{x V_F} &= -\frac{1}{\rho}\frac{\partial \rho}{\partial y} + G_y + f_y - b_y - \frac{R_{SOR}}{\rho V_F}(v - v_w - \partial v_s) \\
\frac{\partial w}{\partial t} + \frac{1}{V_F}\left\{u A_x\frac{\partial w}{\partial x} + v A_y R\frac{\partial w}{\partial y} + w A_z\frac{\partial w}{\partial z}\right\} &= -\frac{1}{\rho}\frac{\partial \rho}{\partial z} + G_z + f_z - b_z - \frac{R_{SOR}}{\rho V_F}(w - w_w - \partial w_s)
\end{aligned} \tag{4}$$

where t is the time, G_x, G_y and G_z are accelerations due to gravity, f_x, f_y, and f_z are viscous accelerations, and b_x, b_y, and b_z are the flow losses in porous media, u_w, v_w, and w_w are velocity of the source component, u_s, v_s, and w_s are velocity of the fluid at the surface of the source.

k-ε RNG turbulence model was suggested by Chie and Wahab (2019) [20] as the viscous model to simulate the flow kinematics around the intake structure. They tested the performance of four turbulence models, namely the standard k-epsilon (k-ε), k-ε renormalized group (RNG), k-omega (k–ω), and large eddy simulation (LES) models and concluded that k-ε RNG model can provide good accuracy and reduce the computational costs compared to the LES model. Furthermore, the k-ε RNG model [21] extend the capabilities of k-ε model to provide better coverage of low intensity turbulence flows and flow in areas with strong shear.

2.2. Model Setup

An existing seawater intake structure was used as a basis for the intake model. The intake structure is fully submerged underwater with a submergence of 2.2 m from mean sea level (MSL) and is extracting water at a rate of 25.43 m³/s. The existing structure is partially buried in the seabed (~4.8 m below the existing seabed level) and the water depth during MSL is approximately 10 m.

The intake structure is composed of 2 major components: velocity cap and intake tower. The velocity cap size of 9.2 m × 5.2 m has 3.1 m horizontal shelf (ℓ_{vc}) and 2.6 m vertical opening (H_{vc}) to convey the flow into the intake tower. The intake tower is rectangular in shape with an outer large and long of 6.2 m × 10.2 m. The intake tower opening size is 5 m large × 9 m long. The total height of intake tower is 9.98 m. Figure 3 shows the overall dimensions of the intake structure.

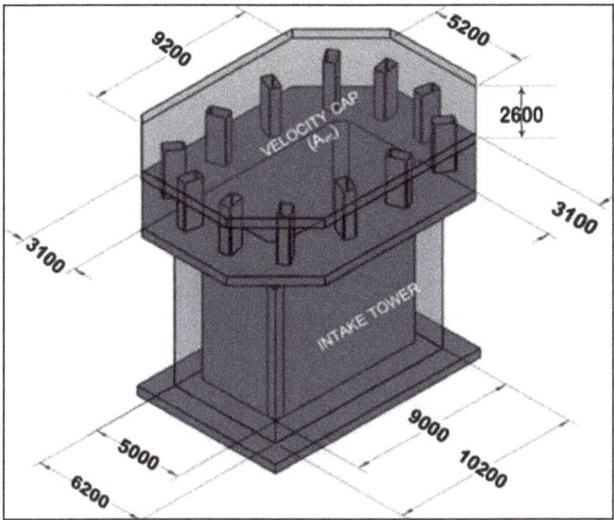

Figure 3. The overall dimensions (in unit mm) of the intake model.

A model with domain sizes of 41 m × 45m × 18 m is constructed for this study. The measured bathymetric data is imported into the model using a universal terrain representation raster format to provide a realistic bed level for modelling. The 3D raster bathymetric map is shown in Figure 4. The intake structure was incorporated into the model bathymetry to mimic the actual site condition.

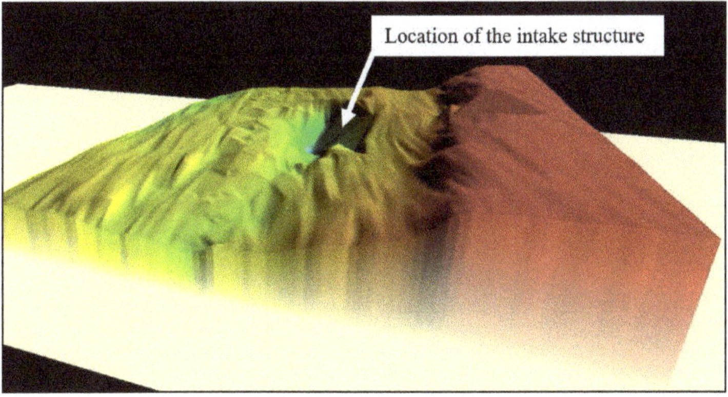

Figure 4. 3D bathymetric map used for model setup.

The boundary condition (BC) of the model domain is defined as velocity, pressure, wall, and mass momentum source. The kinetic energy and dissipation rate are calculated based upon the computational formulas of turbulence quantities at the velocity boundary. The published tidal level [22] is prescribed with pressure-type boundary. The wall boundary is defined at seafloor with no tangential velocities. The model domain is setup with single incompressible fluid with free surface. Fluid fraction (F) = 0 correspond to void region, in which a uniform atmospheric pressure is applied. Mass momentum sinks were added to the intake structure outlet to withdraw water from the model domain. The model boundary conditions are illustrated in Figure 5. One downside regarding the methodology is that the wave effects are not considered in this study. Further data collection would be needed to determine exactly how the wave effects affects the engineering design criteria for the intake structure.

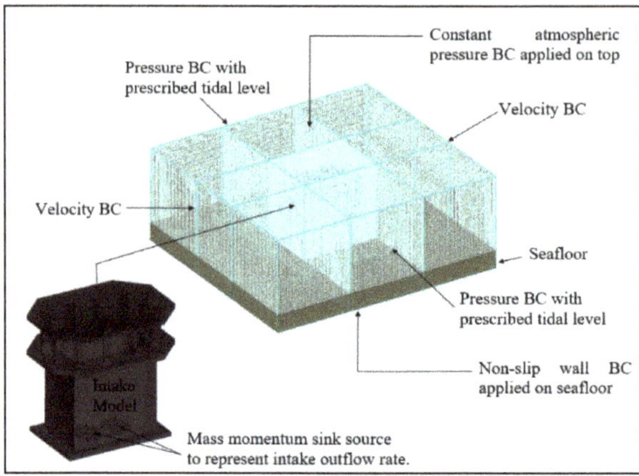

Figure 5. Model boundary conditions.

2.3. Mesh Sensitivity Study

Mesh sensitivity study was performed for three computational grids that consisting of 355,296 (M1), 792,028 (M2), and 1,682,826 (M3) nodes. It is suggested by ASME (2009) [23] that the grid refinement should be conducted systematically, and it is highly recommended that the grid refinement factor, $r = h_{coarse}/h_{fine}$, (where h is the grid size) should be greater than 1.3 for most practical problems. Table 1 summarizes the grid information for the tested grid. An extraction rate of 25.43 m^3/s is added as mass momentum source, and the intake velocity is recorded at four intake windows. Mesh independent solutions are achieved when the differential between two intake velocities (M1-M2 and M2-M3) is less than +/−0.01 m/s. Table 2 tabulated the comparison of intake velocity for M1, M2, and M3 grids. The results show that the maximum velocity differences between M1 and M2 grid is 0.03 m/s. By increasing the grid resolution, the results show a reduction in velocity differences. The maximum velocity differences between M2 and M3 is generally less than 0.01 m/s. This indicates that the M2 grid has reached a solution value that is independent of the grid resolution. Therefore, the optimum grid is M2 with resolution ranging between 0.22 m and 0.4 m.

Table 1. Tested gird for mesh sensitivity study.

Grid Info \ Tested Grid	Coarse Grid (M1)	Medium Grid (M2)	Fine Grid (M3)
Total grid cells	355,296	792,028	1,682,826
Min. grid size (h)	0.29	0.22	0.17
Max. grid size (h)	0.52	0.40	0.31
Grid refinement factor (r)		$h_{M1}/h_{M2} = 1.3$	$h_{M2}/h_{M3} = 1.3$

Table 2. Comparison of intake velocity for M1, M2, and M3 grids.

Tested Grid	No. Grid Cells	Intake Velocity (m/s)			
		Short Window 1	Differential (M2−M1) (M3−M2)	Short Window 2	Differential (M2-M1) (M3-M2)
Coarse Grid (M1)	355,296	0.21		0.23	
Medium Grid (M2)	792,028	0.24	0.03	0.24	0.01
Fine Grid (M3)	1,682,826	0.23	−0.01	0.24	0.00
Tested Grid	No. grid cells	Long Window 1	Differential (M2-M1) (M3-M2)	Long Window 2	Differential (M2-M1) (M3-M2)
Coarse Grid (M1)	355,296	0.26		0.24	
Medium Grid (M2)	792,028	0.25	−0.01	0.25	0.01
Fine Grid (M3)	1,682,826	0.26	0.01	0.25	0.00

2.4. Model Validation

Secondary velocity measurements were adopted for model validation. The tide at the Penang Strait is semi-diurnal with two high tides and low tides in a tidal day with comparatively little diurnal inequality. The measurements, covering both high and low tides, were captured by three Acoustic Doppler Current Profilers (ADCP1, ADCP2 and ADCP3). ADCP1, ADCP2, and ADCP3 were respectively located at 0.5 m, 2 m, and 1 m from the velocity cap as illustrated in Figure 6. The high resolution FLOW3D CFD model was used to reproduce velocity at these locations. The simulated velocity was validated with the measured data, and the results are graphically shown in Figure 7. It is plausible that slightly higher error percentages will be obtained when the validation results are based on field surveys where data was collected in an uncontrolled environment. Department of Irrigation and Drainage (DID) Malaysia suggested the root mean square percentage error (% RMSE) between the measured and simulated data shall be less than 20% for any coastal hydraulic study and impact assessment [24]. Table 3 summarizes the discrepancies between the simulated and measured data with % RMSE, squared of Pearson product-moment correlation coefficient (r^2) and mean absolute error (MAE). The RMSE percentage of all ADCP locations are less than 10%, and r^2 are well above 0.9. This demonstrates that the model is capable and suitable to be used for the investigation of flow field around the intake structure.

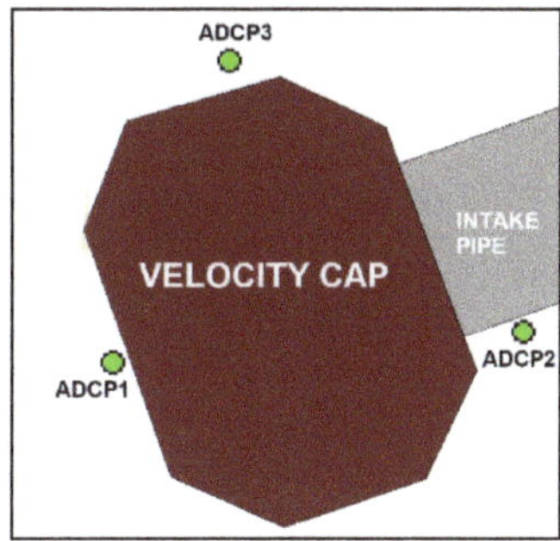

Figure 6. The measurement locations of ADCP1, ADCP2, and ADCP3.

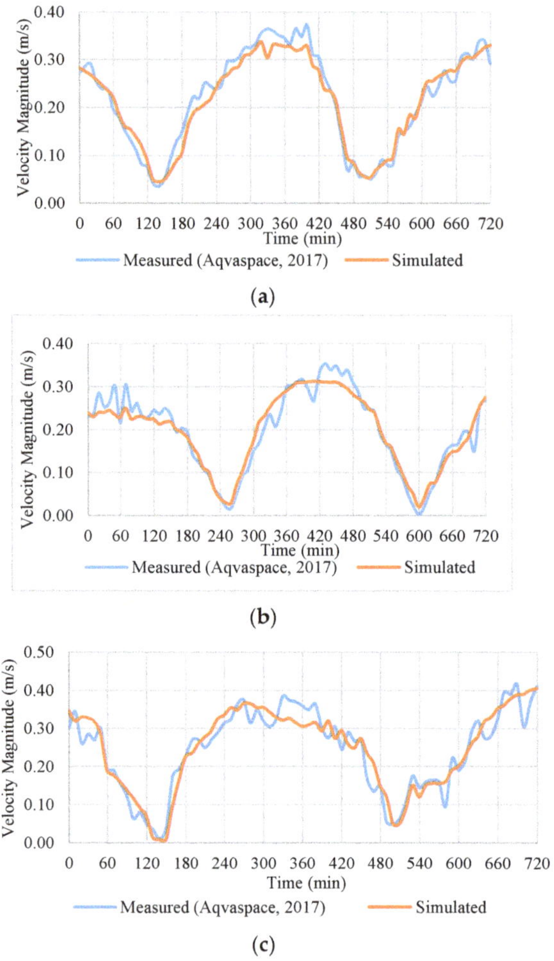

Figure 7. Comparison of velocity magnitude from numerical model with measurement data at (**a**) ADCP1 (**b**) ADCP2, and (**c**) ADCP3.

Table 3. Comparison of velocity magnitude between simulated and measured data.

Station	%RMSE	r^2	MAE (m)
ADCP1	6.81	0.98	0.02
ADCP2	7.16	0.97	0.02
ADCP3	9.63	0.94	0.03

3. Results and Discussion

The validated model was used to study the intake velocity at the velocity cap. Figure 8 illustrates the flow field around the existing intake structure during flood, ebb, and slack current conditions. During the flood and ebb currents, it can be clearly perceived that the flow field around the intake structure is mainly influenced by the tidal current. The intake structure increases the flow velocity upstream but decreases the flow velocity downstream. The slow flow region is observed leeward of the structure, which could benefit sheltering effects for fishes [25,26]. The tidal flow passing through the intake window is affected by the compressive interference of the intake suction, which generate velocity gradients around the intake windows. However, the velocity gradients around the intake windows are mild during tidal flow and potentially stimulate an avoidance response in fishes, which aids to reduce marine life impingement.

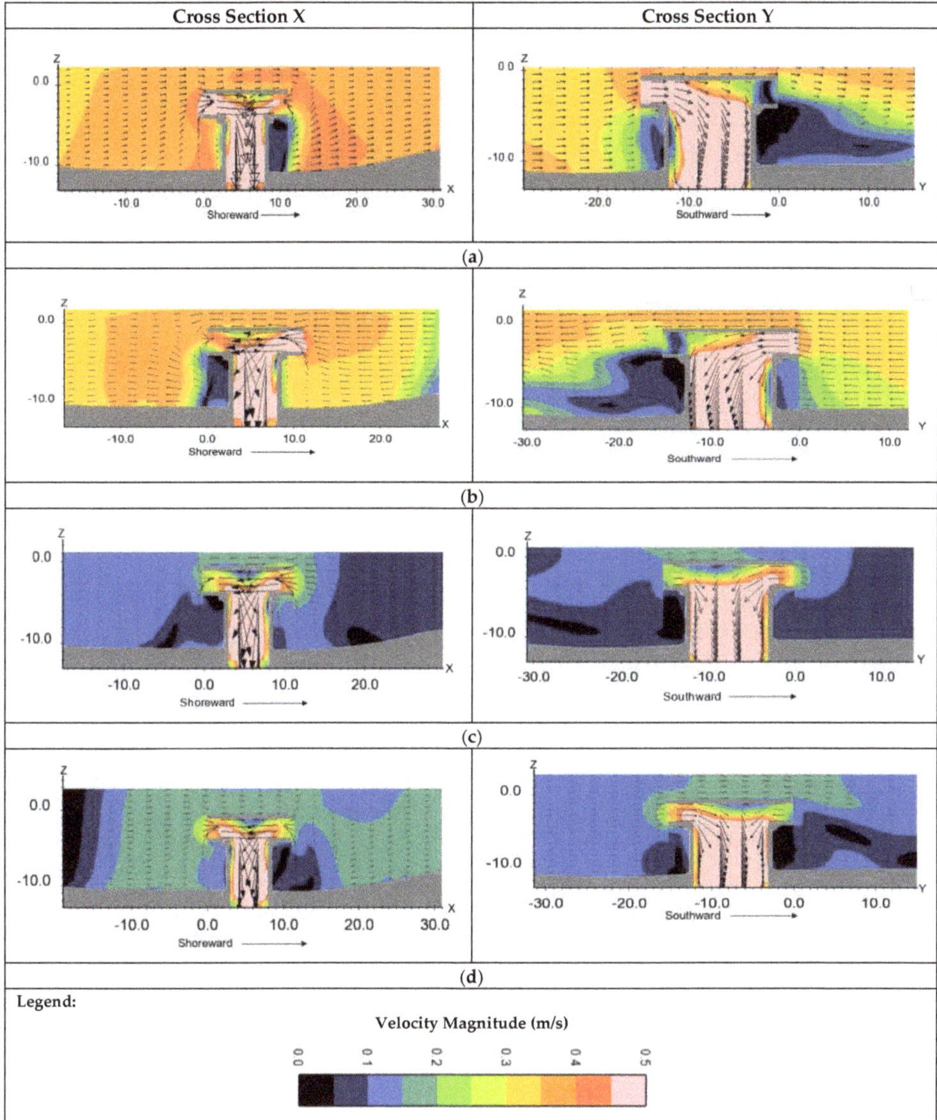

Figure 8. Simulation of flow field around the existing intake structure during flood, ebb, and slack currents (**a**) Flood Current; (**b**) Ebb Current; (**c**) Slack Ebbing Current; (**d**) Slack Flooding Current.

The weakest tidal currents, which happen during slack tides, occur during the transition from flood to ebb currents, and vice versa. At this point of time, the tidal currents decrease rapidly to nearly stagnant and the flow field around the intake structure is mainly attributed by the intake suction. The velocity gradients around the intake windows are strong and could potentially cause the marine life impingement and entrainment. Therefore, slack current condition (with tidal current = 0) will be adopted in this study to derive the fundamental relationship between intake velocity and design parameters of velocity cap.

The design of the velocity cap is influenced by the parameters as illustrated in Figure 9. Basic terminology must first be established to clearly evaluate the intake velocity in this study. For this study, the term "intake velocity" is represented by V_0—the velocity magnitude measured at the vertical opening of the velocity cap. The velocity magnitude was computed with Equation (5) where u, v, and w. are fluid velocities in the Cartesian coordinate directions (x, y, z) computed from governing equation of FLOW3D.

$$V_0 = \sqrt{u^2 + v^2 + w^2} \tag{5}$$

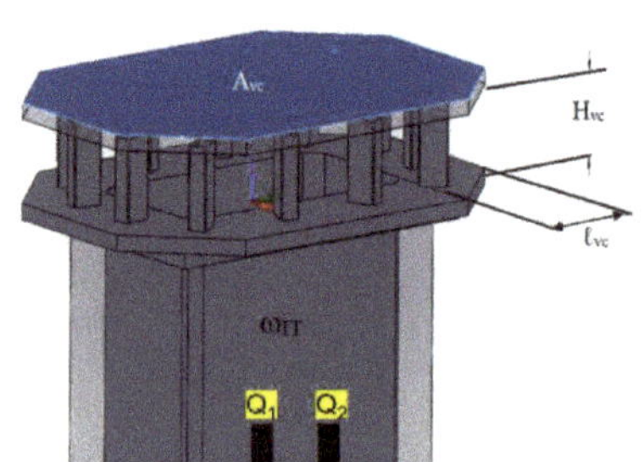

Design parameters for velocity cap:

A_{vc} = Area of the velocity cap (m²)

H_{vc} = Vertical opening (m)

ℓ_{vc} = Horizontal shelf (m)

$\omega_{vc} = A_{vc} \times H_{vc}$ = Size of velocity cap (m³)

ω_{IT} = Size of intake tower (m³)

$Q = Q_1 + Q_2$ = Total intake rate (m³/s)

Figure 9. Design parameters for velocity cap.

To analyze V_0 in the following sections, V_0 profiles were extracted along the vertical openings (H_{vc}) at the middle of the four windows (two long windows at L/2 and two short windows at W/2, as shown in Figure 2). The intake velocity profile at each window was analysed to determine the average V_0. The term "intake velocity ratio (V_r)" is used in this study to represent the dimensionless velocity that considers the ratio of intake velocity (V_0) to average suction velocity at the intake pipe (V_{pipe}) ($V_r = V_0/V_{pipe}$). V_{pipe} is determined by using Equation (6), where Q is the intake rate and A_{pipe} is the pipe area that is calculated with Equation (7). V_r can be used to evaluate the intake structural performance. Lower V_r indicates better structural performance in reducing intake velocy, and vice versa. The use of the dimensionless intake velocity ratio (V_r) is more appropriate and makes the application more generalizable to various intake rates.

$$V_{pipe} = \frac{Q}{A_{pipe}} \tag{6}$$

$$A_{pipe} = \frac{\pi D_{pipe}^2}{4} \tag{7}$$

3.1. Fundamental Relationships

3.1.1. Effect of Horizontal Shelf (ℓ_{vc})

A total of five test cases were conducted to evaluate the effect of ℓ_{vc} on V_r. H_{vc} and ω_{IT} in all test cases are respectively set at 5.5 m and 450 m³. A range of intake rates (Q = 45, 50, and 60 m³/s) are considered in this section. The analyzed simulation results are presented in Figure 10 as a scatter plot. The best fit equation and 95% confidence interval (CI) are presented. The results of a regression analysis revealed that V_r and ℓ_{vc} have a negative non-linear relationship. V_r can be reduced by increasing the length of ℓ_{vc}. The relationship between V_r and ℓ_{vc} can be expressed with the power function:

$V_r = 0.15\,\ell_{vc}^{-0.63}$. All simulated data are lies within the 95% CI. There is a 95% probability that the best fit equation between V_r against ℓ_{vc} lies within the confidence interval.

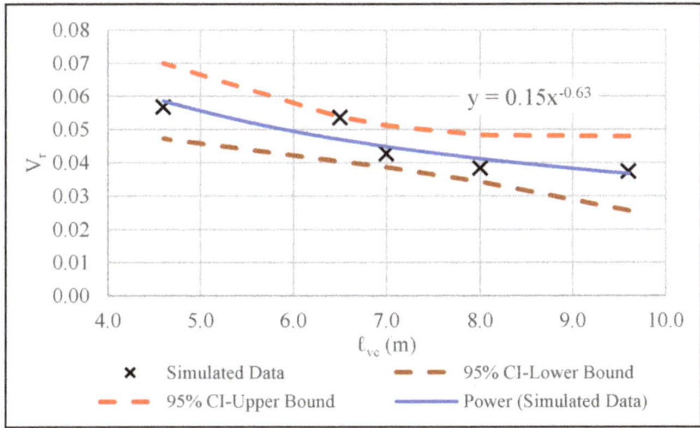

Figure 10. Relationship between V_r against ℓ_{vc}.

3.1.2. Effect of Vertical Opening (H_{vc})

The investigation continued by evaluating the influence of H_{vc} to V_r, with 17 test cases. ℓ_{vc}, A_{vc}, and ω_{IT} in all test cases are respectively set at 3.1 m, 152 m^2, and 450 m^3. Intake rate ranging between 6 and 30 m^3/s are considered in this section. The analyzed simulation results are presented in Figure 11. The results of regression analysis revealed that V_r and H_{vc} have a negative non-linear relationship. V_r can be reduced by the increasing H_{vc}. The relationship between V_r and H_{vc} can be expressed by the equation: $V_r = 0.4H_{vc}^{-0.9}$. All simulated data are lies within the 95% confidence interval bands. The best fit equations have 95% confidence interval.

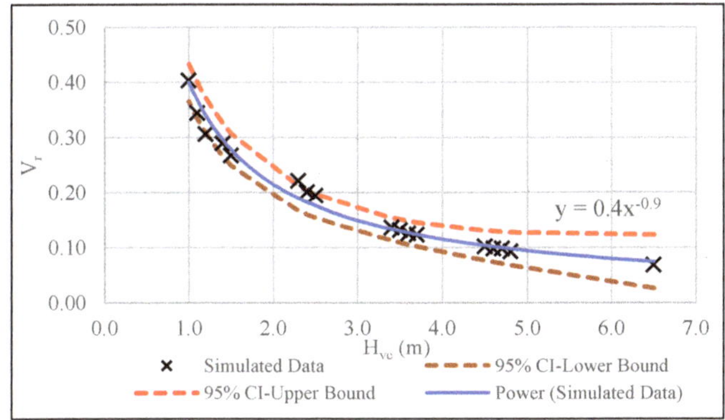

Figure 11. Relationship between V_r against H_{vc}.

3.1.3. Effect of Velocity Cap Size (ω_{vc})

The cumulative influence of the velocity cap design parameters (A_{vc}, H_{vc}, and ℓ_{vc}) to V_r is investigated by using the term "velocity cap size". The velocity cap size (ω_{vc}) is calculated by $A_{vc} \times H_{vc}$, where A_{vc} depends upon the velocity cap geometry and is influenced by ℓ_{vc}. A total of 40 test cases with the following structural design range were adopted in the simulations:

- Intake rate (Q): 6–60 m^3/s

- Vertical opening (H_{vc}): 1.0–6.5 m
- Horizontal shelf (ℓ_{vc}): 3.1–9.6 m
- Area of the velocity cap (A_{vc}): 152–520 m^2
- Size of velocity cap (ω_{vc}): 152–3122 m^3
- Size of intake tower (ω_{IT}): 450 m^3

The simulated results are plotted as scatter plot with best fit equation and 95% Confidence interval (CI) bands, as shown in Figure 12. The results of a regression analysis revealed that V_r and ω_{vc} are fitted by the nonlinear power function with a negative relationship. With the increasing ω_{vc}, V_r is reduced, and the intake performance is increased. However, it is crucial to note that the influence of ω_{vc} to V_r is reduced significantly when $\omega_{vc} > 1000$ m^3.

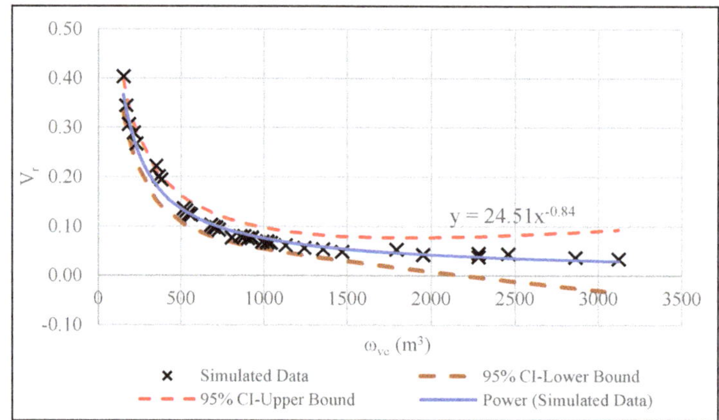

Figure 12. Relationship between V_r against ω_{vc}.

3.2. Engineering Design Criteria

The previous findings indicated the potential application of ℓ_{vc}, H_{vc}, and ω_{vc} to control the intake velocity. In this section, the combined effects of ℓ_{vc}, H_{vc}, ω_{vc}, and ω_{IT} on V_r are investigated. The volume size (ω) is used in this study to make the application more generalizable to the intake tower that is mostly circular in shape. Based on the normal engineering practice [1], the height of intake tower is mostly ranging between 5 and 9 m for constructability considerations (i.e., seabed bathymetry and distance of the structure from the shore). The term "volume ratio (ω_r)", where $\omega_r = \omega_{vc}/\omega_{IT}$, used in this section is a dimensionless term to represent the ratio of velocity cap size to intake tower size. The term "Window opening ratio (O_r)", where $O_r = H_{vc}/\ell_{vc}$, is a dimensionless term to represent the ratio of vertical opening to horizontal shelf. A total of 115 test cases with the following structural design range were adopted in the simulations:

- Intake rate (Q): 6–60 m^3/s
- Vertical opening (H_{vc}): 1.0–6.5 m
- Horizontal shelf (ℓ_{vc}): 3.1–9.6 m
- Area of the velocity cap (A_{vc}): 152–520 m^2
- Size of velocity cap (ω_{vc}): 152–3122 m^3
- Size of intake tower (ω_{IT}): 240–450 m^3

Figure 13 demonstrates the influence of ω_r to V_r. The results of a regression analysis highlighted a strong negative non-linear relationship between the V_r and ω_r. The best fit equation between V_r and ω_r is $0.16\omega_r^{-0.81}$ with 95% confidence interval. Majority of the simulated data lies within the upper and lower bounds. It is noted that V_r is reduced by the increasing volume ratio ω_r. However, it is important to note that the influence of ω_r to V_r is insignificant when $\omega_r > 4$. Figure 14 demonstrates the combined effects of O_r and ω_r to V_r. The results of a regression analysis indicated a strong negative non-linear relationship between V_r, O_r and ω_r, with the best fit equation of $V_r = 0.12(O_r\omega_r)^{-0.65}$. A set of design parameter formulations were derived and is summarized in Table 4.

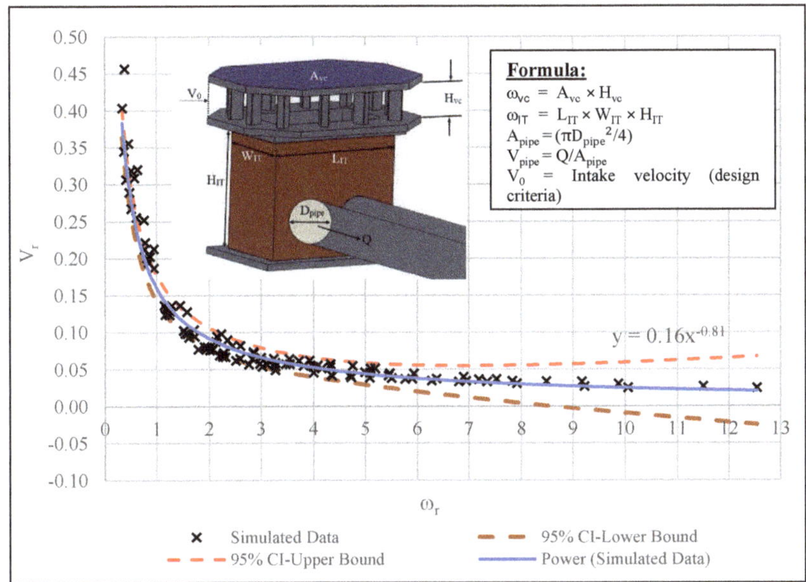

Figure 13. Relationship between V_r against ω_r.

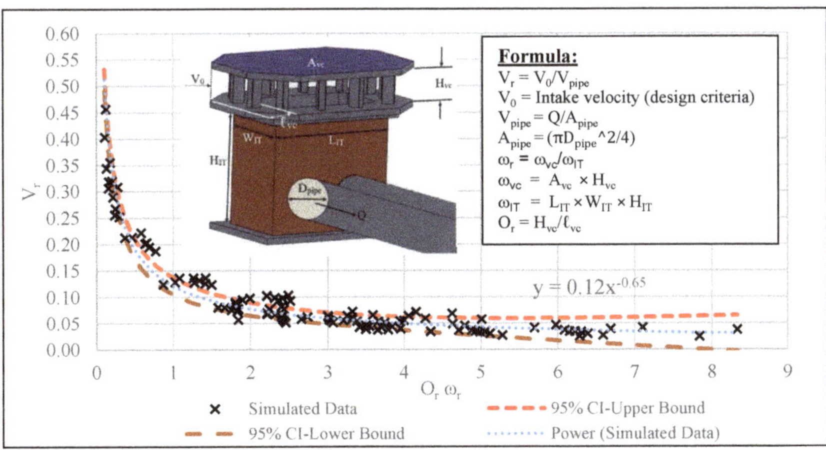

Figure 14. Relationship between V_r against $O_r\omega_r$.

Table 4. Engineering design criteria for seawater intake structure.

Design Parameter	Symbol	Design Formulation
Intake velocity ratio	V_r	V_0/V_{pipe} where V_0 is the design criteria for intake velocity
Pipe velocity (m/s)	V_{pipe}	Q/A_{pipe} where Q = intake rate (m^3/s) and $A_{pipe} = \pi D_{pipe}{}^2/4$
Intake opening ratio	O_r	$0.36 \times V_r^{-0.31}$. where $O_r = H_{vc}/\ell_{vc}$
Volume ratio	ω_r	$0.11 \times V_r^{-1.23}$ where $\omega_r = \omega_{vc}/\omega_{IT}$

Note: The formulations are in SI unit.

The main design consideration for a seawater intake structure is the impact on fish impingement at intakes, resulting from intake velocities (V_0). The Environmental Protection Agency (EPA) of the United States (US) has promulgated final ruling under the Clean Water Act to restrict the through-screen velocity (comparable to the term "intake velocity" used in this study) of a water intake structure to a maximum 0.15 m/s, and specify that the intake structures should be equipped with a velocity cap. The intake velocity performance standard proposed by USEPA is specifically referred to as the "design intake velocity"—which can be used to evaluate intake design prior to construction. The regulation of intake velocity by USEPA can thus be a good design reference to indicate a low potential for detrimental impacts from impingement.

4. Conclusions

This study has shown that intake design can greatly influence the intake velocity, and by optimizing the design parameters, it is greatly reducing the intake velocity. Particular attention is paid to seawater intake structures, which typically consists of velocity cap and intake tower. The research findings are generally summarized into the following areas:

(a) This study has demonstrated the usefulness of velocity cap in mitigating intake velocity. The key design parameters for velocity cap is velocity cap size (ω_{vc}), which comprises of the horizontal shelf (ℓ_{vc}) and vertical opening (H_{vc}). The recommended intake opening ratio (O_r) shall be $0.36V_r^{-0.31}$, where $O_r = H_{vc}/\ell_{vc}$ and $V_r = V_0/V_{pipe}$. V_0 is the velocity at the intake window and V_{pipe} is the suction velocity at the intake pipe.

(b) With increased velocity cap size, the intake velocity is reduced. The volume ratio (ω_r) between the velocity cap (ω_{vc}) and intake tower (ω_{IT}) is recommended at $0.11V_r^{-1.23}$.

(c) The primary environmental impact caused by intake operation is marine life impingement resulting from the intake velocity. Therefore, it is reasonable to regulate intake velocity as an environmental safety factor. However, to the author's best knowledge, most of the country has no single regulation used to enforce the intake performance for environment protection purposes. Thus, the maximum allowable intake value of 0.15 m/s, used by the United States as their national screening value for the permissible regulation of intake velocity, shall be used as reference to establish engineering design criteria for seawater intake structures. This value has been widely utilized to indicate that an intake structure has low potential adverse environmental impact.

Author Contributions: L.H.C. conducted the simulation, analyzed the results, and wrote the paper. A.K.A.W. supervised and reviewed the paper and provided suggestions for the improvement of the paper. All authors have read and agreed to the published version of the manuscript.

Funding: The authors express their appreciation to Ministry of Education (MOE) Malaysia that funded the research via Fundamental Research Grant Scheme (FRGS) to Research Management Center (RMC) of Universiti Teknologi Malaysia (UTM) under the vote number R.J130000.7809.4F979.

Acknowledgments: The authors thank the editors and two anonymous reviewers for their constructive comments to improve this paper. The authors also express sincere gratitude to the research assistants and engineers of Center of Coastal and Ocean Engineering of UTM for all the assistance. Further, this research would not have been possible without the data provided by Aqvaspace Sdn Bhd. The authors thank the surveyors who assisted in the field measurements.

Conflicts of Interest: The authors declare no conflict of interest.

References

1. Pita, E. Seawater intakes for desalination plants: Design and construction. In Proceedings of the International Conference on Desalination, Environment and Marine Outfall Systems, Muscat, Oman, 13–16 April 2014.
2. Voutchkov, N. Design and construction of open intakes. In *Sustainable Desalination Handbook: Plant Selection, Design and Implementation*; Gude, V.G., Ed.; Elsevier: Oxford, UK, 2018; pp. 201–226.
3. Lattemann, S.; Höpner, T. Environmental impact and impact assessment of seawater desalination. *Desalination* **2008**, *220*, 1–15. [CrossRef]
4. Pankratz, T. An overview of seawater intake facilities for seawater desalination. In *The Future of Desalination in Texas*; Biennial Report on Water Desalination, Texas Water Development: Austin, TX, USA, 2004; Volume 2, pp. 1–12.
5. WateReuse Association. White paper on desalination plant intakes: Impingement and entrainment impacts and solutions. *Desalin. Comm.* **2011**. [CrossRef]
6. Environmental Protection Agency (EPA). National pollution discharge elimination system—Final regulations to establish requirements for cooling water intake structures at existing facilities and amend requirements at Phase I facilities: Final rule. *Fed. Regist.* **2014**, *79*, 48300.
7. Turnpenny, A.W.H.; Keeffe, N.O. *Screening for Intake and Outfalls: A Best Practice Guide*; Science Report SC030231; Environment Agency: Bristol, UK, 2005.
8. Pankratz, T. Overview of intake systems for seawater reverse osmosis facilities. In *Intakes and Outfalls for Seawater Reverse-Osmosis Desalination Facilities: Innovations and Environmental Impacts*; Missimer, T.M., Jones, B., Maliva, R.G., Eds.; Springer: New York, NY, USA, 2015; pp. 3–17.
9. Pita, E.; Sierra, I. Seawater Intake Structures. In Proceedings of the International Symposium on Outfall Systems, Mar del Plata, Argentina, 15–18 May 2011.
10. Environmental Protection Agency (EPA). *Requirements Applicable to Cooling Water Intake Structures for New Facilities under Section 316(b) of the Act*; US Electronic Code of Federal Regulations: Washington, DC, USA, 2001; Volume 40.
11. Electric Power Research Institute (EPRI). *Technical Evaluation of the Utility of Intake Approach Velocity as an Indicator of Potential Adverse Environmental Impact under Clean Water Act Section 316(b)*; Final Report; EPRI Inc.: Palo Alto, CA, USA, 2000; p. 1000731.
12. National Marine Fisheries Service (NMFS). *Fish Screening Criteria for Anadromous Salmonids*; U.S. Department of Commerce, National Oceanic and Atmospheric Administration: Washington, DC, USA, 1997.
13. Pearce, R.O.; Lee, R.T. Some design considerations for approach velocities at juvenile salmonid screening facilities. *Am. Fish. Soc. Symp.* **1991**, *10*, 237–248.
14. Schuler, V.J.; Larson, L.E. Improved fish protection at intake systems. *J. Environ. Eng. Div. ASCE* **1975**, *101*, EE6.
15. Moideen, R.; Ranjan Behera, M.; Kamath, A.; Bihs, H. Effect of girder spacing and depth on the solitary wave impact on coastal bridge deck for different airgaps. *J. Mar. Sci. Eng.* **2019**, *7*, 140. [CrossRef]
16. Gomes, A.; Pinho, J.L.S.; Valente, T.; Antunes do Carmo, J.S.; Hegde, A. Performance assessment of a semi-circular breakwater through CFD modelling. *J. Mar. Sci. Eng.* **2020**, *8*, 226. [CrossRef]
17. Karami, H.; Farzin, S.; Sadrabadi, M.T.; Moazeni, H. Simulation of flow pattern at rectangular lateral intake with different dike and submerged vane scenarios. *J. Water Sci. Eng.* **2017**, *10*, 246–255. [CrossRef]
18. Ruether, N.; Singh, J.M.; Olsen, N.R.B.; Atkinson, E. 3D computation of sediment transport at water intakes. In Proceedings of the Institution of Civil Engineers: Water Management; Thomas Telford Ltd.: London, UK, 2005; Volume 158, pp. 1–8.
19. Aqvaspace Sdn. Bhd. *Marine Data Collection around Perai CCGT Power Plant, Penang*; Survey report: AQSP/RPT/07-2017/G&P/DTC/10096; Aqvaspace Sdn. Bhd.: Selangor, Malaysia, 2017; pp. 1–54, Unpublished Report.
20. Chie, L.H.; Abd Wahab, A.K.; Yapandi, F.K.M. Performance of different turbulence models in predicting flow kinematics around and open offshore intake. *SN Appl. Sci.* **2019**, *1*, 1266.
21. Yakhot, V.; Smith, L.M. The renormalization group, the e-expansion and derivation of turbulence models. *J. Sci. Comput.* **1992**, *7*, 35–61. [CrossRef]
22. Royal Malaysian Navy. *Tide Table Malaysa: Volume 1*; National Hydrographic Centre: Klang, Selangor, Malaysia, 2017.

23. American Society of Mechanical Engineers (ASME). *Standard for Verification and Validation in Computational Fluid Dynamics and Heat Transfer*; ASME: New York, NY, USA, 2009.
24. Department of Irrigation and Drainage (DID). *Guidelines for Preparation of Coastal Engineering Hydraulic Study and Impact Evaluation*; Department of Irrigation and Drainage: Kuala Lumpur, Malaysia, 2001.
25. Liu, Y.; Zhao, Y.P.; Dong, G.H.; Guan, C.T.; Cui, Y.; Xu, T.J. A study of the flow field characteristics around star-haped artificial reefs. *J. Fluids Struct.* **2013**, *39*, 27–40. [CrossRef]
26. Liu, T.L.; Su, D.T. Numerical analysis of the influence of reef arrangements on artificial reef flow fields. *Ocean Eng.* **2013**, *74*, 81–89. [CrossRef]

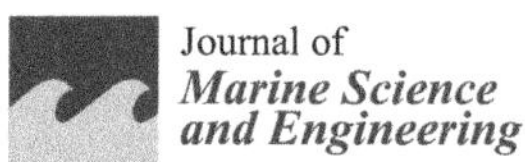

Journal of
Marine Science and Engineering

Article

Investigation of the Starting-Up Axial Hydraulic Force and Structure Characteristics of Pump Turbine in Pump Mode

Zhongyu Mao [1], Ran Tao [1,2], Funan Chen [1], Huili Bi [1], Jingwei Cao [1], Yongyao Luo [1], Honggang Fan [1] and Zhengwei Wang [1,*]

[1] State Key Laboratory of Hydroscience and Engineering, Department of Energy and Power Engineering, Tsinghua University, Beijing 100084, China; maozy14@mails.tsinghua.edu.cn (Z.M.); randytao@cau.edu.cn (R.T.); cfn18@mails.tsinghua.edu.cn (F.C.); bihuili2014@mail.tsinghua.edu.cn (H.B.); caojw18@mails.tsinghua.edu.cn (J.C.); luoyy@tsinghua.edu.cn (Y.L.); fanhg@tsinghua.edu.cn (H.F.)
[2] College of Water Resources and Civil Engineering, China Agricultural University, Beijing 100083, China
*   Correspondence: wzw@mail.tsinghua.edu.cn

Abstract: During the starting up of the pump mode in pump turbines, the axial hydraulic force acting on the runner would develop with the guide vane opening. It causes deformation and stress on the support bracket, main shaft and runner, which influence the operation security. In this case, the axial hydraulic force of the pump turbine is studied during the starting up of pump mode. Its influences on the support bracket and main shaft are investigated in detail. Based on the prediction results of axial hydraulic force, the starting-up process can be divided into "unsteady region" and "Q flat region" with obviously different features. The mechanism is also discussed by analyzing pressure distributions and streamlines. The deformation of the support bracket and main shaft are found to have a relationship with the resultant force on the crown and band. A deflection is found on the deformation of the runner with the nodal diameter as the midline in the later stages of the starting-up process. The reason is discussed according to pressure distributions. The stress concentration of the support bracket is found on the connection between thrust seating and support plates. The stress of the runner is mainly on the connection between the crown and the blade's leading-edge. This work will provide more useful information and strong references for similar cases. It will also help in the design of pump turbine units with more stabilized systems for reducing over-loaded hydraulic force, and in the solving of problems related to structural characteristics.

Keywords: axial hydraulic force; stress; deformation; pump turbine; starting-up

Citation: Mao, Z.; Tao, R.; Chen, F.; Bi, H.; Cao, J.; Luo, Y.; Fan, H.; Wang, Z. Investigation of the Starting-Up Axial Hydraulic Force and Structure Characteristics of Pump Turbine in Pump Mode. *J. Mar. Sci. Eng.* **2021**, *9*, 158. https://doi.org/10.3390/jmse9020158

Academic Editor: José A. F. O. Correia
Received: 11 January 2021
Accepted: 1 February 2021
Published: 5 February 2021

Publisher's Note: MDPI stays neutral with regard to jurisdictional claims in published maps and institutional affiliations.

1. Introduction

Pumped storage power stations are crucial in electric power systems. They have two main modes—the power generating (turbine mode) and pump-storing (pump mode) modes—with the ability to quickly start up and shut down. During the peak period of electricity demand, the pump turbine operates in turbine mode and converts the potential energy of water in upstream to electrical energy. During the off-peak period, it operates in pump mode and stores the excess energy via pumping water into the upstream reservoir.

The pump turbine, designed as reversible in modern times, is the key component of pumped storage power stations [1]. It operates under complex conditions and suffers varying hydraulic force on the runner, shaft and support bracket. Pump turbines are usually designed and installed in vertical-axis style. The operating stability and security becomes very sensitive to axial hydraulic force [2]. The total axial force that shaft systems suffer includes the axial hydraulic force, the weight of the runner and the weight of the shaft system. Thus, the axial force is an important technical requirement and also affects the design of thrust bearing. The desirable condition of axial force is upward but slightly less than the runner-shaft weight. The total axial force will be downward to ensure the unit is stable. However, a desirable condition is usually difficult to achieve. Especially, the axial force strongly and complexly

111

changes in transient conditions such as starting-up, load rejection, stalling, etc. If the upward axial hydraulic force is too strong to exceed the unit weight, the unit lifting happens and causes accidents [3–6]. On the contrary, if the downward axial hydraulic force is too strong, especially with strong pulsation, the support bracket will strongly deform or even be damaged due to plastic deformation or fatigue failure [7].

Investigating the axial hydraulic force characteristics and influences is a popular topic in pump turbine engineering cases and other hydro-turbine cases. The mechanism of axial hydraulic force is currently well known. It is caused by the pressure difference between the inner and outer surfaces of the runner's crown and the band and between the runner blade's suction and pressure sides [8–11]. Usually, the crown and band leakages are filled with high-pressure flow, which is higher than the inner flow of the runner [4,12]. The pressure of the blade pressure side is higher than that of the suction side. However, the total axial hydraulic force is empirically unpredictable because the influence factor is very complicated. The condition parameters of the pump turbine, including head, flow rate and rotation speed, impact the flow regime and pressure distribution [13–15]. Liu et al. [16] presented an analytical method to calculate the axial hydraulic force, considering the leakage size and the angular flow velocity in high-pressure leakages. They also pointed out that the turbine axial force is mainly produced by crown and band leakages. The leakage-induced axial force is much larger than that induced by other runner parts. Based on Genetic Algorithm, large-quantity experiment and computational fluid dynamics(CFD) simulation, Zhao et al. [17] put forward a prediction formulation of axial hydraulic force with condition parameters, runner diameter and empirical coefficients. The axial force of the centrifugal pump simulated by the CFD method was compared with the test result of Zhou et al. [18]. The results indicated that an appropriate impeller rear shroud radius is able to significantly reduce axial force. Based on experiment and simulation, the wear ring radial clearance was found to influence the axial thrust. The solutions of hydraulic axial thrust reduction of pumps were then presented [19,20]. In terms of the load-rejection of generating mode, the axial force variation was found to have a strong relationship with flow rate [21]. Li et al. [22] found that the amplitude of the force depends on the operating conditions and the guide vane openings. For example, the axial force is prominent in the common operation of turbine mode while the radial force is dominant in runaway and shutdown processes.

The structure of the pump turbine mainly includes the support bracket, main shaft, rotor generator and runner. The support bracket is a fixed component while the other components are rotational. As the structural support, the weight of the unit and the hydraulic force of the runner load on the support bracket [23,24]. It should be considered as a fluid–structure interaction (FSI) problem in order to understand the influence of the runner's hydraulic force on structures [25]. Based on FSI, many studies were focused on the stress on the runner. Considering the interaction between the hydraulic force of the flow field and the deformation of the structure field, two-way FSI is used in structural analysis of the runner and the accuracy is compared with experimental studies [26–29]. However, in terms of turbine FSI issues, the deformation of the runner structure is much less than the flow characteristic length. In that case, Zhou et al. and Xiao et al. [30,31] studied the fatigue and stress of runner at off-design operating points via one-way FSI. They found that the stress caused by hydraulic force is one of the main reasons of the runner fatigue failures and cracks. In terms of the other essential components of pump turbines, Luo et al. [32–34] focused on the stress of the rotor bracket of the generator and the piston rod of the blade. Improvements in design are made based on stress analyses. However, the hydraulic force was usually simplified as a resultant force in the research on shafts and support brackets [35–37]. It cannot accurately reflect the influence of hydraulic force on unit structure due to the non-uniform force distributions.

In this study, based on one-dimensional (1D) hydraulic transient simulation in pipeline and 3D CFD simulation in pump turbines, the axial hydraulic force of the pump turbine in the starting up of pump mode is studied. The influence of axial force on pump turbine structure is also researched. The reliability of CFD simulation is verified by comparison with the prototype on-site test. The influence of the operation parameters on axial hydraulic force and the mechanism are analyzed and explained in detail. Based on one-way FSI, the structural strength including deformation and stress of the support bracket, main shaft and runner is calculated. The relationship between axial hydraulic force and structural strength is well discussed. This research will provide more useful information and strong references for similar cases. It helps the design of pump turbine units for a better hydraulic and structural performance.

2. Numerical Method

2.1. Method of 3D Turbulent Flow Simulation

Given the incompressibility of water, the Reynolds averaged Navier-Stokes (RANS) equations were used to calculate the 3D flow field in the pump turbine. The continuity equation and momentum equation are [38]:

$$\frac{\partial u_i}{\partial x_i} = 0 \tag{1}$$

$$\frac{\partial u_i}{\partial t} + u_j \frac{\partial u_i}{\partial x_j} = f_i - \frac{1}{\rho}\frac{\partial p}{\partial x_i} + v\nabla^2 u_i \tag{2}$$

where u is flow velocity, f is body force, ρ is density, p is pressure, v is kinematic viscosity, t is time and x is the coordinate component.

In the RANS method, the instantaneous component is decomposed into its time-averaged component and fluctuating component. To close the equations, the turbulence model is used to empirically model the fluctuating component. The SST $k - \omega$ transient model which is an eddy viscosity model is applied in this study [39]. It can simulate both the shear flow and adverse pressure gradient accurately and is particularly useful in engineering simulations. The RANS equation with SST $k - \omega$ model can be written as:

$$\frac{\partial(\rho k)}{\partial t} + \frac{\partial(\rho u_i k)}{\partial x_i} = P - \frac{\rho k^{\frac{3}{2}}}{l_{k-\omega}} + \frac{\partial}{\partial x_i}\left[(\mu + \sigma_k \mu_t)\frac{\partial k}{\partial x_i}\right] \tag{3}$$

$$\frac{\partial(\rho \omega)}{\partial t} + \frac{\partial(\rho u_i \omega)}{\partial x_i} = C_\omega P - \beta \rho \omega^2 + \frac{\partial}{\partial x_i}\left[(\mu_l + \sigma_\omega \mu_t)\frac{\partial \omega}{\partial x_i}\right] + 2(1 - F_1)\frac{\rho \sigma_{\omega 2}}{\omega}\frac{\partial k}{\partial x_i}\frac{\partial \omega}{\partial x_i} \tag{4}$$

where $l_{k-\omega}$ is the turbulence scale in which $l_{k-\omega}=k^{1/2}\beta_k\omega$, μ is dynamic viscosity, term P is the production term, C_ω is the coefficient of the production term, F_1 is the blending function, σ_k, σ_ω and β_k are model constants.

Based on the CFD commercial software CFX, the high resolution was used for discretization schemes in this paper.

The 3D flow simulation was processed with the steady simulation in CFX. The one-dimensional (1D) hydraulic transient simulation of unsteady flow in pipe was processed to consider the transient effect in starting-up. As the boundary conditions of 3D flow field, the 1D hydraulic transient simulation was based on the continuity equation and momentum equation.

$$U\frac{\partial H}{\partial x} + \frac{\partial H}{\partial t} - U\sin\alpha + \frac{a^2}{g}\frac{\partial U}{\partial x} = 0 \tag{5}$$

$$g\frac{\partial H}{\partial x} + U\frac{\partial U}{\partial x} + \frac{\partial U}{\partial t} + \frac{fU|U|}{2D} = 0 \tag{6}$$

where H is piezometric head, U is average velocity, g is gravity acceleration, f is Darcy-Weisbach friction factor; α is pipeline slope; D is diameter of pipe; a is speed of pressure pulse [40].

2.2. Method of Structural Simulation

The structural simulation was proceeded using commercial software ANSYS. Based on the structural static equilibrium equation, the stress and deformation are calculated using the finite element method (FEM). The equilibrium equation is [41]:

$$[K]\{d\} = \{F\} \tag{7}$$

where $[K]$ is the stiffness matrix of the system, $\{d\}$ is the vector with the nodal displacement and $\{F\}$ is the vector of force loaded on structure.

Via the displacement $\{d\}$ solved by Equation (7), the static stress σ can be calculated by [42,43]:

$$\sigma = [D_s][B_s]\{d\} \tag{8}$$

where $[D_s]$ is the elastic matrix based on Young's modulus and Poisson's ratio for the material and $[B_s]$ is the strain–displacement matrix based on the element shape functions.

Universally, the equivalent von Mises stress is applied in engineering to analyze the stress characteristics of the structure. The equivalent von Mises stress σ_c can be calculated using the fourth strength theory:

$$\sigma_c = \sqrt{\frac{1}{2}\left[(\sigma_1 - \sigma_2)^2 + (\sigma_2 - \sigma_3)^2 + (\sigma_3 - \sigma_1)^2\right]} \tag{9}$$

where σ_1, σ_2 and σ_3 are the first, second and third principal stress.

3. Computational Model and Boundary Conditions

3.1. Flow Field of Pump Turbine

In this study, the model is a prototype reversible pump turbine in a pumped storage power station. In order to acquire the boundary conditions of the pump turbine, the hydraulic system has been modelled in 1D. As shown in Figure 1, the 1D system includes the pipeline, reservoir, gate shaft, pump turbine and tank. The operate process studied in this paper is the starting-up process of the pump mode. Because the reservoir level has principle influence on hydraulic force, the most common situation is considered to be the initial conditions with the maximum lower reservoir level of 294 m and the minimum upper reservoir of 741 m. The length of upper pipeline is about 1160.9 m and the length of lower pipeline is about 1108.5 m.

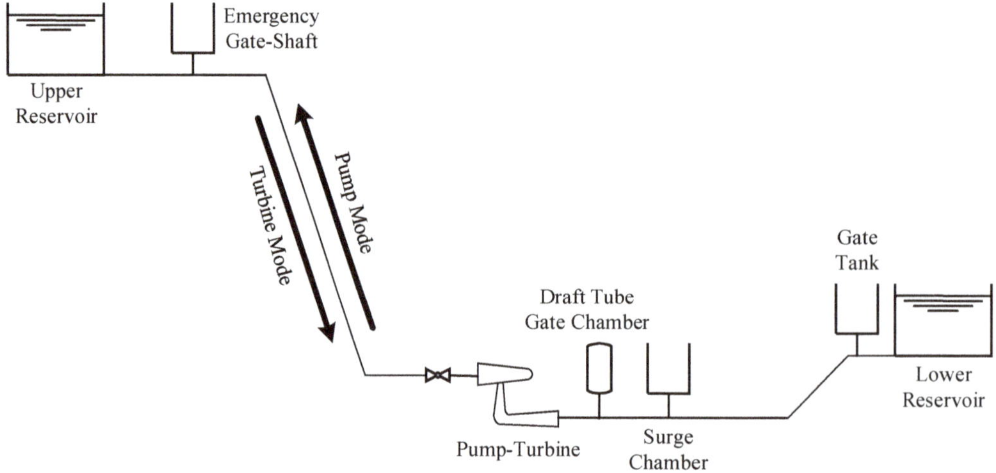

Figure 1. Schematic map of the hydraulic system of the studied pumped storage power station.

The parameters of pump turbine are listed in Table 1. The 3D flow profile of the pump turbine, shown in Figure 2, mainly consists of volute, stay vane, guide vane, runner and draft tube. To consider the hydraulic force on the outside surface of the runner, the flow field of the runner's crown and shroud leakages, and pressure-balancing chamber, are also modelled. Figure 3 shows the details of the runner and leakage. The hydraulic force of runner and leakage flow acts on the fluid–structure interface including the blade suction side (BSS), blade pressure side (BPS), crown outside surface (COS), crown inside surface (CIS), band outside surface (BOS) and band inside surface (BIS). These interfaces are illustrated in Figure 3.

Table 1. Parameters of pump turbine.

Parameter	Value
Rated head H_r [m]	430
Rated rotation speed n_r [r/min]	428.6
Rated power P_r [MW]	300
Rated flow rate Q_r [m^3/s]	68
Diameter on runner pressure side D_{hi} [m]	4.16

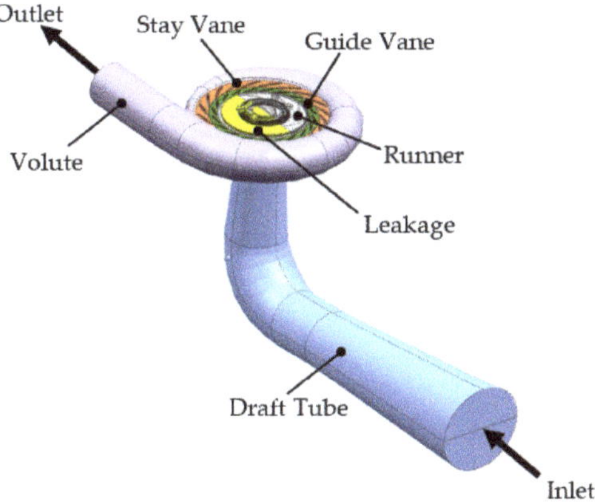

Figure 2. Overview of the 3D flow profile of the pump.

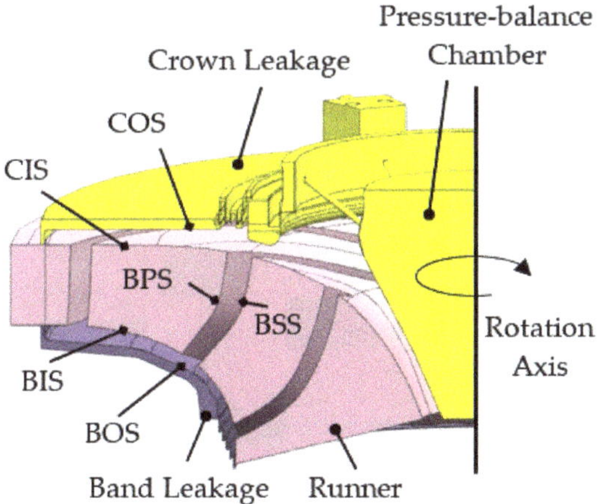

Figure 3. Details of runner and leakage.

While the runner keeps rotating at the rated speed, the air, which is pressurized into the runner chamber to reduce start-up torque, has been already released, instead of filling water. Then, the guide vane opens gradually from 0 degrees to the maximum opening angle. This process is called "starting-up" in pump mode. According to the guide vane opening law, the starting point is when the guide vane opening is 0 degrees and the starting-up calculation duration in this paper is 30 s. The time step of 1D hydraulic transient simulation is 0.005 s.

The 3D flow field during starting-up is calculated using CFD steady simulation at selected typical time points (STP), which is illustrated in Figure 4. The boundary conditions of 3D CFD simulation is based on 1D hydraulic transient results of the pipe at the inlet and outlet of the pump turbine. The draft tube inlet is set as the mass flow rate inlet boundary. The volute outlet is set as the static pressure outlet boundary. The rotation speed is the rated speed. The guide vane opening rule, mass flow rate and static pressure at volute outlet are acquired from 1D hydraulic transient simulation, as shown in Figure 4. All the parameters are at relative values of $Q^* = Q/Q_r$, $H_{out}^* = H_{out}/H_r$, $A^* = A/A_{max}$ where Q_r, H_r are the rated flow rate and rated head, H_{out} is the pressure at the volute outlet, A_{max} is the maximum guide vane opening. There is a fluctuation of H_{out} at the initial stage of starting-up, which is caused by the water hammer effect in the pipeline with the guide vane opening.

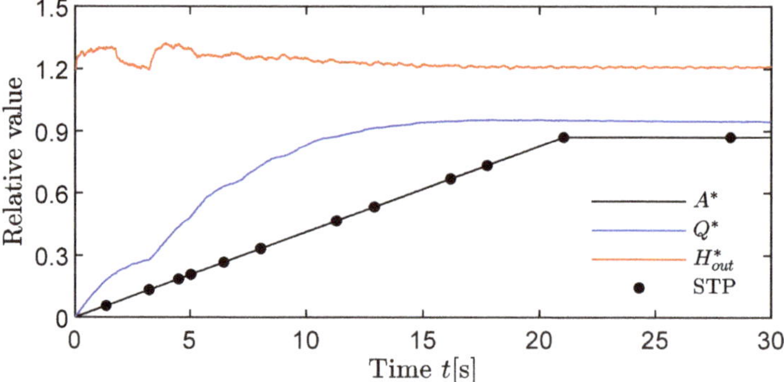

Figure 4. Boundary conditions acquired from 1D hydraulic transient simulation.

The fluid medium is considered as incompressible in this case. The runner walls in leakages are set as counter-rotating. The other solid walls are set as no-slip wall type boundaries. The runner domain and FSI of leakage, COS and BOS are set as rotational. The other domain and wall are set as stationary. Interfaces between stationary domains are set as general connection. Interfaces between stationary and rotational domains are set as frozen-rotor type for good data transfer ability. The convergence criterion is set as the root-mean-square (RMS) residual of continuity equation, and momentum equation is set as less than 1×10^{-4}.

3.2. Structural Field of Pump Turbine Unit

In the structural field simulation, the structural stress and deformation of pump turbine unit are focused and simulated based on the finite element method (FEM). The structural model is composed of the support bracket, shaft, motor and runner. The rotating components including the shaft, motor and runner are regarded as one. Figure 5 shows the 3D structural model with corresponding boundary conditions. The material of the pump turbine unit is steel; the properties that refer to the prototype pump turbine are listed in Table 2. The total weight of rotating components is about 550 t. The coordinate system is shown in Figure 5 with a downward +z (axial) direction. The bracket has, in total, eight supporting arms, fixed by concrete foundations at the arm end (fixed support 1) and connected with the generator stator at the bottom (fixed support 2). The diameter of the basic support part is 4 m and the diameter with eight supporting arms is 10 m. The thrust bearing is simplified by spring element (the element type is COMBIN14), connecting the thrust collar and thrust seating of the support bracket. The upper bearing, lower bearing and turbine bearing are simplified by the bearing element (the element type is COMBI214), constraining the radial motion of the shaft. The stiffness coefficients k of thrust bearing and guide bearing are listed in Table 3 [37]. Hydraulic force loads on the fluid–structure interfaces of runner based on one-way FSI. The hydraulic force is acquired from 3D flow field simulation and is mapped onto the fluid–structure interface of the runner structure using the profile-preserving method. The rotation speed of the rotating components is n_r. The gravity and centrifugal force of the rotating components are fully considered in this study.

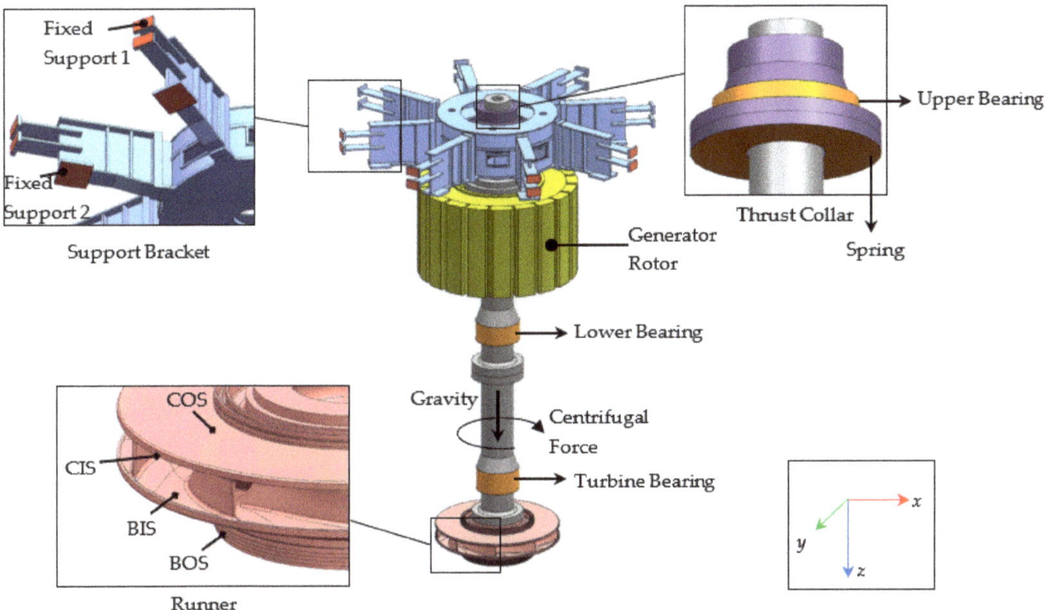

Figure 5. Model and boundary conditions of structural field of pump turbine unit.

Table 2. Pump turbine unit material properties.

Density ρ [kg/m^3]	Young's Modulus E [Pa]	Poisson's Ratio μ [-]
7850	2.1×10^{11}	0.3

Table 3. Bearing stiffness coefficients k.

	Upper Bearing	Lower Bearing	Turbine Bearing	Thrust Bearing
Stiffness coefficients k [N/m]	2.0×10^9	2.0×10^9	1.5×10^9	2.5×10^9

4. Mesh and Independence Check

4.1. Mesh of Flow Field

The flow field of the pump turbine is discretised using tetrahedral-hexahedral hybrid mesh elements to balance the computational cost and simulation accuracy. The schematic map of mesh is shown in Figure 6. It is worth noting that the guide vane flow domain is remeshed with consistent element size and rules as the guide vane opening is changing. The y^+ value is also checked by adjusting the near-wall mesh height. To apply the automatic wall functions, y^+ is finally controlled within 30~300.

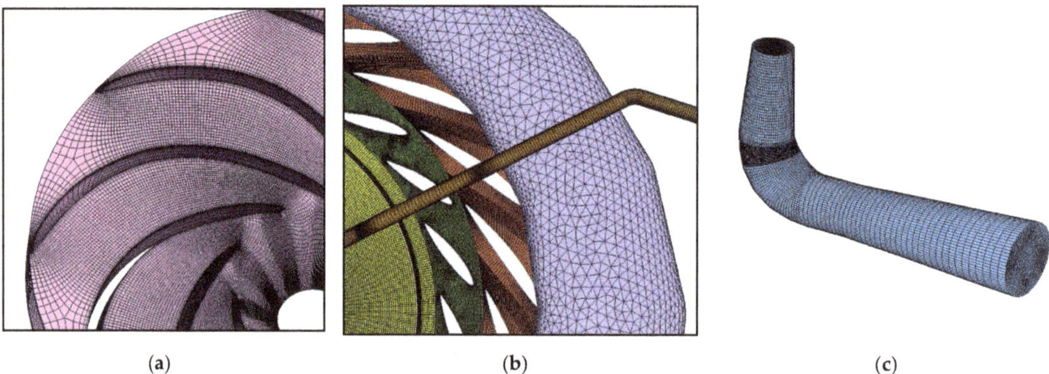

(a)   (b)   (c)

Figure 6. Schematic map of the mesh for flow field. (**a**) Runner; (**b**) Leakage, guide vane, stay vane and volute; (**c**) Draft tube.

In order to validate the mesh independence and the simulation accuracy, the comparison of pressure on typical locations between the prototype pump turbine test data and simulation is conducted, as pressure distribution is the key point of hydraulic force. Four pressure sensors were arranged on the prototype pump turbine at the location marked in Figure 7, working during the starting-up in pump mode. The sampling frequency of the pressure sensor is 800 Hz. For comparison with simulation results at STP, the frequency of test pressure shown in Figure 8 is 4 Hz, which is averaged from the original data with 800 Hz. The relative pressure coefficient C_p is defined as:

$$C_p = \frac{P}{\rho g H_r} \tag{10}$$

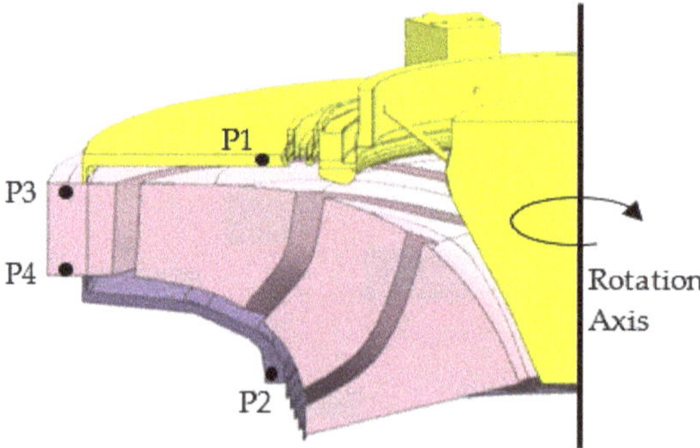

Figure 7. Pressure measurement points. P1: crown leakage before seal; P2: band leakage before seal; P3: upper vaneless space; P4: lower vaneless space.

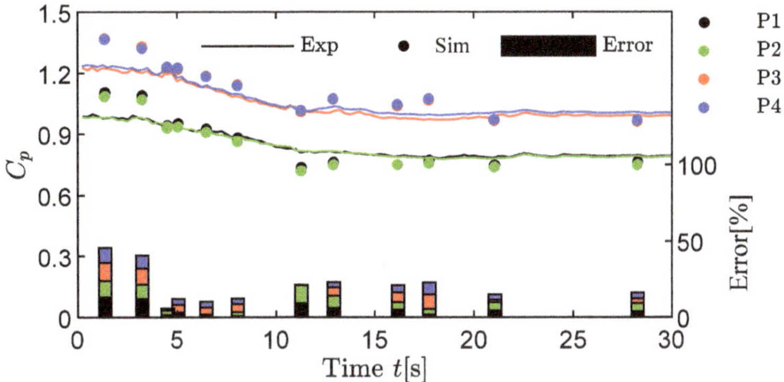

Figure 8. Comparison of pressure between prototype test and simulation on typical locations.

As shown in Figure 8, the pressure on typical locations shows good agreement between the test and simulation, with errors below 10.7%. Larger errors mainly exist in the initial stage of the starting-up process as both the guide vane opening and mass flow rate are very small. Furthermore, relative head $H^* = H/H_r$ and relative pressure at draft tube inlet $H_{in}^* = H_{in}/H_r$ between 1D hydraulic transient and 3D CFD steady simulation are compared in Figure 9. The H^* and H_{in}^* predicted by CFD matches well with the 1D-predicted results. Therefore, the mesh is sufficient for the hydraulic force simulation based on steady state CFD simulation. The final mesh of the flow field in this study has about 6.89 million nodes and 9.91 million elements. The mesh detail of each component is, respectively, listed in Table 4.

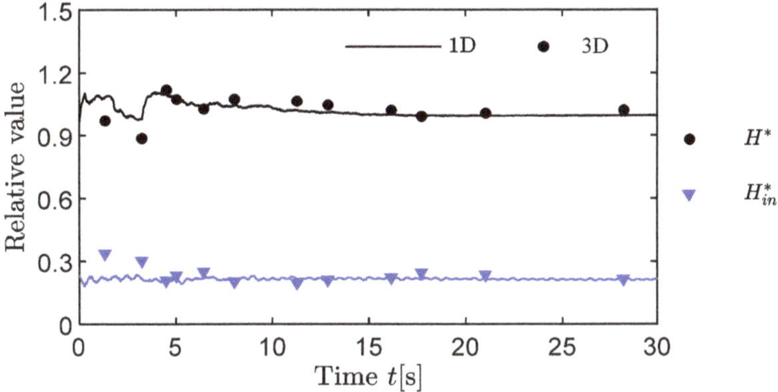

Figure 9. Comparison of head between 1D hydraulic transient simulation and 3D CFD simulation.

Table 4. Mesh details of the flow field.

Component	Nodes
Volute	150,796
Stay Vane	417,420
Guide Vane	750,688
Runner	1,831,959
Draft Tube	174,411
Leakages	2,947,430
Balancing-Pipe	615,984
Total	6,888,688

4.2. Mesh of Structural Field

Figure 10 shows the mesh of structural field including support bracket, shaft and runner. The support bracket is meshed by tetrahedral mesh elements. The shaft and runner are meshed by hexahedral mesh elements. The element type in ANSYS simulation is SOLID185. The tress concentration often occurs at the "T-shape" connection and at the corners. These regions are extremely sensitive to the mesh quality [44]. Therefore, as shown in detail in Figure 10, mesh in special regions like the corners of support plates, the connections between the blade and the crown, and the connection between the blade and the runner's band, are locally refined. According to the maximum stress on local refinement zones of the runner and support bracket, the mesh independence check is conducted with four schemes, as listed in Table 5. The four schemes have different element sizes, especially in sensitive regions. As shown in Figure 11, the changes of maximum stress at typical sites are monitored to be less than 2% in the check. The final mesh scheme has 0.92 million nodes and 1.17 million elements. The mesh details of each component are listed in Table 6.

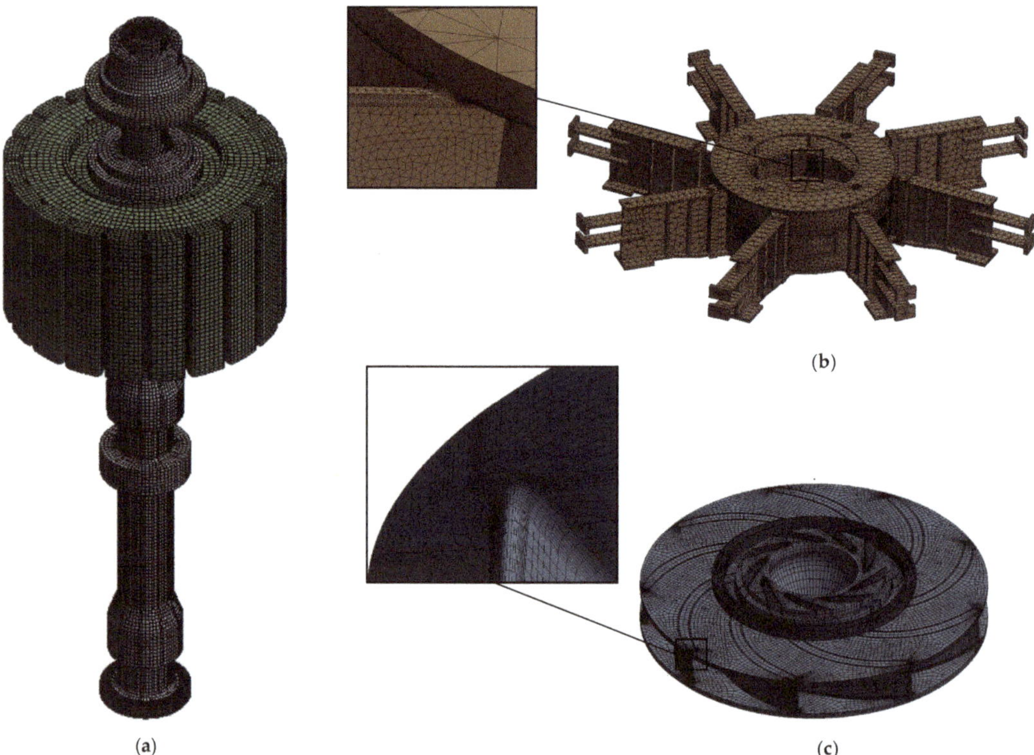

Figure 10. Schematic map of the structure field meshes. (**a**) Shaft and Generator rotor; (**b**) Support Bracket; (**c**) Runner.

Table 5. Mesh details for mesh independence check.

	Mesh1	Mesh2	Mesh3	Mesh4
Nodes	635,012	827,539	915,961	1,972,471
Elements	774,907	1,009,839	1,169,321	2,640,587

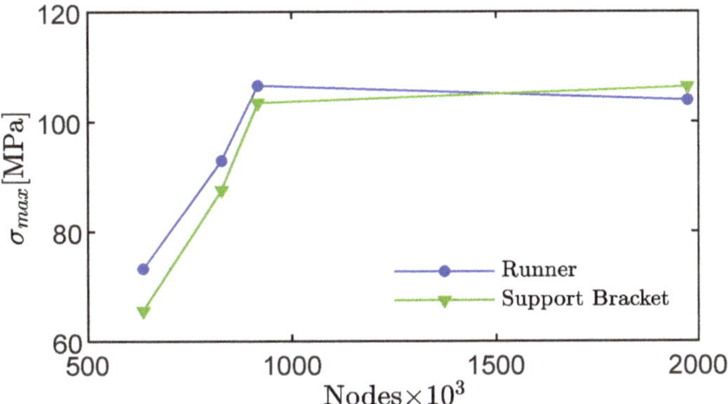

Figure 11. Mesh independence check.

Table 6. Mesh details of the final structure field.

Component	Shaft and Generator Rotor	Support Bracket	Runner	Total
Nodes	262,888	121,470	540,639	915,961

5. Results and Analysis

5.1. Axial Hydraulic Force

5.1.1. Characteristic and Development

Based on the CFD simulation at the STP, the development of total axial hydraulic force acting on runner during the starting up process is analyzed, as presented in Figure 12. The relative axial hydraulic force is defined as:

$$F_z^* = \frac{F_z}{m_t g} \tag{11}$$

where m_t is the weight of rotating components including the shaft system and runner.

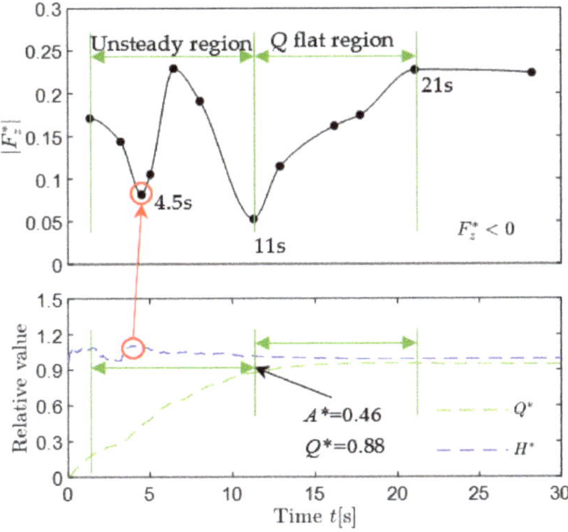

Figure 12. Schematic map of the structure field meshes.

In this paper, the downward axial force is defined as positive according to the definition of the +z direction. $|F_z^*|$ is adopted to stress the magnitude of axial hydraulic force. To find out the mechanism of axial hydraulic force and its relationship with flow parameters of the turbine, the relative flow rate Q^* and relative head H^* from 1D hydraulic simulation are also shown in Figure 12 as a reference.

During starting-up, $|F_z^*|$ is upward and changes between approximately 0.05 and 0.23. There, $t = 0$ corresponds to the onset of the opening up of the guide vane. In 0~4.5 s, $|F_z^*|$ gradually decrease and reach the first local valley. During this period, H^* reaches the maximum with fluctuation. Before $t = 11$ s, the guide vane has opened for 46% and Q^* increases to 0.88. $|F_z^*|$ suffers the largest fluctuation from 0.05 to 0.23. It can be found that the axial hydraulic force, head and flow rate are very fluctuant until the guide vane opens to 46%, which is about half of the maximum opening. This period is defined as the "unsteady region" where the phenomenon will be discussed in later sections. Then, Q^* increases flatly when the guide vane is continually opening, while H^* is almost constant. This period is defined as the "Q flat region". It is worth stressing the point that $|F_z^*|$ has a sharp increase and is strongly positively related to flow rate in this "Q flat region".

In order to identify the axial hydraulic force on specific locations of the runner, the components of $|F_z^*|$ are plotted in Figure 13. It may be helpful to have a clear knowledge of the variation mechanism of axial hydraulic force during the starting up process. As shown in Figure 13a, the directions of F_z^* on COS and BIS are downward and those on CIS and BOS are upward. It is very unstable for axial hydraulic force on the surface of the crown and band in the "unsteady region" while it is almost flat in the "Q flat region". For the runner's crown, $|F_z^*|$ on COS is larger than CIS, so the resultant axial force acting on crown is downward, as shown in Figure 13b. For the runner's band, the resultant axial force is upward as $|F_z^*|$ on BOS is larger than BIS. It can be estimated that pressure in the leakages is higher than that in the runner. Hence, the axial hydraulic force on the outside surface of the runner is larger than on its inside surface. With the opening of the guide vane, the resultant force on the crown and band decreases with the increasing of the flow rate. The other point observed in Figure 13b is that the resultant force on the crown is smaller than that on the band. Figure 13c provides the $|F_z^*|$ value on the blade with the comparison against that on the crown and the band. $|F_z^*|$ on the blade is found downward because pressure on blade pressure side is always higher than that on the suction side. In the "unsteady region", $|F_z^*|$ on the blade increases with the opening of the guide vane because the differential pressure on the blade is increasing. In contrast, it decreases in the "Q flat region" as the guide vane is continually opening and the flow rate is flatly increasing. For the resultant force on the crown and the band, some fluctuations can be found in the "unsteady region", as well as a slight increase in the "Q flat region".

In summary, the process of pump mode's starting-up should be divided to two parts—the "unsteady region" and the "Q flat region"—when discussing the characteristics of axial hydraulic force. The dividing point of two periods is that the guide vane opens to around half. In the two periods, obviously different axial hydraulic force characteristics and their relationships with flow parameters can be found. By analyzing the local components, it is found that the magnitude of axial hydraulic force on the runner's local surface is 4~10 times the weight of the pump turbine unit, but in the opposite direction. The total axial hydraulic force is generated due to the counteraction among all the force components. The total $|F_z^*|$ value is much smaller than these local axial force values. Therefore, because of the combined influence of local axial force, the resultant force develops according to a complicated law.

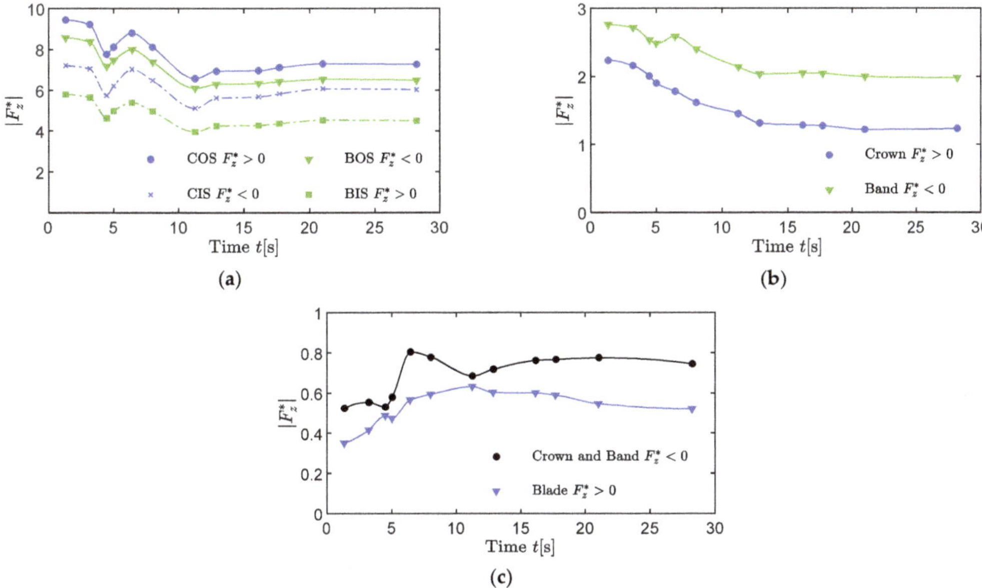

Figure 13. Axial hydraulic force components during pump mode's starting up. (**a**) Inside and outside surface of runner; (**b**) Crown and band; (**c**) Crown and band, blade.

5.1.2. Mechanism Discussion

Pressure distribution in the runner and leakages is the key factor influencing axial hydraulic force. In order to discuss the mechanism of axial hydraulic force, Figure 14 provides the pressure coefficient C_p distribution on the cross-section view in runner domain and leakages at typical selected time points, $t = 3$ s, 4 s, 6 s, 11 s, 16 s, 21 s.

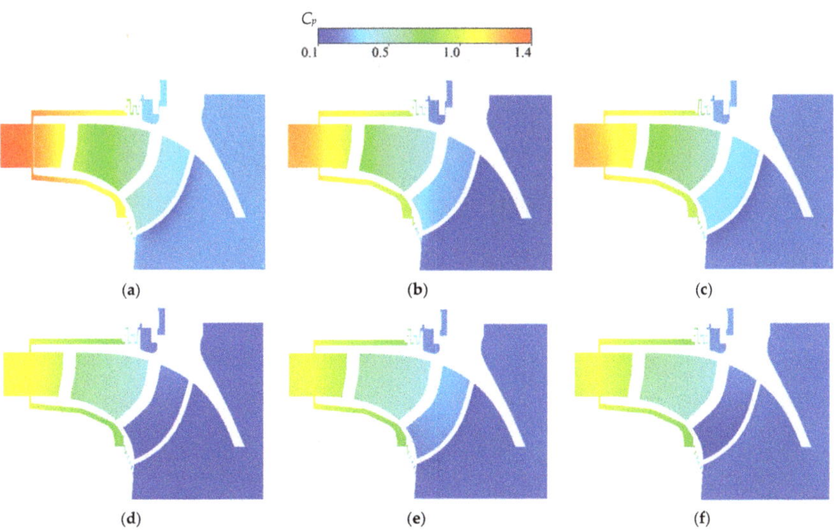

Figure 14. Pressure distribution on axial cross-section of runner and leakages during starting-up at $t = 3$ s, 4 s, 6 s, 11 s, 16 s, 21 s. (**a**) $t = 3$ s; (**b**) $t = 4$ s; (**c**) $t = 6$ s; (**d**) $t = 11$ s; (**e**) $t = 16$ s; (**f**) $t = 21$ s.

During the starting-up process, pressure reaches the lowest level at $t = 11$ s, accompanied with the smallest axial hydraulic force. Before reaching this point (in the "unsteady region"), unstable pressure development can be observed in the runner and leakages. At $t = 4$ s, the pressure level becomes lower than before and after. This is the time with the first valley value, as shown in Figure 12. Hence, the valley values of axial hydraulic force are, along with the lower pressure level in the "unsteady region", at $t = 4$ s and 11 s. Moreover, in this period, the pressure in crown and band leakages are apparently higher than that in the runner. In contrast, there is a small difference of pressure between the runner and leakages from $t = 11$ s to 21 s. Pressure in the crown and band leakages remain almost unchanged in the "Q flat region". At the same time, pressure in the runner near the draft tube increases with the flow rate.

Figure 15 shows the streamlines and relative velocity coefficient C_v in the mid-span of the guide vane at typical selected time points of $t = 3$ s, 6 s in the "unsteady region" and $t = 16$ s, 21 s in the "Q flat region". The relative velocity coefficient C_v is defined as:

$$C_v = \frac{v}{\frac{\pi n D}{60}} \tag{12}$$

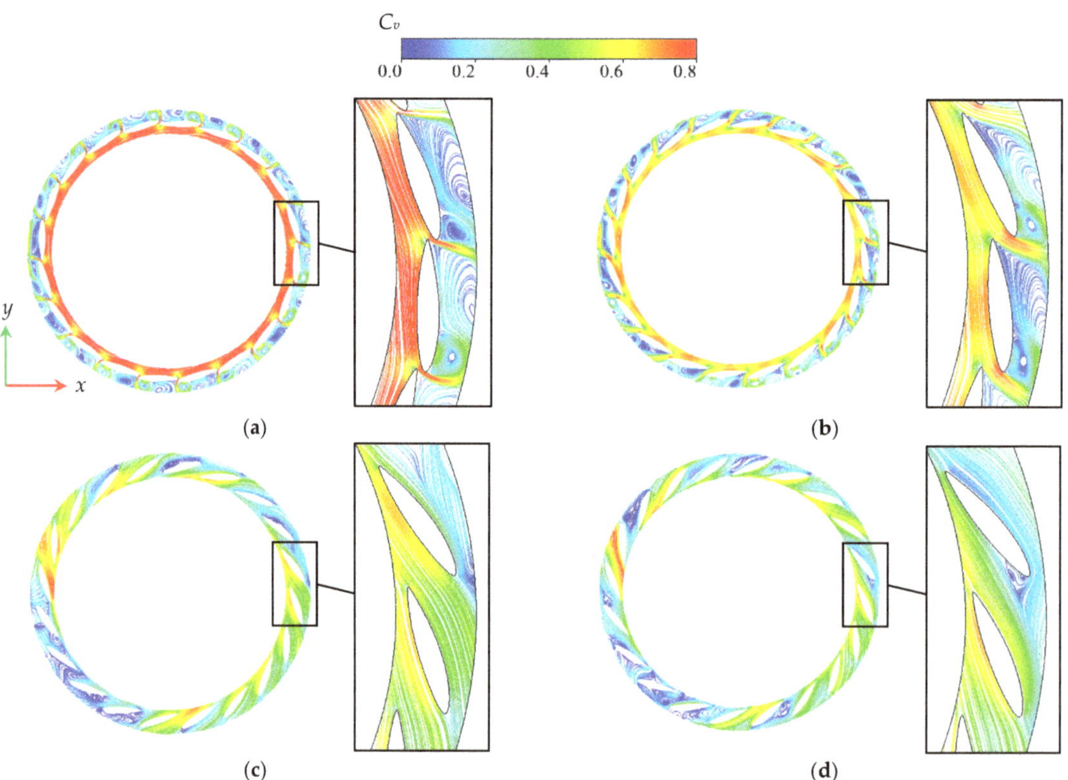

Figure 15. Streamlines and relative velocity coefficient C_v in the mid-span of guide vane during starting-up at $t = 3$ s, 6 s, 16 s, 21 s. (**a**) $t = 3$ s; (**b**) $t = 6$ s; (**c**) $t = 16$ s; (**d**) $t = 21$ s.

In the "unsteady region", the guide vane opening is relatively small. The high-speed flow from runner is blocked in front of the guide vane and forms an obvious jet flow between two guide vane blades. Due to the disturbance of the jet, the twin-vortex flow structure can be seen in the vaneless region between the guide vane and the stay vane. This twin-vortex flow structure causes strong local blockage in the guide vane. Therefore, the flow blockage leads to high-pressure in the runner and leakages. The twin-vortex causes the flow pattern to be significantly more turbulent and unstable. This is the reason why axial hydraulic force fluctuates in this period. With the guide vane opening becoming larger in the "Q flat region", the flow pattern in the guide vane becomes well-behaved with no obvious separation and vortex. The axial hydraulic force develops stably with a positive correlation against flow rate. It is worth noting that the flow pattern distribution in the guide vane is asymmetric in the later period during starting-up. This phenomenon will be discussed in the following sections.

5.2. Structural Characteristic of Unit

Based on the FEM method, structural simulation of pump turbine unit, including the support bracket, shaft and runner, is conducted during pump mode's starting-up process. The hydraulic force on the runner and leakages are obtained from the CFD results above. It is loaded on the fluid–structure interface based on the one-way FSI method. The Von-Mises stress σ and axial deformation D, which influence the operation safety of the pump turbine, are the main parameters in the structural simulation.

5.2.1. Deformation

The deformation D_{max} distribution on the main shaft and support bracket are almost constant during starting-up, as shown in Figures 16 and 17 at $t = 21$ s. In this paper, the downward deformation is positive. The D_{max} of the main shaft is 1.3 mm at $t = 21$ s and the location is at the top of shaft. It is noted that the end of the shaft, which connects the runner, has the smallest deformation. This means that the shaft is pushed upward because of centrifugal force. The D_{max} of the support bracket is 0.35 mm at $t = 21$ s and the location is at the thrust seating, which bears the axial hydraulic force and the self-weight of rotating components. The deformation decreases as the radius increases since the ends of the support arms are fixed.

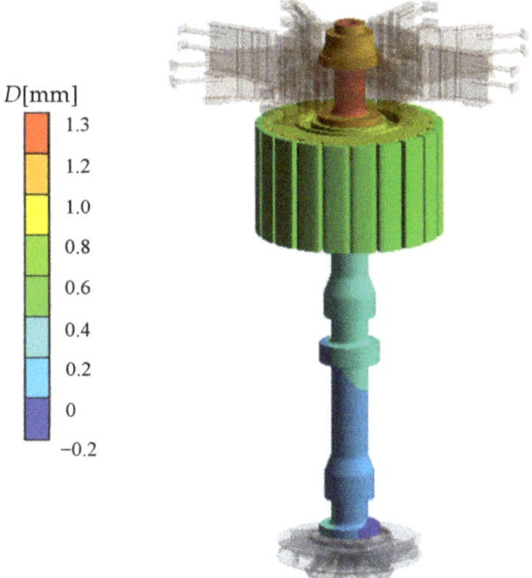

Figure 16. Deformation of main shaft at $t = 21$ s.

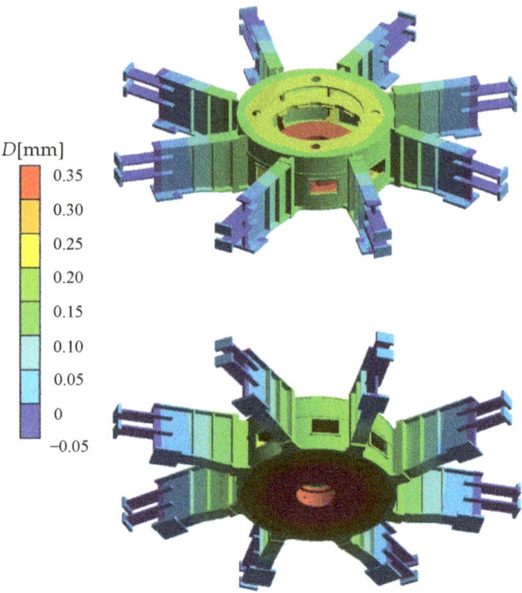

Figure 17. Deformation of support bracket at $t = 21$ s.

Figure 18 shows the D_{max} variation of the main shaft and support bracket during pump mode's starting-up process. To emphasize the effect of hydraulic force, the deformation without hydraulic force is plotted as a reference. The resultant force on the crown and band is also shown as a reference. The total axial force F_{zT} acting on the shaft and bracket can be calculated as:

$$F_{zT} = F_{zH} + F_{zW} \tag{13}$$

where F_{zH} is the hydraulic axial force and F_{zW} is the self-weight of rotating components.

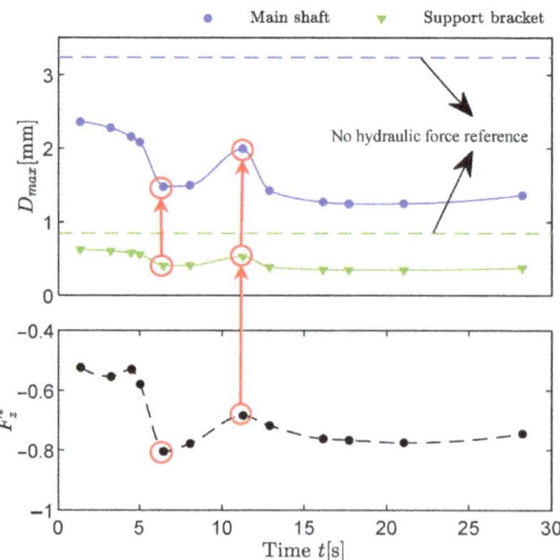

Figure 18. Maximum deformation of main shaft and support bracket during starting-up.

Based on Equation (7), there is the relationship that $D \propto F_{zT}$. Since F_{zW} is constant, it is the case that $D \propto F_{zH}$. The finding that the axial deformation of the main shaft and support bracket do not relate to the total axial hydraulic force is unexpected. However, it relates to the resultant axial hydraulic force on the crown and band (F_{zCB}). As Figure 18 shows, when hydraulic force has not been loaded on the fluid–structure interface, the D_{max} of the shaft and bracket is 3.2 mm and 0.85 mm. The hydraulic force makes D_{max} decrease, as a result of the upward axial hydraulic force. At the initial point of starting-up, D_{max} is the largest, with values of 2.35 mm for the shaft and 0.63 mm for the bracket. D_{max} of the shaft and bracket have similar tendencies during starting-up. They decrease before $t = 6$ s and reach the first valley value, which corresponds to the first valley of F_{zCB}. Then, they increase and reach the peak value at $t = 11$ s which corresponds to the first peak of F_{zCB}. In the "Q flat region", the D_{max} of the shaft and bracket decrease and correspond to the F_{zCB} with a similar tendency. It indicates that the resultant axial hydraulic force on the crown and band plays a principal role that affects the deformation of the shaft and bracket. However, the axial force on the blade has only a slight effect. This is a new breakthrough understanding because total axial hydraulic force was, previously, usually regarded as important. Now, the influence of the resultant axial hydraulic force on the runner's crown and band should be specially focused.

Unlike the shaft and bracket, a developing distribution of the runner axial deformation during starting-up is shown in Figure 19. Initially, the D distribution of the runner is radial-symmetric. The D_{max} is 1.65 mm and the location is at the outer edge of the crown. The D_{min} is at the bottom of the band (shown in Figure 19a) due to the effect of centrifugal force. However, at $t = 6$ s, there are different degrees of deformation between the two sides of the runner. As shown in Figure 19b,c, the maximum is at one side of the outer edge of the crown while the minimum is at the other side, at the bottom of the band. In the later period of the starting-up process, the runner is obviously deflected with the nodal diameter as midline. D_{max} and D_{min} are, respectively, at the two sides outer edge of crown, as shown in Figure 19d. Considering the operation of the pump turbine, it is important to find out this phenomenon. This is because the deformation on the crown and band may change the size of leakages and then influence the leakage flow field.

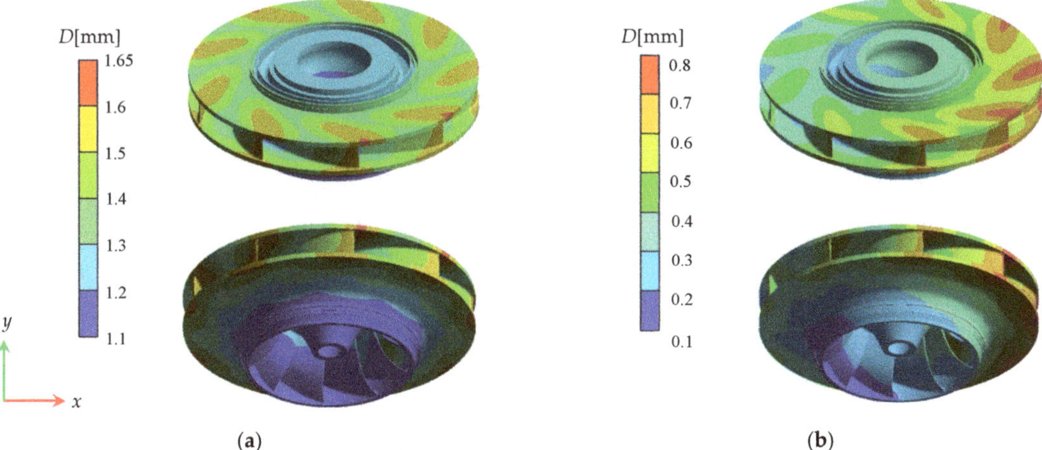

(a)

(b)

Figure 19. *Cont.*

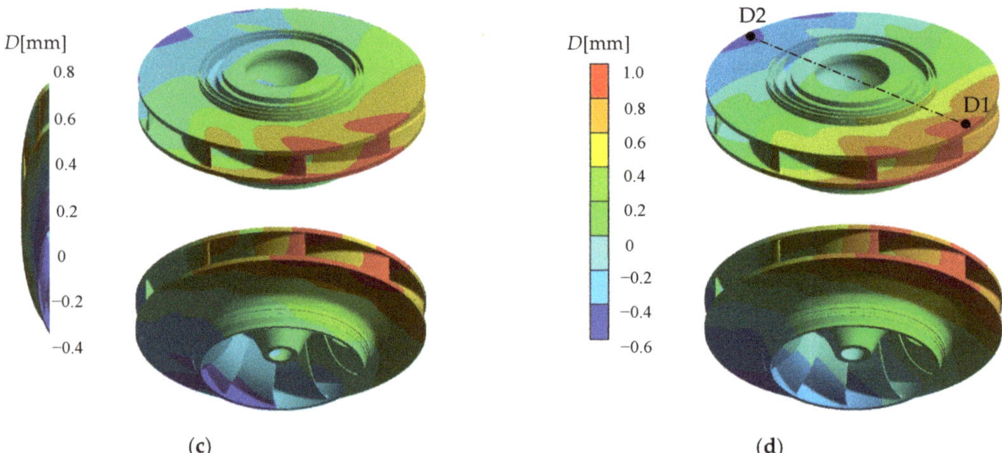

(c) (d)

Figure 19. Deformation of main shaft during starting-up at t = 3 s, 6 s, 16 s, 21 s. (**a**) t = 3 s; (**b**) t = 6 s; (**c**) t = 16 s; (**d**) t = 21 s.

In order to understand the runner deflection, based on the location of D_{max} and D_{min} at t = 21 s (marked as D1 and D2 in Figure 19d), the deformation development is provided in Figure 20. During starting-up, the maximum location is at D1 all the time. The variation of D_{max} of the runner has a similar tendency to those of the shaft and bracket as a result of the superposition of deformation. The largest D_{max} is 1.7 mm at t = 11 s. After that, the D_{max} is at lower values of around 0.75~1 mm. In terms of D2, it is very close to D1 at the beginning. With the increase in guide vane opening, the difference between D1 and D2 also increases. Figure 21 provides the development of $\Delta D = D1 - D2$ which provides a visualized illustration. According to the asymmetric streamlines in the guide vane region, as shown in Figure 15, a possible explanation is that the increasing of the flow rate leads to a more serious asymmetric flow field distribution as the volute are asymmetric. This may induce the large runner deflection phenomenon.

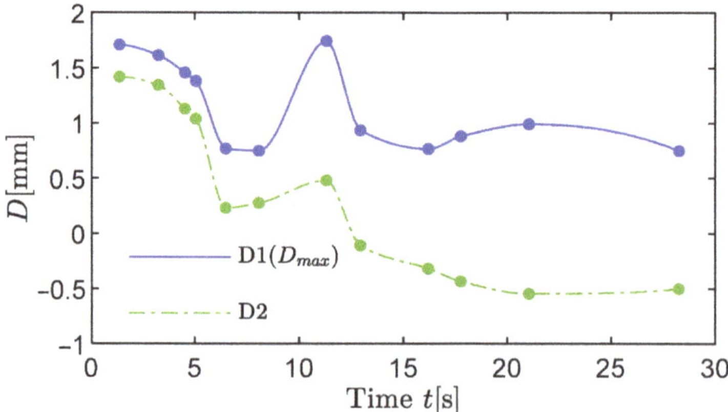

Figure 20. Deformation of runner's typical location during starting-up.

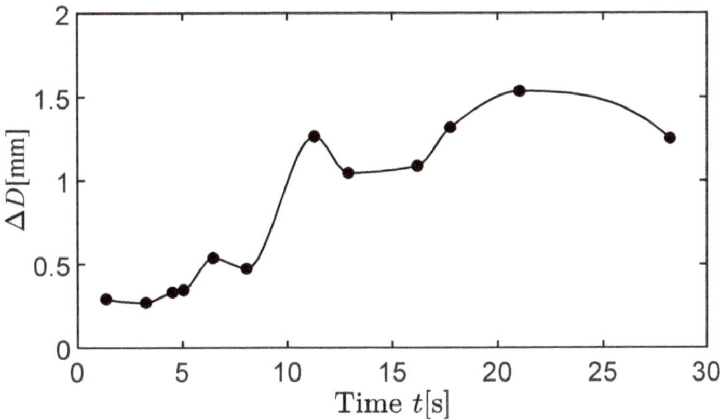

Figure 21. ΔD of runner during starting-up.

To illustrate the reason of runner deflection, the pressure distribution of COS and BOS are found to be asymmetric in the later part of starting-up, when $t = 21$ s, as shown in Figure 22. It is obvious that the pressure distribution at 135° direction of both COS and BOS are smaller than that of the opposite direction. As the direction of axial hydraulic force of COS and BOS is opposite and the effects would be counteracted against each other, the pressure distribution on every 45° line (as marked in Figure 22a) of COS and BOS is provided in Figure 23 for comparison. R^* is the relative radial distance where $R^* = 2R/D_{hi}$. In COS and BOS, C_p of PL4 is the largest while that of PL2 is the smallest. It can be compared to the difference between the PL4 and PL2 of COS, which is larger than that of BOS, which means that COS plays the principal role in affecting the runner deflection. In this case, it can be summarized that the downward hydraulic force on the D1 side is larger than that on the D2 side. Therefore, the deflection of runner can be well explained.

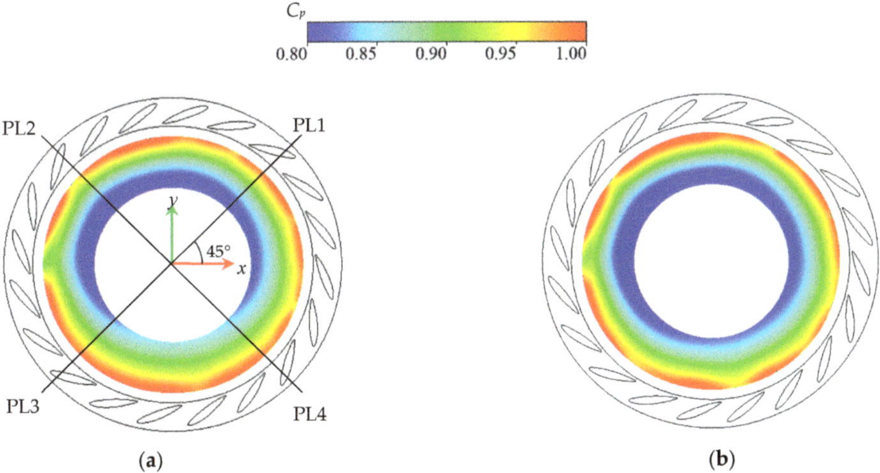

Figure 22. Pressure distribution of COS and BOS at $t = 21$ s. (a) COS; (b) BOS.

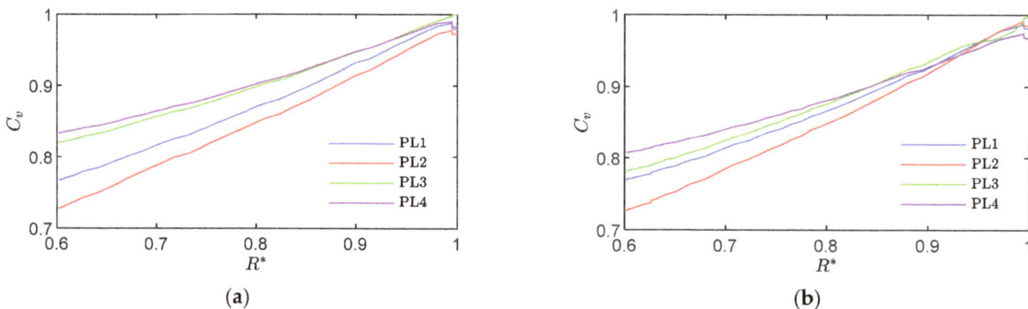

Figure 23. Pressure distribution on typical lines of COS and BOS at $t = 21$ s. (**a**) COS; (**b**) BOS.

5.2.2. Von-Mises Stress

Unlike the deformation of the runner, the Von-Mises stress distribution and the σ_{max} location on the runner are almost constant during starting-up. It is also almost constant on the support bracket. The stress of the main shaft, which is very small, is not discussed in detail in this case. Figures 24 and 25 show the distribution of Von-Mises stress on the support bracket and runner at $t = 21$ s. The maximum stress of the support bracket concentrates on the connection between the thrust seating and the support plates. The maximum stress value is about 90.2 MPa at $t = 21$ s. However, the strength of other sites is strong enough as the stress is only around 10 MPa. For the runner, the stress concentration occurs on the connection between the crown and the leading-edge of the blade. The σ_{max} of the runner is about 105.8 MPa at $t = 21$ s and that of the other regions is less than 60 MPa.

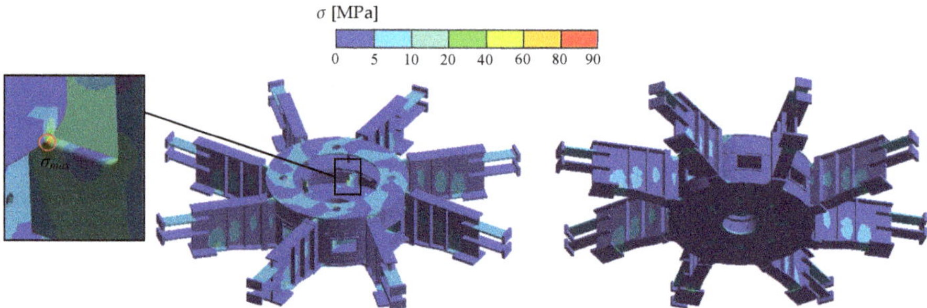

Figure 24. Von-Mises stress of support bracket at $t = 21$ s.

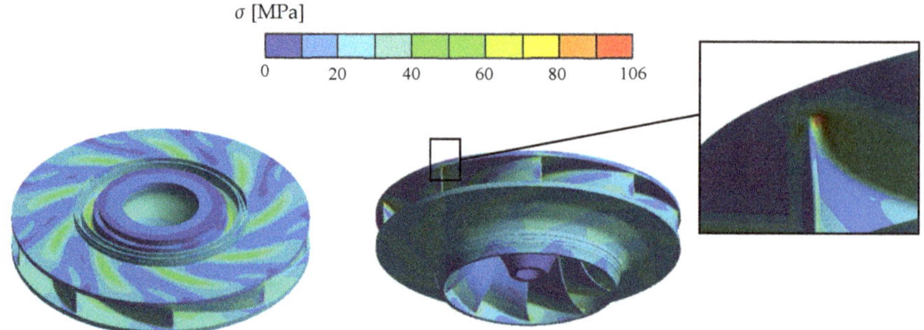

Figure 25. Von-Mises stress of runner at $t = 21$ s.

Figure 26 shows the σ_{max} development of the support bracket and the runner during starting-up. For the support bracket, the σ_{max} development has a similar trend with deformation. It is principally influenced by the resultant force on the crown and band. The largest σ_{max} of the support bracket is 172.8 MPa at the beginning of starting-up. As guide vane opens completely, σ_{max} reaches the smallest value of 90.2 MPa. It indicates that the beginning of starting-up should be focused when considering the strength and safety of the support bracket. Differently, the σ_{max} of the runner reaches the peak value of 134.1 MPa at $t = 11$ s. It is obvious that the stress of the runner is not only influenced by the hydraulic force on the crown and band but also affected by the blade force. In this paper, the axial hydraulic force is mainly focused. The radial and circumferential force components and their influence on structural stress will be discussed in future research.

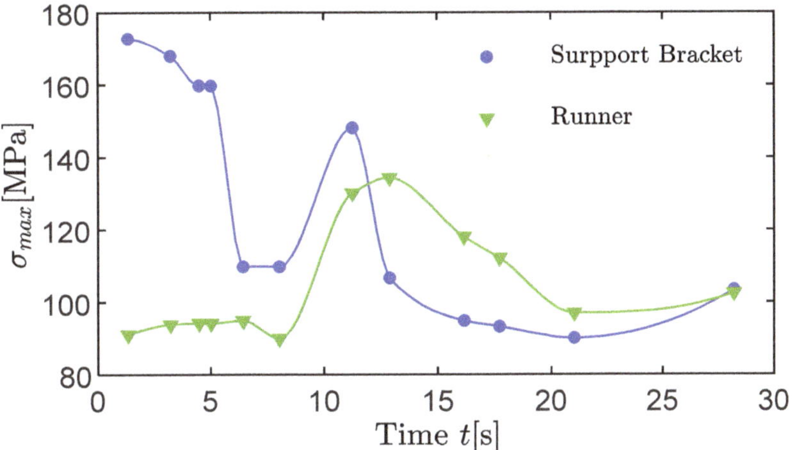

Figure 26. Maximum Von-Mises stress of support bracket and runner during starting-up.

6. Conclusions

This paper studied the axial hydraulic force characteristics on the runner in a pump turbine case during the starting up of the pump mode. Conclusions can be drawn as follows:

1. According to the features of axial hydraulic force, the pump mode's starting-up process can be divided into two parts—those of the "unsteady region" and the "Q flat region". In the "unsteady region", the axial hydraulic force and its components are obviously fluctuant. In the "Q flat region", the axial hydraulic force shows a strong positive relationship with flow rate. The dividing point of these two regions is, approximately, at the half-opening of the guide vane. The components of axial hydraulic force are in different directions at different positions. Therefore, the total axial hydraulic force is formed by the counteraction among force components. It is the main reason for the complexity of axial hydraulic force characteristics.

2. The pressure distribution in the runner and leakages and the streamline in the guide vane region enabled the identification of the mechanism of axial hydraulic force development. In the "unsteady region", the pressure is obviously unstable. Pressure in the runner's crown and band leakages is apparently higher than in the runner. A twin-vortex flow structure can be observed in the vaneless region between the guide vane and the stay vane with strong flow blockage. In the "Q flat region", the pressure in the crown and band leakages remains almost unchanged, while the flow regime in the guide vane is well-behaved. It is worth noting that the flow pattern distribution in the guide vane is asymmetric in the later period of the starting-up process.

3. The maximum deformation of the main shaft is located at the top of shaft. The maximum deformation of the support bracket is on the thrust seating. The finding that

the resultant axial hydraulic force on the crown and band plays a principal role in affecting the deformation of the shaft and bracket is unexpected. However, the axial force on the blade has just a slight effect. The deformation on the runner is radially symmetric at the beginning of starting-up, while the runner clearly deflects with the nodal diameter at the midline in the later period. The reason is found to be the asymmetric pressure distribution of COS and BOS. Among them, COS plays the principal role.

4. The maximum stress on the support bracket concentrates on the connection between the thrust seating and the support plates. The maximum stress on the runner is on the connection between the crown and the blade inlet edge. The σ_{max} development of the support bracket has a similar tendency with the development of deformation. The stress on the runner is found not only to be influenced by hydraulic force on the crown and band, but also to be affected by the blade force.

Finally, this paper will be helpful in realizing the axial hydraulic force of the pump turbine during the starting-up of the pump mode and can provide support for the design of structural components. In the further works in the future, the transient characteristics of axial hydraulic force, vibration and structural dynamic stress should be analyzed and discussed in order to facilitate further improvement in more actual engineering cases.

Author Contributions: Conceptualization, Y.L. and Z.W.; Data curation, R.T.; Formal analysis, Z.M. and H.F.; Resources, R.T. and H.B.; Software, Z.M. and F.C.; Supervision, R.T. and Y.L.; Validation, F.C. and J.C.; Visualization, J.C.; Writing—Original draft, Z.M.; Writing—Review & editing, Z.M. All authors have read and agreed to the published version of the manuscript.

Funding: This research was funded by National Key R&D Program of China, grant number 2016YFC0401905 and National Natural Science Foundation of China, grant number 51909131.

Institutional Review Board Statement: Not applicable.

Informed Consent Statement: Not applicable.

Data Availability Statement: Not applicable.

Acknowledgments: This study is financially supported by the National Key R&D Program of China No. 2016YFC0401905 and the National Natural Science Foundation of China No. 51909131. Fujian Xianyou Pumped Storage Power Station provided strong technical support to the work, especially with help in the on-site prototype tests.

Conflicts of Interest: The authors declare no conflict of interest.

References

1. Zhang, Y.; Wu, Y. A review of rotating stall in reversible pump turbine. *Proc. Inst. Mech. Eng. Part C J. Mech. Eng. Sci.* **2016**, *231*, 1181–1204. [CrossRef]
2. Zhang, J. Study on Thrust Support Types of Vertical Axis Hydraulic Turbine Generating Unit. *Large Electr. Mach. Hydraul. Turbine* **2014**, *1*, 52–56.
3. Huang, D.; Yuan, B.; Yu, Z. Treatment of Machine Lifting for Kaplan Turbine. *Water Resour. Power* **2009**, *27*, 161–162.
4. Wu, G.; Zhang, K.W.; Dai, Y.F. Influences of the Runner Gap and Seal Construction on the Safety in Operation of Francis Water Power Sets. *Large Electr. Mach. Hydraul. Turbine* **2005**, *1*, 44–48+52.
5. Li, Z.; Bi, H.; Karney, B.; Wang, Z.; Yao, Z. Three-dimensional transient simulation of a prototype pump-turbine during normal turbine shutdown. *J. Hydraul. Res.* **2017**, *55*, 520–537. [CrossRef]
6. Li, Z.; Bi, H.; Wang, Z.; Yao, Z. Three-dimensional simulation of unsteady flows in a pump-turbine during start-up transient up to speed no-load condition in generating mode. *Proc. Inst. Mech. Eng. Part A J. Power Energy* **2016**, *230*, 570–585. [CrossRef]
7. Ji, X.-Y.; Li, X.-B.; Su, W.-T.; Lai, X.; Zhao, T.-X. On the hydraulic axial thrust of Francis hydro-turbine. *J. Mech. Sci. Technol.* **2016**, *30*, 2029–2035. [CrossRef]
8. Faria, M.T.C.; Paulino, O.G.; De Oliveira, F.H.; Barbosa, B.H.G.; Martinez, C.B. Influence of Mechanical Draft Tube Fish Barrier on the Hydraulic Thrust of Small Francis Turbines. *J. Hydraul. Eng.* **2010**, *136*, 924–928. [CrossRef]
9. Kazakov, Y.A.; Pelinskii, A.A. Experimental investigation of the axial force in a submersible, electric well pump. *Chem. Pet. Eng.* **1970**, *6*, 262–263. [CrossRef]
10. Tao, R.; Xiao, R.; Liu, W. Investigation of the flow characteristics in a main nuclear power plant pump with eccentric impeller. *Nucl. Eng. Des.* **2018**, *327*, 70–81. [CrossRef]

11. Wang, C.; Shi, W.; Zhang, L. Calculation Formula Optimization and Effect of Ring Clearance on Axial Force of Multistage Pump. *Math. Probl. Eng.* **2013**, *2013*, 1–7. [CrossRef]

12. Wu, G.; Zhang, K.; Dai, Y.; Sun, J. Influences of the leakage rate of low specific speed Francis runner on phenomenon of the lifting hydroelectric generator set. *J. Hydroelectr. Eng.* **2004**, *4*, 106–111.

13. Meng, L.; Zhang, S.P.; Zhou, L.J.; Wang, Z.W. Study on the Pressure Pulsation inside Runner with Splitter Blades in Ultra-High Head Turbine. In Proceedings of the IOP Conference Series: Earth and Environmental Science, Montreal, QC, Canada, 22–26 September 2014; Volume 22, p. 32012.

14. Zhang, Y.; Zheng, X.; Li, J.; Du, X. Experimental study on the vibrational performance and its physical origins of a prototype reversible pump turbine in the pumped hydro energy storage power station. *Renew. Energy* **2019**, *130*, 667–676. [CrossRef]

15. Xia, L.; Cheng, Y.G.; Yang, Z.; You, J.; Yang, J.; Qian, Z. Evolutions of Pressure Fluctuations and Runner Loads During Runaway Processes of a Pump-Turbine. *J. Fluids Eng.* **2017**, *139*, 091101. [CrossRef]

16. Liu, D.; You, G.; Wang, F.; Zhang, J. Calculation and analysis of axial thrust acting on turning wheel of flow-mixing reversible hydraulic turbines. *J. Hohai Univ.* **2004**, *32*, 557–561.

17. Zhao, X.; Lai, X.; Gou, Q.; Zhu, L.; Tang, J. Research on the calculation formula of axial hydro-thrust of Francis turbine. *China Rural Water Hydropower* **2015**, *5*, 172–175.

18. Zhou, L.; Shi, W.; Li, W.; Agarwal, R. Numerical and Experimental Study of Axial Force and Hydraulic Performance in a Deep-Well Centrifugal Pump with Different Impeller Rear Shroud Radius. *J. Fluids Eng.* **2013**, *135*, 104501. [CrossRef]

19. Gantar, M.; Florjancic, D.; Sirok, B. Hydraulic Axial Thrust in Multistage Pumps—Origins and Solutions. *J. Fluids Eng.* **2002**, *124*, 336–341. [CrossRef]

20. Zhao, W.G.; He, M.Y.; Qi, C.X.; Li, Y.B. Research on the effect of wear-ring clearances to the axial and radial force of a centrifugal pump. In Proceedings of the IOP Conference Series: Materials Science and Engineering, Beijing, China, 19–22 September 2013; Volume 52, p. 072015.

21. Li, X.; Mao, Z.; Lin, W.; Bi, H.; Tao, R.; Wang, Z. Prediction and Analysis of the Axial Force of Pump-Turbine during Load-Rejection Process. *IOP Conf. Series Earth Environ. Sci.* **2020**, *440*, 052081. [CrossRef]

22. Li, J.-W.; Zhang, Y.; Liu, K.-H.; Xian, H.-Z.; Yu, J.-X. Numerical simulation of hydraulic force on the impeller of reversible pump turbines in generating mode. *J. Hydrodyn.* **2017**, *29*, 603–609. [CrossRef]

23. Tang, W.; Zheng, X. Group Theory Method for Upper Bracket-stator System Vibration Analysis. *Large Electr. Mach. Hydraul. Turbine* **1999**, *3*, 10–14.

24. Luo, Y.; Wang, Z.; Chen, G.; Lin, Z. Elimination of upper bracket resonance in extremely high head Francis hydro-generators. *Eng. Fail. Anal.* **2009**, *16*, 119–127. [CrossRef]

25. Trivedi, C.; Cervantes, M.J. Fluid-structure interactions in Francis turbines: A perspective review. *Renew. Sustain. Energy Rev.* **2017**, *68*, 87–101. [CrossRef]

26. Wang, W.Q.; He, X.; Zhang, L.X.; Liew, K.; Guo, Y. Strongly coupled simulation of fluid-structure interaction in a Francis hydroturbine. *Int. J. Numer. Methods Fluids* **2009**, *60*, 515–538. [CrossRef]

27. Wang, W.; Yan, Y. Strongly coupling of partitioned fluid–solid interaction solvers using reduced-order models. *Appl. Math. Model.* **2010**, *34*, 3817–3830. [CrossRef]

28. Schmucker, H.; Flemming, F.; Coulson, S. Two-Way Coupled Fluid Structure Interaction Simulation of a Propeller Turbine. *Int. J. Fluid Mach. Syst.* **2010**, *3*, 342–351. [CrossRef]

29. Dompierre, F.; Sabourin, M. Determination of turbine runner dynamic behaviour under operating condition by a two-way staggered fluid-structure interaction method. *Ser. Earth Environ. Sci.* **2010**, *12*, 012085.

30. Xiao, R.; Wang, Z.; Luo, Y. Dynamic stresses in a francis turbine runner based on fluid-structure interaction analysis. *Tsinghua Sci. Technol.* **2008**, *13*, 587–592. [CrossRef]

31. Zhou, L.; Wang, Z.; Xiao, R.; Luo, Y. Analysis of dynamic stresses in Kaplan turbine blades. *Eng. Comput.* **2007**, *24*, 753–762. [CrossRef]

32. Luo, Y.; Wang, Z.; Chen, L.; Wu, J. Finite Element Analysis Design of a Split Rotor Bracket for a Bulb Turbine Generator. *Adv. Mech. Eng.* **2013**, *5*, 428416. [CrossRef]

33. Luo, Y.; Wang, Z.; Zeng, J.; Lin, J. Fatigue of piston rod caused by unsteady, unbalanced, unsynchronized blade torques in a Kaplan turbine. *Eng. Fail. Anal.* **2010**, *17*, 192–199. [CrossRef]

34. Luo, Y.; Wang, Z.; Zhang, J.; Zeng, J.; Lin, J.; Wang, G. Vibration and fatigue caused by pressure pulsations originating in the vaneless space for a Kaplan turbine with high head. *Eng. Comput.* **2013**, *30*, 448–463. [CrossRef]

35. Momčilović, D.; Odanović, Z.; Mitrović, R.; Atanasovska, I.; Vuherer, T. Failure analysis of hydraulic turbine shaft. *Eng. Fail. Anal.* **2012**, *20*, 54–66. [CrossRef]

36. Urquiza, G.; García, J.C.; González, J.G.; Castro, L.; Rodríguez, J.A.; Basurto-Pensado, M.A.; Mendoza, O.F. Failure analysis of a hydraulic Kaplan turbine shaft. *Eng. Fail. Anal.* **2014**, *41*, 108–117. [CrossRef]

37. Zhai, L.; Luo, Y.; Wang, Z.; Liu, X. Failure Analysis and Optimization of the Rotor System in a Diesel Turbocharger for Rotor Speed-Up Test. *Adv. Mech. Eng.* **2014**, *6*, 476023. [CrossRef]

38. Terentiev, L. *The Turbulence Closure Model Based on Linear Anisotropy Invariant Analysis*; VDM Verlag Dr. Müller: Saarbrücken, Germany, 2008.

39. Menter, F.R.; Kuntz, M.; Langtry, R. Ten years of industrial experience with the SST turbulence model. *Turbul. Heat Mass Transf.* **2003**, *4*, 625–632.
40. Bi, H.; Fan, H.; Cao, B.; Xu, Y. Research on the Performance of Check Valve at a Long Distance Pumping System. In Proceedings of the ASME-JSME-KSME 2019 8th Joint Fluids Engineering Conference, San Francisco, CA, USA, 28 July–1 August 2019.
41. Wang, H.; Qin, Q.-H. *Methods of Fundamental Solutions in Solid Mechanics*; Elsevier: Amsterdam, The Netherlands, 2019.
42. Megson, T.H.G. *Structural and Stress Analysis*; Butterworth-Heinemann: Oxford, UK, 2019.
43. Yu, M. *Engineering Strength Theory*; High Education Publication: Beijing, China, 1999; Volume 150.
44. Cook, R.D.; Saunders, H. *Concepts and Applications of Finite Element Analysis*; John Wiley & Sons: Hoboken, NJ, USA, 2007.

Journal of
*Marine Science
and Engineering*

Article

Effect of Cutting Ratio and Catch on Drag Characteristics and Fluttering Motions of Midwater Trawl Codend

Wei Liu [1], Hao Tang [1,2,3,4,5,*], Xinxing You [6], Shuchuang Dong [7], Liuxiong Xu [1,2,3,4,5] and Fuxiang Hu [6]

1   College of Marine Sciences, Shanghai Ocean University, Shanghai 201306, China;
    d190200047@st.shou.edu.cn (W.L.); lxxu@shou.edu.cn (L.X.)
2   National Engineering Research Center for Oceanic Fisheries, Shanghai 201306, China
3   Key Laboratory of Oceanic Fisheries Exploration, Ministry of Agriculture and Rural Affairs,
    Shanghai 201306, China
4   The Key Laboratory of Sustainable Exploitation of Oceanic Fisheries Resources, Shanghai Ocean University,
    Ministry of Education, Shanghai 201306, China
5   Scientific Observing and Experimental Station of Oceanic Fishery Resources, Ministry of Agriculture and
    Rural Affairs, Shanghai 201306, China
6   Faculty of Marine Science, Tokyo University of Marine Science and Technology, Minato, Tokyo 108-8477,
    Japan; yuukinsei@gmail.com (X.Y.); fuxiang@kaiyodai.ac.jp (F.H.)
7   Institute of Industrial Science, The University of Tokyo, 5-1-5 Kashiwanoha, Chiba, Kashiwa 277-8574, Japan;
    dongsc@iis.u-tokyo.ac.jp
*   Correspondence: htang@shou.edu.cn; Tel.: +86-21-61900309

Abstract: The codend of a trawl net is the rearmost and crucial part of the net for selective fish catch and juvenile escape. To ensure efficient and sustainable midwater trawl fisheries, it is essential to better understand the drag characteristics and fluttering motions of a midwater trawl codend. These are generally affected by catch, cutting ratio, mesh size, and twine diameter. In this study, six nylon codend models with different cutting ratios (no cutting, 6:1, 5:1, 4:1, 7:2, and 3:1) were designed and tested in a professional flume tank under two conditions (empty codends and codends with catch) and five current speeds to obtain the drag force, spatial geometry, and movement trend. As the cutting ratio of empty codends decreased, the drag force decreased, and the drag coefficient increased. The unfolding degree of codend netting and the height of empty codends were found to be directly proportional to the current speed and inversely proportional to the cutting ratio. The positional amplitude of codend with cutting ratio 4:1 was the smallest for catch. The drag force of codends with catch increased as the current speed increased, and first decreased and then increased as the cutting ratio decreased. To ensure the best stability and minimum drag force of the codend, it is recommended to use the 4:1 cutting ratio codend.

Keywords: cutting ratio; codend; hydrodynamic characteristics; fluttering motions; the Fourier series

Citation: Liu, W.; Tang, H.; You, X.; Dong, S.; Xu, L.; Hu, F. Effect of Cutting Ratio and Catch on Drag Characteristics and Fluttering Motions of Midwater Trawl Codend. *J. Mar. Sci. Eng.* **2021**, *9*, 256. https://doi.org/10.3390/jmse9030256

Academic Editor: Yuriy Semenov

Received: 24 January 2021
Accepted: 24 February 2021
Published: 28 February 2021

1. Introduction

Trawling plays an important role in marine fishing, accounting for approximately 35% of the world's catches [1,2]. Midwater trawling involves pulling a fishing net horizontally through the water behind one or two vessels. This fishing equipment is normally designed according to the behavior of target species to achieve more selective netting, and to reduce the nets' drag to decrease the fuel consumption of fishing vessels, thus improving fishing efficiency and increasing economic benefits [3,4]. The shape of a midwater trawl is almost similar to an elliptical cone when dragged; the scientific conicity and outline make the midwater trawl smooth and stable, guiding the fish better and with uniform force [5,6]. The conicity and contour of a trawl are obtained by the cutting ratio. However, to ensure that all of the nets of the trawl are at the same inclined angle with a smooth contour, the cutting ratio generally increases from the wing to the codend, causing the trawl

135

contour line to form a convex curve inwardly when the nets are connected. Then, the curve becomes a straight line by water impact [3].

The codend is connected to the narrow end of a tapered trawl, and is most often an elliptical cone shape due to the cutting ratio [7]. The codend is an essential part of the trawl, because its function is to store large fish by catching them while releasing the juvenile fish [8–10]. However, as the catch accumulates, the codend shape changes, with the front mesh of the codend closing and the rear contour of the codend bulging [11]. To ensure consistent trawl performance, the increase in codend drag as the catch builds up must be considered. Moreover, as the codend is connected to the end of trawl, the effect of shadowing on the codend cannot be ignored [12]. Druault and Germain [13] reported that to optimize the codend efficiency in terms of catchability and its effect on energy consumption, it must have a high static stability, which is difficult due to the significant influence of hydrodynamic turbulence flow. Furthermore, they mentioned that the knowledge of flow instability is important to better understand the force acting on a codend and to implement a selected device.

The selectivity, drag force, and stability of the movement of the codend are significantly affected by the hydrodynamic turbulent boundary layer, vortex, and wake flow developing around the fishing equipment [14]. However, it is difficult to determine the hydrodynamic force on each part of the trawl net because of the fluid–structure interactions that induce large deformation and oscillation of the net, modifying the drag force instantaneously [15,16]. In recent decades, considerable progress has been made in understanding the flow field around a trawl net in general, and a codend in particular, using experimental and numerical approaches. Bouhoubeiny et al. [14] evaluated the flow around rigid and oscillating codend models. These preliminary analyses showed that a symmetrical vortex exists behind the rigid codend and possible turbulent flow interactions with the fluttering codend structure without a complete study of the PIV (Particle Image Velocimetry) database. Recently, Druault and Germain [13] used the PIV method to evaluate the flow field distribution around a codend and found that, while dragging the trawl, vortices are alternately generated behind the codend. The shedding of the vortex generates vertical pressure on the codend and causes it to oscillate. Due to the local hydrodynamic effects (fluctuating velocities, vortex shedding wake, etc.), warp tension variations, movements of the deformable structure itself, and the geometry of rigid structure can cause the codend to oscillate. Moreover, each excitation mechanism can occur simultaneously and interact with each other, thereby increasing the complexity of characterizing the fluid–structure interaction [17]. However, these investigations explain the oscillation of the codend, but do not provide a method to minimize the oscillation amplitude or provide a model to fit the oscillation trend. Madsen et al. [18] evaluated the behavior of six different codends at full scale in a flume tank of SINTEF and found that each codend oscillates considerably when loaded with fish, and the standard codend is the most stable of those six codends. Thus, he considered that adjusting the net type and combination mode of the codend can minimize its oscillation amplitude.

The codend motion and catch build-up in the codend can influence catch quality because they can cause epidermal damage [12,19]. Therefore, it is important to improve the codend performance. As shown by the literature review, there are also differences in the cutting ratio of the codend to suit target species. Kumazawa et al. [20], Zhou et al. [21,22], and Yao et al. [23] investigated the drag characteristics of the midwater trawl, but different cutting ratios of codend structures exist. To date, it has not been determined whether the cutting ratio can be adjusted to reduce the drag force and improve the stability. In addition to influencing the effect of the cutting ratio on attack angle and contour, it also creates a special geometrical shape with the codend that, in turn, influences the hydrodynamic performance of the codend and consequently its catchability, stability, and selectivity. Therefore, in this study, the effects of cutting ratio on the net shape and drag force of empty codends and codends with catch were analyzed. Six codends with different cutting ratios and in two states, with or without catch, were selected. Model tests were performed under five conditions to compare the net shape, drag force, and motion of these codends. The Fourier series was used to fit the positional and drag force oscillations of the codend with catch, providing information to improve the hydrodynamic performance and stability of the codend.

2. Material and Methods

2.1. Codend Design

The Antarctic krill trawl (codend length = 30 m) is designed for the target vessel, F/V "Long Teng" of China National Fisheries Corp. This net codend has a mesh size of 144 mm and bar diameter of 16 mm. The model trawl codend scales are: length scale = 1/20 (λ = 20), small-scale ratio λ' = 5, flow velocity scale = 1/2.24, according to Tauti's law.

Six models of codends constructed using nylon material and with different cutting ratios were used based on the model codend (Net 1). These codends were constructed by assembling four pieces of netting with a diamond mesh size of 30 mm and a twine diameter of 3 mm. Each piece of netting was joined by part 1 (20 × 25 mesh) and part 2 (20 × 25 mesh), in addition to cutting part 2 into six cutting ratios (no cutting, 6:1, 5:1, 4:1, 7:2, and 3:1). The codend parameters and joining methods are shown in Table 1 and Figure 1.

Table 1. Specifications for each netting panel of codend.

Codend		Twine Materials	Bar Length (mm)	Bar Diameter (mm)	Cutting Ratio	Cutting Sequence (Subscripts Represent Cycle Index)
Part 1		nylon	15	3	No cutting	$[N]_{22}$
	Net 1	nylon	15	3	No cutting	$[N]_{25}$
	Net 2	nylon	15	3	6:1	N $[NBNNNBN]_4$
Part 2	Net 3	nylon	15	3	5:1	$[NBNNBN]_5$
	Net 4	nylon	15	3	4:1	N $[NBNBN]_6$
	Net 5	nylon	15	3	7:2	NB$[NBNBNBNBN]_3$ BNN
	Net 6	nylon	15	3	3:1	$[NBNB]_8$ N

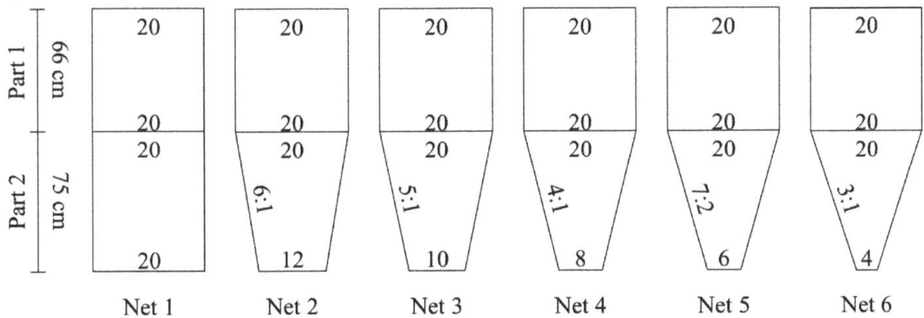

Figure 1. Design sketch of six codends.

2.2. Experimental Setup

Figure 2 shows the experimental apparatus used for measuring the drag force and net shape of the codend. The experiments were conducted in a flume tank at the Tokyo University of Marine Sciences and Technology (TUMST). The test section of the flume tank is 9.0 m in length, 2.2 m in width, and 1.6 m in depth, containing ≈150 tons of freshwater. The flow is circulated using four contrarotating impellers through constant-speed hydraulic delivery pumps. The impellers are 1.6 m in diameter and deliver a maximum flow speed of 2 m/s. A side-viewing window on one side of the flume tank allows users to observe the behavior of the codend during testing and to record video. A camera with a frequency of 59 Hz per frame image and a resolution of 1920 × 1080 pixels was used to record the codend behavior. To accurately measure the drag force of the codend and ensure that the drag force of the equipment is lower than that of the codend, a circular rigid frame was used to determine the drag force of the codend, and the drag force of the equipment was finally subtracted from the total drag force. In the experiment, the codend opening was joined around the rigid frame, and the rigid frame was combined with a six-component force (5 kgf, Denshikogyo Co., Japan) instrument (Figure 2b). The data were sampled at 50 Hz. A flow meter was placed 1.2 m directly in front of the six-component force instrument to detect the current speed. The measurements were conducted at five different flow velocities of 0.5, 0.6, 0.7, 0.8, and 0.9 m /s. The water density of the flume tank was 999.8 kg/m^3, and the water temperature was maintained at 17.6–18.4 °C during the experiments.

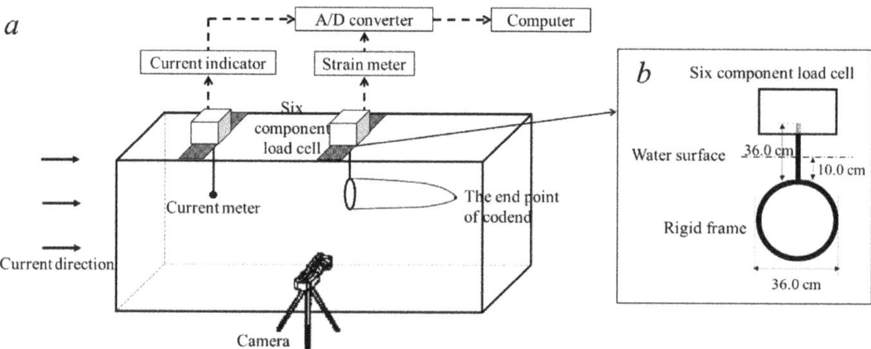

Figure 2. The setup of flume tank experiment (**a**): schematic diagram of experimental set up and apparatus for codend testing; (**b**): schematic diagram of rigid frame).

To measure the drag force and net shape of the codend with catch, a water-injected balloon was used instead of the catch, and a tape measure with an accuracy of 0.1 cm was used to measure the circumference of the balloon at ten different positions. The balloon diameter was calculated as 16.6 ± 0.3 cm, and the density of the water-injected balloon was similar to that of the water in the tank.

2.3. Experimental Procedures

The experiment was divided into two parts with a total of 30 tests. The first part was to measure the drag force and net shape of the empty codend, and the second part was to measure the codend with catch.

Before the experimental measurements of different codends, the drag force of a rigid frame was first measured. The rigid frame was directly combined with the six-component force instrument, making the plane of rigid frame perpendicular to the current direction. This rigid frame handle was immersed in water at 10.0 cm. The drag force of frame was measured at different current velocities, ranging from 0.3 to 1.1 m/s, and set up with an interval of 0.1 m/s. The drag force data measurements of the rigid frame (500 datapoints in total) were sampled for a duration of 10 s, and the average value was calculated using the measurement data.

After the measurement of rigid frame drag force, the second measurement was conducted on the codend without catch. The rigid frame and codend were attached to the six-component instrument and submerged in the water at 10.0 cm, making the rigid frame plane perpendicular to the current direction. The current velocity was adjusted until the codend shape was unfolded and stable. The camera was used directly in front of the observation window to record the net shape of the codend. The drag force and net shape of each codend were recorded for 10 s. Finally, the codend with catch inside (water-injected balloon) was measured, following the same experimental procedure as the first experiment.

The drag coefficient was calculated using Equation (1) as follows:

$$Cx = \frac{2R}{\rho S V^2} \tag{1}$$

where Cx is the drag coefficient, R is the drag force, ρ is the density of water, S is the trawl net opening area, and V is the current speed.

2.4. Data Collection and Analysis

The net shape geometry was obtained from the video camera (Figure 3). The following method was used for image processing in this study: First, a series of images separated by 0.25 s were selected from the recorded video footage; second, graph digitizing software was used to extract the coordinates of characteristic points of the model net based on a plane-coordinate system; finally, a standard bar was used in different locations to calibrate the measurements and consider the effect of camera lens and water refraction. In addition, by maintaining the 50 Hz high-frequency data of drag force for the codend, the high-frequency data were processed to obtain the corresponding 4 Hz data.

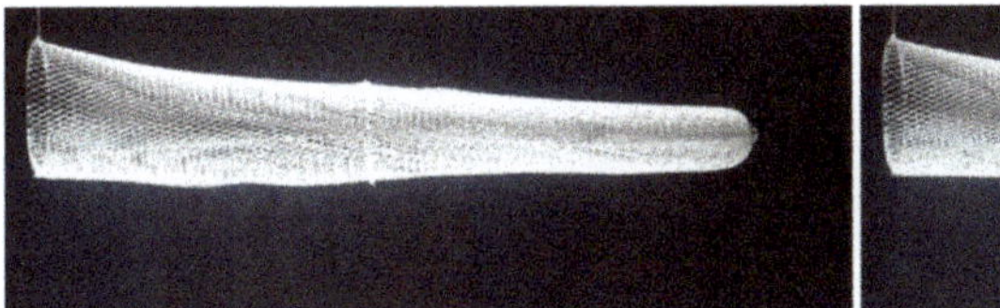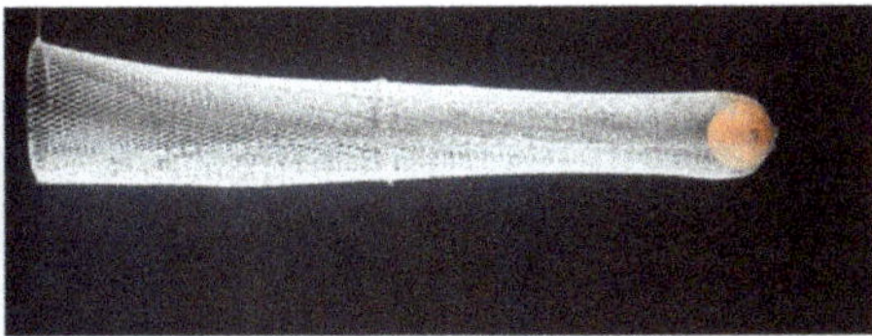

Figure 3. Side-view of the empty codend (**left**) and codend with catch (**right**) in the flume tank.

2.5. Data Fitting and Testing

The Fourier series was used to fit the net motion and drag force oscillation of the codend with catch. The Pearson product-moment correlation coefficient was used to assess the correlation between the fitted and measured values, where 80–100% indicates very strong correlation, 60–80% indicates strong correlation, 40–60% indicates moderate correlation, 20–40% represents weak correlation, and 0–20% indicates very weak correlation or no correlation.

The Fourier series formula can be estimated as follows:

$$f(t) = \frac{A_0}{2} + \sum_{n=1}^{\infty} (A_n \cos\omega_n t + B_n \sin\omega_n t) \tag{2}$$

$$\text{or } f(t) = \frac{A_0}{2} + \sum_{n=1}^{\infty} A_n \sin(\omega_n t + \varphi_n) \tag{3}$$

where t is the time, A_n and B_n are amplitude, ω_n is the angular frequency, φ_n is the initial phase, and n is the series.

The Pearson product-moment correlation coefficient can be calculated using Equation (4):

$$\rho_{XY} = \frac{Cov(X,Y)}{\sqrt{D(X)}\sqrt{D(Y)}} = \frac{E\{[X - E(X)][Y - E(Y)]\}}{\sqrt{D(X)}\sqrt{D(Y)}} \tag{4}$$

where E is the mean, D is the variance, $\sqrt{D(X)}$ and $\sqrt{D(Y)}$ are the standard deviation of variables X and Y, and $E\{[X - E(X)][Y - E(Y)]\}$ is the covariance of variables X and Y, denoted $Cov(X, Y)$, i.e., $Cov(X, Y) = E\{[X - E(X)][Y - E(Y)]\}$.

3. Results

3.1. Frame Drag Force

To verify the effectiveness of the experimental frame, the rigid frame drags were measured at different current speeds (Figure 4). The rigid frame drag force varied from 32.5 to 541.4 g as the current speed varied from 0.3 to 1.1 m/s. The results indicate that the rigid frame drag increased as the current speed increased, and their relationship was exponential (Figure 4). Using nonlinear regression with the data argument predicted from the formula of drag force of rigid frame with different current speeds, the following expression is proposed:

$$R_{frame} = 422.47 \times V_{flow}{}^{2.070}, \left(R^2 = 0.99 \right) \tag{5}$$

where R_{frame} is the drag force of the rigid frame and V_{flow} is the current speed.

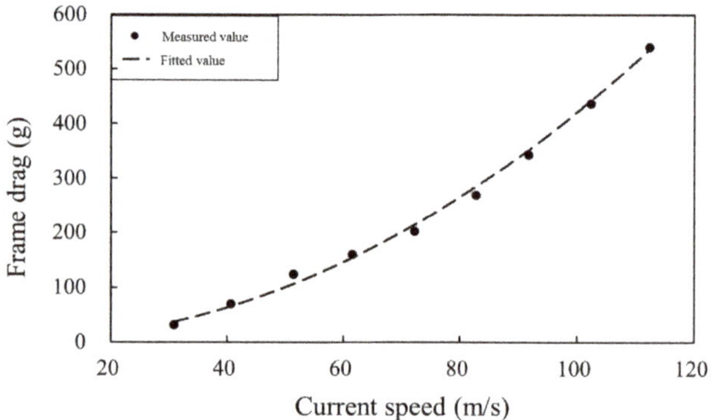

Figure 4. The relationship between frame drag and current speed.

3.2. Empty Codend Profile

As shown in Figure 5, the codend motion increases as the current speed increases. The codend without cutting droops the most at a current speed ≤0.6 m/s. However, at a current speed of ≥0.8 m/s, the central axis of each codend is horizontal. Moreover, the unfolding degree of codend netting and the height of the codend endpoint are directly proportional to the current speed and inversely proportional to the cutting ratio.

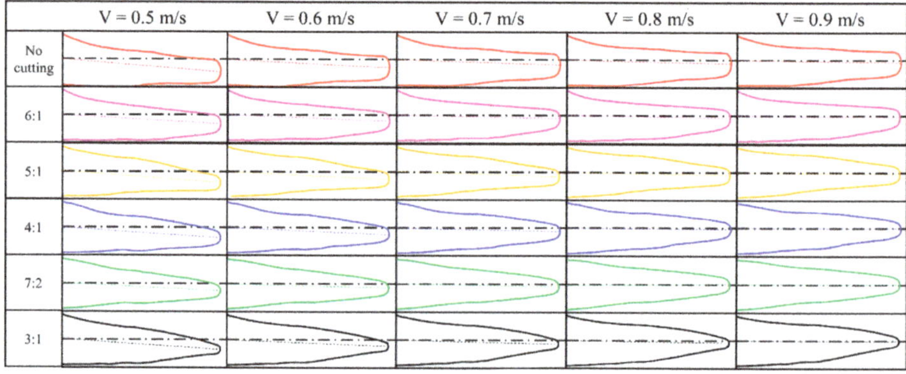

Figure 5. Variation of the profile of empty codends with different cutting ratios at different current speeds.

3.3. Drag Force of Empty Codends

The drag forces on the codends were determined by subtracting the averaged measurements for each current speed on the rigid frame from the averaged measurements for each current speed on the frame and codends; the results are shown in Figure 6. The codends' drag force increased as the current speed and cutting ratio increased. On average, the drag forces of codend without cutting are 9.44% ($\pm$0.94%), 13.25% ($\pm$0.67%), 16.68% ($\pm$0.81%), 16.05% ($\pm$1.04%), and 18.24% ($\pm$0.95%) greater than those of codend with cutting ratios 6:1, 5:1, 4:1, 7:2, and 3:1, respectively.

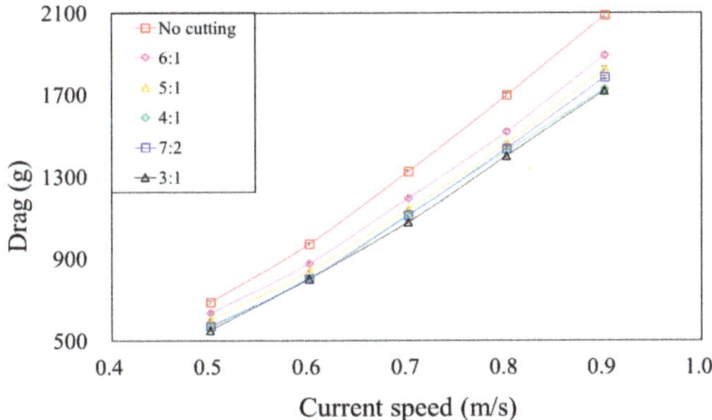

Figure 6. Drag of empty codends with different cutting ratio at 5 current speeds.

On average, the drag coefficients of codend without cutting are 2.99% ($\pm$0.66%), 3.04% ($\pm$0.53%), 3.09% ($\pm$1.44%), 7.72% ($\pm$0.32%), and 9.33% ($\pm$1.19%) smaller than those of codend with cutting ratios 6:1, 5:1, 4:1, 7:2, and 3:1, respectively (Figure 7). The drag coefficient of the codend decreased as the Reynolds number and cutting ratio increased (Figure 7). Moreover, the codends with cutting ratios 6:1, 5:1, 4:1, 7:2, and 3:1 were designed to have a small amount of twine compared with codends without cutting and those used in the midwater fishing industry to evaluate the effect of cutting ratio on the total drag.

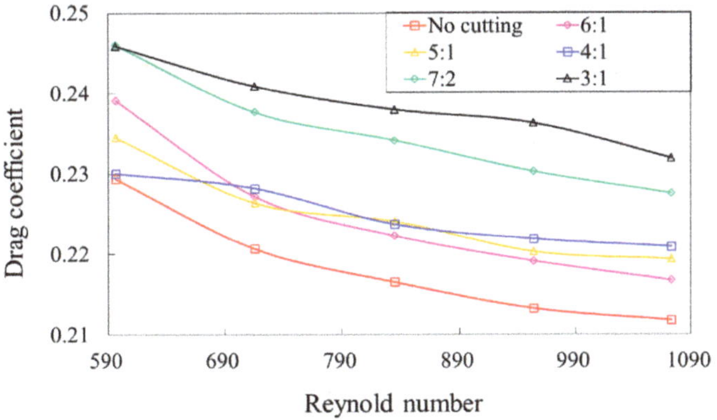

Figure 7. Drag coefficient of empty codends with different cutting ratio at five current speeds.

3.4. Positional Oscillation of Codends with Catch

Evaluation of fluid–structure–catch interaction requires consideration of the geometry and motions of codend structure. Figure 8 shows the time evolution of fluctuating transverse motion of codend structure with catch inside during selected measurements. Clearly, the codend oscillates when it has a catch. However, using the first-order and second-order

Fourier series to fit the positional oscillation, the results show that as the current speed increased, the oscillation cycle of each codend decreased, but the amplitude did not change significantly with a gap of <1.5%. Except for the amplitude of codend without the cutting ratio, the amplitude first decreased (current speed lower than 0.7 m/s) and then increased (current speed higher than 0.7 m/s) as the cutting ratio decreased.

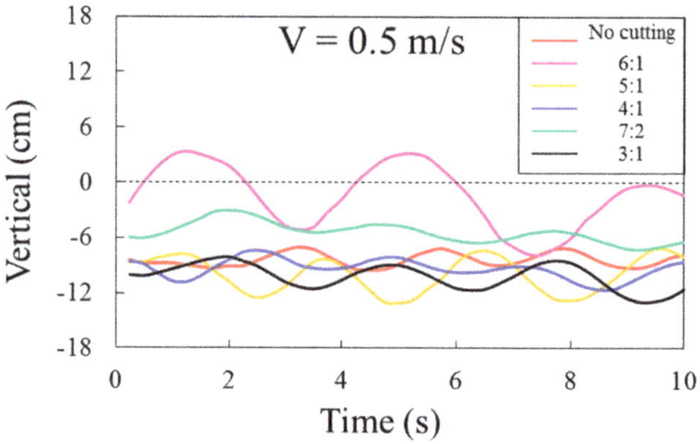

Figure 8. Variations of the end point height for codends with catch in different cutting ratios.

3.5. Use of Fourier Series to Fit the Positional Oscillation of Codends

The Fourier series was used to fit the positional oscillation of the codend, and the Pearson product-moment correlation coefficient was used to test the fitted value (Figure 9, Table 2). The results show that when the first-order Fourier series was used to fit the motion of codend, the fitted value was the same as the measured value with a very strong correlation (43.3%), but others still have an error compared with the measured value because of a moderate correlation of 10%. When the second-order Fourier series was used, the fitted value showed a very strong correlation of 93.3% compared with the measured value, and the remaining (6.7%) had a strong correlation.

Table 2. Correlation of measured value and fitted value of the end point height for codends by the Pearson product-moment correlation coefficient.

					Cutting Ratio			
First-order Fourier Series		**No Cutting**	**6:1**	**5:1**	**4:1**	**7:2**	**3:1**	**Average**
	0.5	79.0%	88.3%	91.9%	71.8%	66.9%	85.2%	80.5%
	0.6	94.5%	91.0%	80.4%	70.6%	75.8%	91.8%	84.0%
Current speed (m/s)	0.7	92.5%	86.6%	96.1%	56.2%	77.8%	86.8%	82.7%
	0.8	92.2%	74.6%	68.0%	70.1%	58.3%	89.9%	75.5%
	0.9	77.8%	62.5%	79.1%	76.0%	53.2%	84.1%	72.1%
Second-order Fourier Series								
	0.5	88.8%	90.5%	94.0%	95.4%	97.4%	92.0%	93.0%
	0.6	96.4%	96.3%	89.3%	89.9%	98.1%	92.2%	93.7%
Current speed (m/s)	0.7	93.0%	96.5%	98.0%	87.3%	94.6%	96.7%	94.3%
	0.8	94.8%	95.5%	78.5%	75.6%	85.6%	93.7%	87.3%
	0.9	90.7%	84.9%	86.7%	84.4%	83.4%	89.1%	86.5%

Note: Percentage values represent the similarity between the Fourier series fitted value and the measured value.

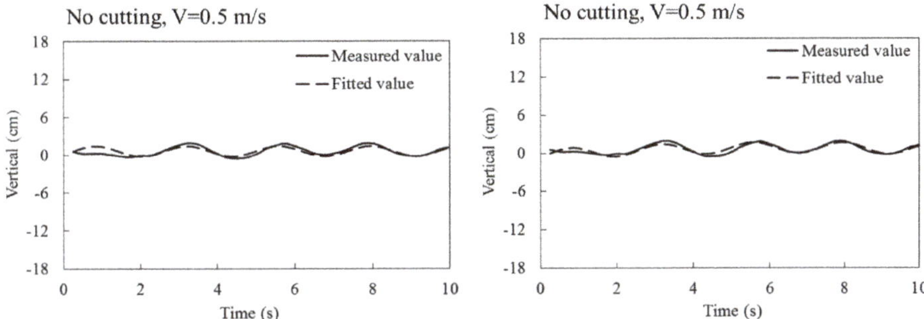

Figure 9. Comparison of the fitted value and the measured value of the end point height for codends with catch using different order Fourier series (an example: (**left**): using the first-order Fourier series; (**right**): using the second-order Fourier series).

3.6. Drag Force of Codends with Catch

The drag force of codends with catch increased as the current speed increased; however, when the cutting ratio decreased, the drag force first decreased and then increased, except the drag force of the 4:1 cutting ratio, which was lower than those of other codends (Figure 10). Furthermore, as shown in Figure 10, the drag force evolution of codends was quasi-periodic oscillatory. The oscillations include high-frequency (50 Hz) and low-frequency (4 Hz). However, no significant difference was observed between the high-frequency and low-frequency oscillation of each codend in the first-order Fourier series. In contrast, when the second-order and third-order Fourier series were used to fit the low-frequency oscillation, the amplitude of each codend drag force increased, and the cycle decreased with the increase in current speed. Moreover, the oscillation cycle of each codend drag force had no significant difference in terms of cutting ratio at the same current speed, but the amplitude initially decreased and then increased as the cutting ratio decreased. The amplitude of codend without the cutting ratio is the highest, and that of 4:1 cutting ratio codend is the smallest, compared to those of other codends.

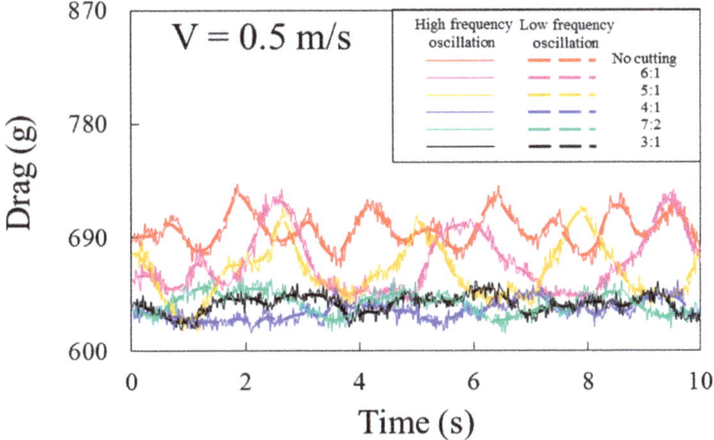

Figure 10. Variations of drag for codends with catch in different cutting ratios.

3.7. Use of Fourier Series to Fit the Drag Force Oscillation of Codends

The Fourier series can be used to fit the drag force oscillation of a codend. However, the accuracy of the fitted value obtained using the second-order Fourier series is different from that obtained using the third-order Fourier series (Figure 11). Using the Pearson product-moment correlation coefficient to test the fitted and measured values (Table 3), the use of second-order Fourier series to fit the measured value has a lower correlation with very strong correlation (56.7%), but the use of third-order Fourier series to fit the drag force of codend with different cutting ratios has also very strong correlation with the increase in measured value to 86.7%.

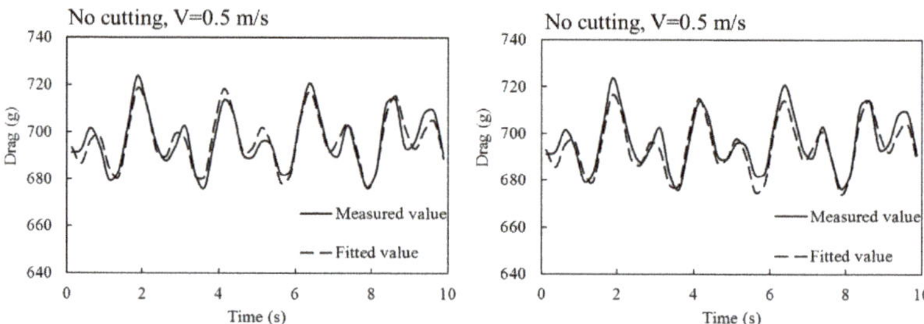

Figure 11. Comparison of the fitted value and the measured value of drag for codends with catch using different order Fourier series (an example: (**left**): using the second-order Fourier series; (**right**): using the third-order Fourier series).

Table 3. Correlation of measured value and fitted value of codend drag by the Pearson product-moment correlation coefficient.

Second-order Fourier Series		No Cutting	6:1	5:1	4:1	7:2	3:1	Average
					Cutting Ratio			
	0.5	94.0%	95.1%	92.9%	81.7%	88.0%	87.8%	89.9%
	0.6	91.3%	81.4%	89.6%	69.9%	69.4%	66.0%	77.9%
Current speed (m/s)	0.7	76.7%	72.8%	92.3%	73.7%	87.5%	50.3%	75.5%
	0.8	82.7%	76.7%	70.2%	72.6%	93.5%	81.0%	79.4%
	0.9	81.5%	70.3%	76.9%	63.9%	85.3%	94.4%	78.7%
Third-order Fourier Series								
	0.5	94.5%	97.4%	95.6%	84.5%	89.8%	92.3%	92.3%
	0.6	92.2%	86.0%	90.9%	80.7%	90.9%	82.5%	87.2%
Current speed (m/s)	0.7	85.9%	81.4%	94.0%	83.4%	93.5%	73.3%	85.3%
	0.8	85.0%	84.6%	72.2%	82.0%	94.4%	88.2%	84.4%
	0.9	86.3%	79.0%	85.2%	74.6%	90.7%	96.7%	85.4%

Note: Percentage values represent the similarity between the Fourier series fitted value and the measured value.

4. Discussion

4.1. Effect of Rigid Frame Drag Force on Codend

The experimental apparatus effect of model nets is a common problem, especially in codend model tests. This study and previous studies by Pichot et al. [24], Bouhoubeiny et al. [14], and Druault and Germain [13] used a self-designed configuration of this rigid frame to accurately measure the hydrodynamic forces of the codend. One of the principal objectives in fishing gear hydrodynamic measurements is to design frames with very lower drag forces unlike those of fishing equipment in general, and nettings or codends in particular. In this case, the average drag force caused by a rigid frame was 17.6% (±1.3%) and 16.1% (±0.7%) of the total compared to the codend without catch and with catch (codend drag + rigid frame drag), respectively. In addition, the frame stability was evaluated under different current speeds and cutting ratios. Indeed, the ratio between drag force caused by a rigid frame without a codend and the total drag increased as current speed and cutting ratio increased. This ratio was 13.32% (±0.42%), 14.51% (±0.54%), 15.05% (±0.49%), 15.60% (±0.46%), 15.47% (±0.35%), and 15.82% (±0.40%) for the codend without cutting ratio and the codends with cutting ratios 6:1, 5:1, 4:1, 7:2, and 3:1, respectively. However, the results reported by Tang et al. [25,26] on netting showed that the frame drag force accounted for a maximum of between 17% and 20% of the total (netting drag + rigid frame drag) using streamlined frames, close to those obtained in this experiment. Therefore, the application of a rigid frame reduced the impact of turbulent or vortex flow on the experimental hydrodynamic force and the geometrical shape of codends. When studying hydrodynamic forces and codend motion at different current speeds, the use of a rigid frames is strongly recommended.

4.2. Effect of Cutting Ratio on the Net Shape and Drag Force of Empty Codends

In this study, the codend did not oscillate significantly under the condition of an empty codend. Unlike the results of previous studies, the results obtained by Pichot [27] and Bouhoubeiny et al. [14] using a rigid codend showed that an asymmetrical vortex was generated at the codend tail due to the obstruction of codend. At the same time, the vortex shedding generated water pressure in the vertical flow direction to the codend and caused an oscillation of this structure. The differences between the results obtained in this study and those obtained by Bouhoubeiny et al. [14] are due to the codend filtration and the current speed, which are not the same. In addition, the difference is also because Bouhoubeiny et al. [14] placed a spherical cap to block the water directly passing through the mesh of the codend at the codend bottom, but allowed the water to flow back and around the outer surface of the spherical cap, thereby generating a symmetrical vortex. In this study, the codend was completely open, facilitating the passage of water through the mesh of the codend and not causing a vortex in the codend that would deviate the fish trajectory. Indeed, the bar of mesh also blocks the water passing through the codend directly [24,28,29], but the pressure generated by the shading effect of the bar was not sufficient to make the entire codend oscillate significantly. Thus, the codend did not significantly oscillate in the camera shot.

As shown in the side view of each codend shape in Figure 5, the lower part of the codend without cutting ratio had obvious stacking at a low current speed, and the stacking nets rippled with the current in the video. These results were also observed in the study carried out by Balash and Sterling [30] on prawn trawl, Bouhoubeiny et al. [31] on the flow measurement around a fishing net, and Tang et al. [32] on the shape measurement of a purse seine. However, according to the drag force results obtained in each empty codend at different current speeds (Figure 6), the drag force of the empty codend positively correlates with the cutting ratio, i.e., a higher cutting ratio led to a greater drag force of codend at the same current speed. Thus, the reason is not only that the cutting ratio was directly proportional to the twine area, but also that the codend with a high cutting ratio was more likely to stack the net under the low current speed condition, thus increasing its drag force. A significant decrease in twine area and drag force was observed due to a decrease in cutting ratio. Indeed, the twine area of the codend without a cutting ratio was 7.81%, 9.96%, 12.20%, 14.52%, and 16.96% greater than that of codends with cutting ratios of 6:1, 5:1, 4:1, 7:2, and 3:1, respectively. As mentioned above, the codends with cutting ratio were designed to contain a small amount of twine compared to codends without a cutting ratio and those used in the midwater trawl fishery industry to demonstrate the effect of the cutting ratio on hydrodynamic force and net shape. In conclusion, according to the experimental results of Balash et al. [33] and Thierry et al. [3] on the effect of net cutting of the prawn trawl body and bottom trawl wing on the trawl performance, and the results of this study, the net cutting of some part of the trawl net allows it to change its shape, which automatically reduces its twine area and drag force while making it more efficient.

This study also demonstrated that the increase in cutting ratio led to the increase in the angle between the flow direction and codend surface (attack angle) during the operation. However, Tang et al. [25] reported that the relationship between the nylon mesh drag coefficient and attack angle ($0°–20°$) has a positive correlation, and the netting drag coefficient increased as the attack angle increased, decreasing as the Reynolds number increased before moving asymptotically toward a constant value. This trend was confirmed by the experiment carried out in this study, demonstrating that the drag coefficient for the empty codend increased as the Reynolds number decreased and attack angle and cutting ratio increased. However, the drag coefficients obtained in this study are different from those obtained in the netting by Tang et al. [25,26,34] and Hosseini et al. [35], whereas they are similar to those obtained on the optimized trawls studied by Balash [30] and Thierry et al. [6]. The main reason for this difference is that the mesh of the codend is in a free state in this experiment, rather than the mesh being tightly bound to the frame. In addition, the net at the bottom of the codend piles up and waves with the current. Moreover, the attack angles are not the same in the whole codend structure, and the shadowing effect of the rigid frame affected the codend drag coefficient. By comparing the effect of the cutting ratio and twine area on the codend drag force, it can be seen that the effect of twine area on the drag force for the codend is greater.

4.3. Effect of Catch Substitutes on the Experiment

The actual catch build-up in a codend produces debris and fall, affecting the outcome and equipment. Some researchers use catch substitutes for model experiments. The catch substitutes are generally divided into two categories, namely multiple small-volume objects and a single large-volume object. The multiple small-volume objects mainly include water-filled table tennis balls or small-volume water-filled bags. Madsen et al. [18] used multiple small-volume water-filled bags to evaluate the effects of mesh types on the hydrodynamics and oscillations of the codend. Multiple small objects in the codend directly block the flow through the mesh, similar to the actual catch when investigating the flow field distribution. However, because small objects have a smooth surface and most of them are spherical with a large space, the overall drag force of the model is less than the actual drag force. In addition, the number of small-volume objects required in the test is mostly large; operating errors can cause some small-volume objects to pass into the tank and damage the instrument. Single large-volume objects mainly include a water-injected balloon or a spherical cap. When Druault and Germain [13] used the PIV technology to study the effect of catch on shadowing and flow field around the codend, a spherical cap and water-injected balloon were used in the place of catches. Such a surrogate also plays a role in blocking the flow directly through the mesh, but it completely blocks the flow. Therefore, the use of a single large volume to replace the catch will result in an overall drag force of the codend that is greater than that actually obtained, and also makes the oscillation of codend more obvious. The drag focus in this study was to analyze the effect of the cutting ratio on the oscillation of the codend. Obvious codend oscillation allowed easy result analysis, so a water-injected balloon was used instead of real catch.

4.4. Effect of Cutting Ratio on the Shape and Drag Force Oscillation of Codends with Catch

A codend with catch has obvious oscillations in position or drag force. Previously, Bouhoubeiny et al. [14], Madsen et al. [18], and Druault and Germain [13] reported that codend oscillations can be caused by the vortices behind the codend. When dragging the trawl, vortex shedding generates vertical pressure on the codend and makes it oscillate. Furthermore, this study and Madsen et al. [18] found that codend oscillation is not limited to the specific direction, but codend oscillation in any direction perpendicular to the codend central axis is possible. The main reason is that the codend motion is free in all directions perpendicular to the central axis. In addition, the oscillating track of the codend is mainly distributed near the central axis according to the results obtained by this experiment and Madsen et al. [18].

The codend drag force oscillation mainly included a high-frequency oscillation and a low-frequency oscillation according to the results obtained in this study and those obtained by Bouhoubeiny et al. [14] and Druault and Germain [13]. In addition, it was also found that the low-frequency oscillation of codend drag force included strong wave oscillation (large amplitude) and weak wave oscillation (small amplitude), and the strong wave and weak wave oscillations appeared alternately. The sum of two oscillating cycles of drag force is approximately equal to one oscillating cycle of the position. Moreover, this experiment and those of Bouhoubeiny et al. [14] and Druault and Germain [13] also confirmed that the oscillation of codend position and drag force oscillation are synchronous, i.e., the alternate cycle of two vortices in the same direction is equal to one codend positional oscillation cycle.

Because of two main factors, namely, the twine area and attack angle, the positional and drag force oscillation amplitudes of large cutting ratios (no cutting, 6:1, and 5:1) are usually greater than those of small cutting ratios (4:1, 7:2, and 3:1). A larger cutting ratio can lead to a larger twine area, making the codend drag force greater. Therefore, the vortex shedding generates more vertical pressure on the codend and increases the amplitude. The periodic variation in the oscillation is mainly due to the periodicity of vortex shedding [13]. The greater the current speed, the greater the vortex shedding speed, and the greater the pressure. This also shows that, as the current speed increases, the oscillation cycle is shorter and leads to a greater amplitude. In addition, according to the orthogonal decomposition analysis, the amplitude decreased as the attack angle and binding force of free direction (vertical current direction) increased.

4.5. Fourier Series Fitting

The Fourier series was used in this study to fit cyclic motion. The higher the series, the closer the fitted value to the measured value, but the calculation increased as the series increased. In this study, the first-order and second-order Fourier series were used to fit the codend positional oscillation. The results show that the value obtained using the second-order Fourier series is close to the measured value, depending on the positional oscillation of the codends with catch taken in a different cutting ratio (Figure 8). This also shows that the codend positional oscillation has a trigonometric function, with the exception of the presence of alternate strong and weak waves of positional oscillation, such as a 6:1 cutting ratio with 0.8 and 0.9 m/s. Therefore, when fitting the codend positional oscillation, it is recommended to use $\geq$ second-order Fourier series to fit the codend positional oscillation.

The Fourier series was used to fit the oscillation of codend drag force which should be greater than that of positional oscillation fitting. As shown in Figure 10, the force oscillation of the codend included high-frequency oscillation and low-frequency oscillations. The low-frequency oscillations included strong wave oscillation and weak wave oscillation. In this study, we only fitted low-frequency oscillation. The results show that although a third-order Fourier series was used to fit the codend drag force oscillations, some of the fitted values were still different from the measured values. Therefore, when using the Fourier series to fit the low-frequency oscillation of the codend drag force, it is recommended to use $\geq$ third-order Fourier series.

Author Contributions: Literature, W.L. and S.D.; figures, W.L. and S.D.; study design, W.L., H.T., X.Y., L.X., and F.H.; data collection, W.L., H.T., X.Y., L.X. and F.H., funding acquisition, H.T. and L.X.; data analysis, W.L. H.T., S.D. and F.H.; writing—original draft, W.L.; writing—review and editing, W.L. and H.T. All authors have read and agreed to the published version of the manuscript.

Funding: This study was financially sponsored by the National Natural Science Foundation of China (Grand No. 31902426), Shanghai Sailing Program (19YF1419800), and Special project for the exploitation and utilization of Antarctic biological resources of Ministry of Agriculture and Rural Affairs (D-8002-18-0097).

Institutional Review Board Statement: Not applicable.

Informed Consent Statement: Not applicable.

Data Availability Statement: Not applicable.

Conflicts of Interest: The authors declare no conflict of interest.

References

1. Watson, R.; Revenga, C.; Kura, Y. Fishing gear associated with global marine catches: II. Trends in trawling and dredging. *Fish. Res.* **2006**, *79*, 103–111. [CrossRef]
2. Anticamara, J.A.; Gelchu, R.; Watson, A.; Pauly, D. Global fishing effort (1950–2010): trends, gaps, and implications. *Fish. Res.* **2011**, *107*, 131–136. [CrossRef]
3. Thierry, N.N.B.; Tang, H.; Achile, N.P.; Xu, L.; Hu, F.; You, X. Comparative study on the full-scale prediction performance of four trawl nets used in the coastal bottom trawl fishery by flume tank experimental investigation. *Appl. Ocean Res.* **2020**, *95*, 102022. [CrossRef]
4. Sala, A.; De Carlo, F.; Buglioni, G.; Lucchetti, A. Energy performance evaluation of fishing vessels by fuel mass flow measuring system. *Ocean Eng.* **2011**, *38*, 804–809. [CrossRef]
5. Park, H.-H.; Cho, B.-K.; Ko, G.-S.; Chang, H.-Y. The gear shape and cross section of sweep at mouth of a bottom trawl. *J. Korean Soc. Fish. Technol.* **2008**, *44*, 120–128. [CrossRef]
6. Thierry, N.N.B.; Tang, H.; Liuxiong, X.; You, X.; Hu, F.; Achile, N.P.; Kindong, R. Hydrodynamic performance of bottom trawls with different materials, mesh sizes, and twine thicknesses. *Fish. Res.* **2020**, *221*, 105403. [CrossRef]
7. Wileman, D.A.; Ferro, R.S.T.; Fonteyne, R.; Millar, R.B. (Eds.) *Manual of Methods of Measuring the Selectivity of Towed Fishing Gears*; ICES Cooperative Research Report No. 215; International Council for the Exploration of the Sea (ICES): Copenhagen, Denmark, 1996; Available online: http://www.vliz.be/imisdocs/publications/266109.pdf (accessed on 27 February 2021).
8. Brčić, J.; Herrmann, B.; Sala, A. Predictive models for codend size selectivity for four commercially important species in the Mediterranean bottom trawl fishery in spring and summer: Effects of codend type and catch size. *PLoS ONE* **2018**, *13*, e0206044. [CrossRef] [PubMed]
9. Sala, A.; Herrmann, B.; De Carlo, F.; Lucchetti, A.; Brčić, J. Effect of codend circumference on the size selection of square-mesh codends in trawl fisheries. *PLoS ONE* **2016**, *11*, e0160354. [CrossRef]
10. Sala, A.; Lucchetti, A.; Buglioni, G. The influence of twine thickness on the size selectivity of polyamide codends in a Mediterranean bottom trawl. *Fish. Res.* **2007**, *83*, 192–203. [CrossRef]
11. O'Neill, F.G.; O'Donoghue, T. The fluid dynamic loading on catch and the geometry of trawl cod–ends. *Proc. R. Soc. A Math. Phys. Eng. Sci.* **1997**, *453*, 1631–1648. [CrossRef]
12. O'Neill, F.; Knudsen, L.; Wileman, D.; McKay, S. Cod-end drag as a function of catch size and towing speed. *Fish. Res.* **2005**, *72*, 163–171. [CrossRef]
13. Druault, P.; Germain, G. Analysis of hydrodynamics of a moving trawl codend and its fluttering motions in flume tank. *Eur. J. Mech. B Fluids* **2016**, *60*, 219–229. [CrossRef]
14. Bouhoubeiny, E.; Germain, G.; Druault, P. Time-resolved PIV investigations of the flow field around rigid cod-end net structure. *Fish. Res.* **2011**, *108*, 344–355. [CrossRef]
15. Liu, L.; Kinoshita, T.; Wan, R.; Bao, W.; Itakura, H. Experimental investigation and analysis of hydrodynamic characteristics of a net panel oscillating in water. *Ocean Eng.* **2012**, *47*, 19–29. [CrossRef]
16. Zheng, Z.; Zhang, N. Frequency effects on lift and drag for flow past an oscillating cylinder. *J. Fluids Struct.* **2008**, *24*, 382–399. [CrossRef]
17. Blevins, R.D.; Saunders, H. Flow Induced Vibration. *J. Mech. Des.* **1979**, *101*, 6. [CrossRef]
18. Madsen, N.; Hansen, K.; Madsen, N.A.H. Behavior of different trawl codend concepts. *Ocean Eng.* **2015**, *108*, 571–577. [CrossRef]
19. Wan, R.; Jia, M.; Guan, Q.; Huang, L.; Cheng, H.; Zhao, F.; He, P.; Hu, F. Hydrodynamic performance of a newly-designed Antarctic krill trawl using numerical simulation and physical modeling methods. *Ocean Eng.* **2019**, *179*, 173–179. [CrossRef]
20. Kumazawa, T.; Hu, F.X.; Watanabe, T.; Kinoshita, H.; Tokai, T. Development of a Pelagic and/or Mid Water Trawl with Canvas Kites. *Fish. Eng.* **2010**, *46*, 197–204. Available online: http://ci.nii.ac.jp/naid/110007580742 (accessed on 27 February 2021).
21. Zhou, A.Z.; Zhang, Y.; Wang, Y.J.; Zhang, X. Comparison on the Hydrodynamic Performance of Mid-Water Trawl 1056 Type for Catching Small Upper-Middle-Class Fish in Middle-East Atlantic and 1040 Type Mid-Water Trawl for Catching Chilean Jack Mackerel. *Fish. Inf. Strategy* **2015**, *2*, 112–118. Available online: http://en.cnki.com.cn/Article_en/CJFDTOTAL-XYYZ201502006.htm (accessed on 27 February 2021).
22. Zhou, A.Z.; Feng, C.L.; Zhang, X.; Wang, Y.J. Influence of Adjustment of Operation Parameters on Small-Mesh Antarctic Krill trawl. *Mar. Fish.* **2016**, *1*, 74–82. Available online: http://www.en.cnki.com.cn/Article_en/CJFDTotal-HTYY201601010.htm (accessed on 27 February 2021).
23. Yao, Y.; Chen, Y.; Zhou, H.; Yang, H. A method for improving the simulation efficiency of trawl based on simulation stability criterion. *Ocean Eng.* **2016**, *117*, 63–77. [CrossRef]
24. Pichot, G.; Germain, G.; Priour, D. On the experimental study of the flow around a fishing net. *Eur. J. Mech. B Fluids* **2009**, *28*, 103–116. [CrossRef]
25. Tang, H.; Xu, L.X.; Hu, F.X. Hydrodynamic characteristics of knotted and knotless purse seine net panels as determined in a fume tank. *PLoS ONE* **2018**, *13*, e0192206. [CrossRef] [PubMed]
26. Tang, H.; Hu, F.; Xu, L.; Dong, S.; Zhou, C.; Wang, X. Variations in hydrodynamic characteristics of netting panels with various twine materials, knot types, and weave patterns at small attack angles. *Sci. Rep.* **2019**, *9*, 1–13. [CrossRef]

27. Pichot, G. Modélisation et Analyse Numérique du Couplage Filet-Écoulement Hydrodynamique dans une Poche de Chalut. Ph.D. Thesis, Université de Rennes, Rennes, France, 6 December 2007. Available online: https://archimer.ifremer.fr/doc/2007/these-3350.pdf (accessed on 27 February 2021).
28. Bi, C.-W.; Balash, C.; Matsubara, S.; Zhao, Y.-P.; Dong, G.-H. Effects of cylindrical cruciform patterns on fluid flow and drag as determined by CFD models. *Ocean Eng.* **2017**, *135*, 28–38. [CrossRef]
29. Bearman, P. Vortex shedding from oscillating bluff bodies. *Annu. Rev. Fluid Mech.* **1984**, *16*, 195–222. [CrossRef]
30. Balash, C.; Sterling, D. Prawn trawl drag due to material properties-an investigation of the potential for drag reduction. *J. Appl. Phys.* **2012**, *46*, 1376–1381. [CrossRef]
31. Bouhoubeiny, E.; Druault, P.; Germain, G. Phase-averaged mean properties of turbulent flow developing around a fluttering sheet of net. *Ocean Eng.* **2014**, *82*, 160–168. [CrossRef]
32. Tang, H.; Xu, L.; Hu, F.; Kumazawa, T.; Hirayama, M.; Zhou, C.; Wang, X.; Liu, W. Effect of mesh size modifications on the sinking performance, geometry and forces acting on model purse seine nets. *Fish. Res.* **2019**, *211*, 158–168. [CrossRef]
33. Balash, C.; David, S.; Jonathan, B.; Giles, T.; Neil, B. Drag characterisation of prawn-trawl bodies. *Ocean Eng.* **2016**, *113*, 18–23. [CrossRef]
34. Tang, H.; Hu, F.; Xu, L.; Dong, S.; Zhou, C.; Wang, X. The effect of netting solidity ratio and inclined angle on the hydrodynamic characteristics of knotless polyethylene netting. *J. Ocean Univ. China* **2017**, *16*, 814–822. [CrossRef]
35. Hosseini, S.A.; Lee, C.-W.; Kim, H.-S.; Lee, J.; Lee, G.-H. The sinking performance of the tuna purse seine gear with large-meshed panels using numerical method. *Fish. Sci.* **2011**, *77*, 503–520. [CrossRef]

Journal of
Marine Science and Engineering

MDPI

Article

Study on Vibration Characteristics of Marine Centrifugal Pump Unit Excited by Different Excitation Sources

Cui Dai [1], Yuhang Zhang [2], Qi Pan [2], Liang Dong [2,*] and Houlin Liu [2]

[1] School of Energy and Power Engineering, Jiangsu University, Zhenjiang 212013, China; daicui@ujs.edu.cn
[2] Research Center of Fluid Machinery Engineering and Technology, Jiangsu University, Zhenjiang 212013, China; 2221911038@stmail.ujs.edu.cn (Y.Z.); 2221811022@stmail.ujs.edu.cn (Q.P.); liuhoulin@ujs.edu.cn (H.L.)
* Correspondence: dongliang@ujs.edu.cn

Abstract: In order to study the vibration mechanism of a marine centrifugal pump unit and explore the contribution of vibration caused by different vibration excitation sources, a marine centrifugal pump with a specific speed of 66.7 was used for research. A numerical calculation model of the flow field and electromagnetic field of the pump unit was established to analyze the frequency spectrum characteristics and contribution of pump unit vibration caused by different excitation sources. Using the modal superposition method, the vibration characteristics of the pump unit caused by fluid excitation and electromagnetic excitation were analyzed. The results show that the main frequency of pump unit vibration caused by fluid excitation was at the $1\times$ blade passing frequency. The main frequency of pump unit vibration caused by electromagnetic excitation was at the $2\times$ utility frequency. The contribution of different excitation sources to the vibration of marine centrifugal pump unit was in the following order: fluid excitation on the inner surface of the pump > electromagnetic excitation > fluid excitation in the impeller.

Keywords: marine centrifugal pump; vibration excitation source; fluid excitation; electromagnetic excitation; numerical simulation

Citation: Dai, C.; Zhang, Y.; Pan, Q.; Dong, L.; Liu, H. Study on Vibration Characteristics of Marine Centrifugal Pump Unit Excited by Different Excitation Sources. *J. Mar. Sci. Eng.* **2021**, *9*, 274. https://doi.org/10.3390/jmse9030274

Academic Editor: Wei-Bo Chen and Claudio Ferrari

Received: 20 January 2021
Accepted: 22 February 2021
Published: 3 March 2021

Publisher's Note: MDPI stays neutral with regard to jurisdictional claims in published maps and institutional affiliations.

1. Introduction

The marine centrifugal pump is one of the important auxiliary devices on a ship, which plays an important role in its operation. On the one hand, the self-excited vibration of fluid during the operation of the marine pump will affect the stable operation of other equipment. On the other hand, the vibration generated by other mechanical equipment or the marine main engine during operation also affects the marine pump. Excessive vibration will not only affect the measurement accuracy of some precision equipment, but also cause harm to people's physical and mental health.

The marine centrifugal pump in the process of operation will show obvious non-stationary properties. Vibration analysis is widely used in centrifugal pump condition monitoring, fault diagnosis, and other fields [1]. Compared with the pressure pulsation sensor, the vibration sensor does not need to contact the fluid and does not destroy the flow characteristics in the centrifugal pump. In addition, the vibration sensor can be placed on the surface of the pump body, which can better reflect the high-frequency characteristics of the centrifugal pump during operation [2,3]. The vibration excitation sources of the marine pump unit can be divided into three types according to the generation mechanism, namely, shafting excitation, fluid excitation, and electromagnetic excitation. Shafting excitation is mainly caused by a rotor fault or rolling bearing fault [4]. Fluid excitation is mainly caused by unsteady fluid excitation in the marine pump. Electromagnetic excitation is mainly caused by the radial electromagnetic force generated by the rotating magnetic field inside the motor. The vibrations caused by the above three excitation sources will be coupled with each other, which will affect the pump unit.

151

Usually, in non-fault cases, the shaft excitation is generally small and can be controlled by active control technology. Compared with shaft excitation, the occurrence mechanism of fluid excitation is more complex. In recent years, with the development of CFD (computational fluid dynamics) technology, more and more scholars have realized that the deep-seated reason for pump vibration caused by fluid excitation is the unsteady flow of fluid in the pump, and they began using numerical calculation methods to analyze the internal flow of the pump [5–7]. At first, most scholars' research on fluid excitation was mainly focused on the analysis of pressure pulsation on the inner wall of the pump [8–10]. They hoped to explore the law of vibration caused by fluid excitation through the analysis of pressure pulsation. With the deepening of research, the calculation methods of vibration caused by fluid excitation have made great progress. Unidirectional coupling and bidirectional coupling are two commonly used methods at present [11]. In the unidirectional coupling method, the pulsating force of the fluid is calculated using the CFD method, and the pulsating force is loaded on the inner wall of the finite element structure. Subsequently, the vibration response of the structure is obtained. Many scholars have conducted in-depth research on the unidirectional coupling method. Ye [12] used the Reynolds-averaged method to calculate the internal flow field of a centrifugal pump. He calculated the vibration response of the pump body through the radial force acting on the inner wall of the volute. However, he ignored the influence of motor structure on the vibration of the pump unit when establishing the structural model. Moreover, pump vibration calculated using the radial force inside the volute as excitation is unreasonable, and there is a big difference between calculation and the actual results. Jiang [13] et al. matched the fluid–solid boundary grid and mapped the pressure pulsation data obtained from fluid numerical calculations to the structural grid, which corresponded to the fluid grid. Jiang's method was more accurate in calculating the structural vibration caused by fluid excitation. Wang [14] calculated the pressure pulsation of the pump chamber, guide vane, impeller, and volute of a centrifugal pump via numerical calculations and compared the results with an experiment. He [15] established a finite element model of a pump unit including a motor. By writing the UDF (user-defined function), he calculated pressure pulsations on the impeller surface at each time step. He integrated these pulse forces as the force of the rotor subjected to the impeller. The force on the rotor was regarded as the excitation source that caused the vibration of the pump unit, and the vibration response of the pump unit was obtained on the basis of the modal analysis results. Following He, Jiang [16] studied the vibration of the bracket through the fluid in the volute and the impeller. The research results showed that the bracket vibration caused by fluid excitation through the impeller was greater than that caused by the volute. Although He and Jiang considered that the vibration of the pump unit would be affected by the motor structure, they did not consider the impact of the vibration generated by the motor on the pump unit. In summary, structural vibration caused by fluid can be quickly calculated using the unidirectional coupling method. However, the unidirectional coupling method only considers the effect of the fluid on the structure, while it ignores the reaction of the structure to the fluid. Therefore, it is widely used in the case of large stiffness of the structure and large volume of the flow field [17,18].

Compared with the unidirectional coupling method, the bidirectional coupling method considers the reaction of the structure to the fluid, which can more truly reflect the real situation and accurately capture the changes in the force and motion laws of the geometry. Pei [19] took a single-stage centrifugal pump as the research object. He calculated the structural response of the pump unit by using the bidirectional fluid–solid coupling method. Zhang [20] used the bidirectional fluid–solid coupling method to calculate the volute vibration of a thick-blade centrifugal pump. Guo [21] calculated the radial force on the impeller by using the bidirectional fluid–solid coupling method and used it as the fluid excitation force to calculate the vibration response on the impeller. Although the bidirectional coupling method considers the interaction between the fluid and the structure more comprehensively, it should be noted that the bidirectional fluid–solid coupling method is more suitable for occasions where the structure has a greater impact on the fluid. Due to the complex structure of the centrifugal pump, a high computational cost is required when using the two-way fluid–solid coupling method, and the calculation is difficult to converge. In addition, the calculation results can be obtained more conveniently by using the unidirectional coupling method because of the high rigidity of the centrifugal pump structure.

In summary, many scholars conducted extensive research on the vibration of pump unit caused by fluid excitation in the study of pump unit vibration, but few scholars considered the influence of motor vibration on the pump unit [22,23]. The vibration generated by the motor can be transmitted to the pump body through the bracket or connecting plate according to different types of pump. Many scholars established numerical calculation models to study the vibration generated by the motor during operation, but they only analyzed the impact of motor vibration on itself, whereas they did not study the impact of motor vibration on the structure of the pump [24–26]. Motors can generate electromagnetic fields through coils, the electromagnetic force inside the electromagnetic field can be calculated by finite element analysis, and then the electromagnetic force can be mapped to the inner wall of the structure to calculate the vibration response of the motor.

Since the transfer vibration between the motor and the pump in a marine centrifugal pump unit has higher coupling than that of a horizontal pump, the vibration of the motor must be taken into account in the study of the vibration characteristics of a marine pump unit.

2. Numerical Calculation Model and Strategy

2.1. Flow Field Calculation Model and Calculation Method

A marine centrifugal pump with a specific speed of 66.7 was used in this paper for research. The main design parameters were as follows: flow rate, $Qd = 25$ m^3/h; rated head, $H = 35$ m; rated speed, $n = 2950$ r/min. The structural parameters of the marine pump are shown in Table 1.

Table 1. Main structural parameters of marine pump.

Components	Geometric Parameters	Symbol	Value
	Inlet diameter (mm)	D_1	65
	Exit diameter (mm)	D_2	165
Impeller	Exit width (mm)	b_2	7
	Blade wrap angle (°)	φ	110
	Blade numbers	z	6
	Basic circle diameter (mm)	D_3	170
Volute	Inlet width (mm)	b_3	20
	Exit diameter (mm)	D_d	50

The whole flow field computational domain includes the inlet elbow, fluid field in the impeller, pump chamber, fluid field in the volute, and outlet extended section. In order to achieve a stable state, the inlet extended section with a length of four times the diameter of the pipe is set before the inlet elbow, which is also conducive to the stability of the internal flow field of the marine pump. The whole model is shown in Figure 1.

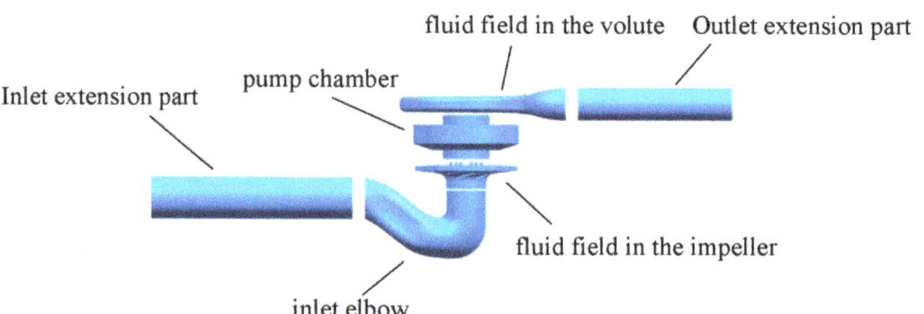

Figure 1. Numerical calculation model of fluid in a marine centrifugal pump.

In this paper, ICEM-CFD (The Integrated Computer Engineering and Manufacturing code for Computational Fluid Dynamics) was used to mesh the fluid domain. In order to ensure the accuracy of the calculated pressure pulsations, the boundary layer grid in the fluid domain was refined. In order to avoid the influence of grid density on the calculation results of the flow field, five sets of meshes with different numbers were used for the mesh-dependence test in this paper. The results are shown in Table 2. Considering the difference between the head from the numerical calculation and the design value, and considering the time required for numerical calculation, Scheme 3 was finally determined for subsequent numerical calculations.

Table 2. Scheme of mesh-dependence test.

Scheme	Number of Grids	Number of Nodes	Head (m)
1	1,647,157	1,474,148	34.5
2	2,457,849	2,287,414	35.2
3	2,914,979	2,741,943	35.5
4	3,278,458	3,024,785	35.5
5	3,715,756	3,546,854	35.6

The y+ values of the stationary domain surface and the moving domain surface are shown in Figure 2. Figure 2a shows the y+ values of the inner wall of the volute and the pump chamber. Figure 2b shows the y+ values of the inner surface of the impeller. It can be seen from Figure 2 that the y+ values of all walls are less than 12. Most of the y+ values of the inner wall of the volute and pump chamber are less than 6, and most of the y+ values of the inner wall of the impeller are below 8, which can better ensure the accuracy of the wall flow information calculation

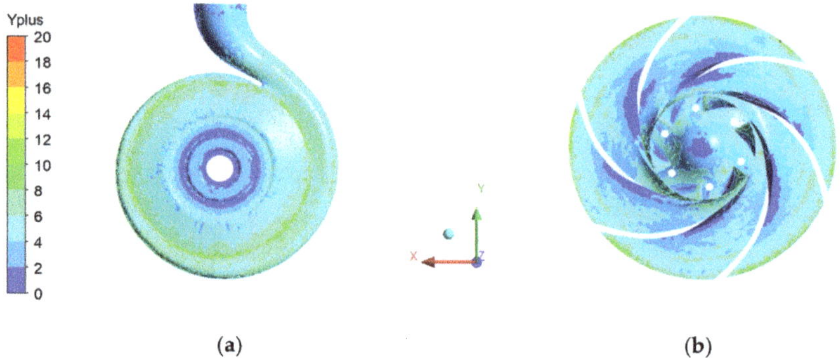

Figure 2. y+ distribution of the volute, pump chamber, and impeller: (**a**) static domain; (**b**) moving domain.

In this paper, the standard k–ε turbulence model was adopted, and the pressure boundary inlet condition and mass flow outlet boundary condition were adopted. In order to ensure that the calculated fluid excitation had high time resolution, the time step was set to $\Delta T = 0.565 \times 10^{-4}$ s, which is the time required for the impeller to rotate $1°$. When the flow field is stable, the time-domain information of pressure pulsation on the surface of the stationary domain and moving domain can be derived. The pressure pulsation data for 10 stable cycles of impeller rotation were used as the fluid excitation source, and the total duration was set to 0.25s.

2.2. Electromagnetic Field Calculation Model and Calculation Method

The three-phase asynchronous motor model Y132S1-2 was selected in this paper. The parameters of the motor are shown in Table 3. Due to the imprecise shortcomings in the calculation of the three-dimensional motor model and the symmetrical structure of the motor, many scholars used the two-dimensional model to analyze the internal electromagnetic field of the motor.

Table 3. Geometric parameters of the motor.

Voltage (V)	380	Pole Number	2
Rated speed (rpm)	2950	Phase number	3
Frequency (Hz)	50	Connection method	Delta connection
Stator outer diameter (mm)	210	Stator inner diameter (mm)	116
Rotor outer diameter (mm)	114	Rotor inner diameter (mm)	74
Stator slot number	30	Rotor slot number	26

Altair Flux was used for parametric modeling of the above motors in this paper. This model uses element PLANE53 for meshing. The electromagnetic field calculation domain included the stator iron core, rotor iron core, stator tooth, rotor tooth, and motor shaft. The electromagnetic field calculation model is shown in Figure 3.

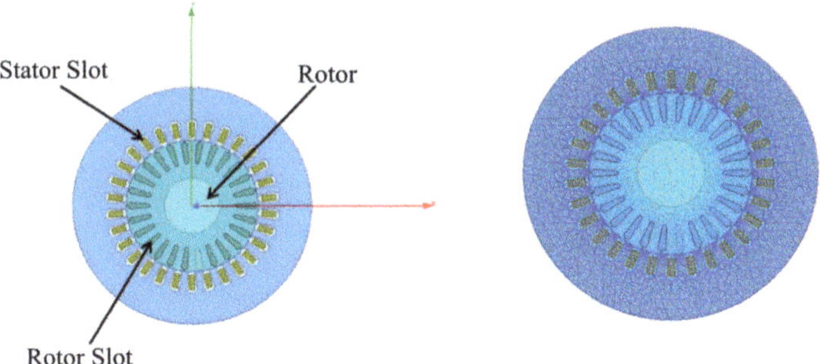

Figure 3. Electromagnetic field calculation model.

2.3. Structure Calculation Model and Calculation Method

The structure model of the marine pump unit in this paper included the motor, bracket, pump body, inlet pipe, base, and rotor system.

ANSYS Workbench was used to establish a finite element analysis model of the pump unit in this paper. The three-dimensional (3D) model of the pump unit is shown in Figure 4. The rotor and pump body adopted element SOLID187, and the surface of the structure in contact with the flow field and electromagnetic field adopted element SURF154. The finite element analysis model is shown in Figure 5. In order to reduce the computation time, the details which had less influence on the structure were simplified when building the structure model. In addition, since the bearing position was the focus of this study, the rotor was simplified when the model was established. The grids of the rotor system were hexahedral grids, and the meshes were refined.

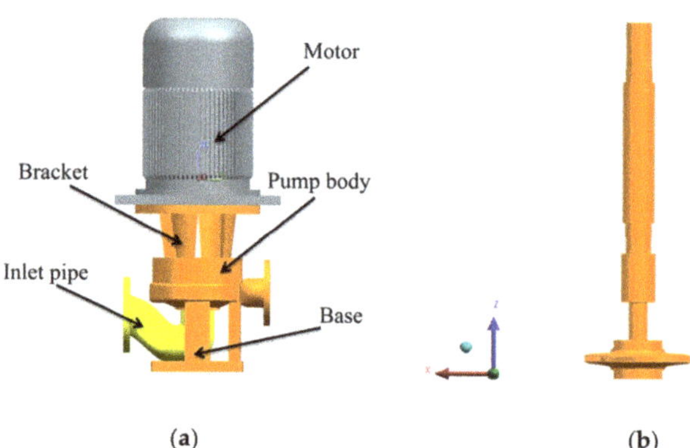

Figure 4. The three-dimensional (3D) model of the pump unit: (**a**) pump unit structure; (**b**) rotor system.

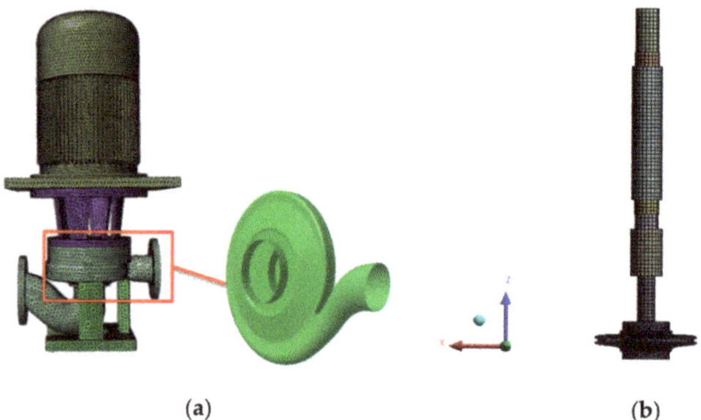

Figure 5. Finite element model of pump unit: (**a**) pump unit structure; (**b**) rotor system.

Before vibration calculation, a modal analysis of the marine pump was needed. The LMS Virtual Lab software was used in this paper to analyze the modal response of the centrifugal pump unit. In the modal analysis, the bolt holes of the pump unit base were set as rigid surface constraints, and the constraints of the inlet and outlet flanges were released. Figure 6 shows the natural frequency of the centrifugal pump unit less than 1000 Hz obtained through modal analysis.

According to the results of modal analysis, the calculated pressure pulsation on the fluid grid was mapped to the structural grid to calculate the vibration.

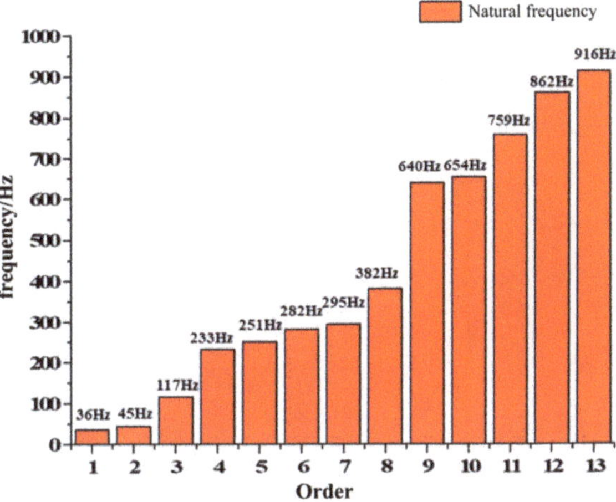

Figure 6. The natural frequency of the pump unit.

2.4. Test Object and External Characteristic Experiment

The vibration test was carried out on the closed test bench of the marine pump of Jiangsu University. This test used a three-axis vibration acceleration sensor model INV9832. The sampling frequency was 6.4 kHz and the sampling time was 30 s. The arrangement of vibration measuring points is shown in Figure 7.

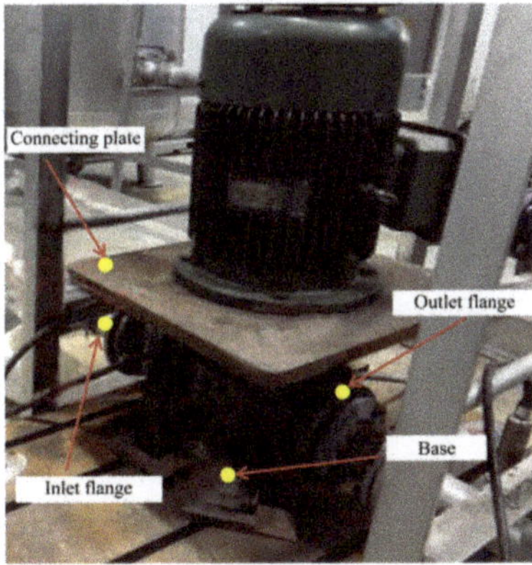

Figure 7. Vibration measuring point location.

In order to verify the reliability of the numerical simulation results, the external characteristic curve of marine pump was compared with that of the numerical simulation. The comparison result of the external characteristic curve is shown in Figure 8. Under the rated flow, the numerical simulation result of the head differed by 4.43% from the test value, and the efficiency differed from the test value by 1.90%, which shows that the adopted numerical simulation calculation method was reliable.

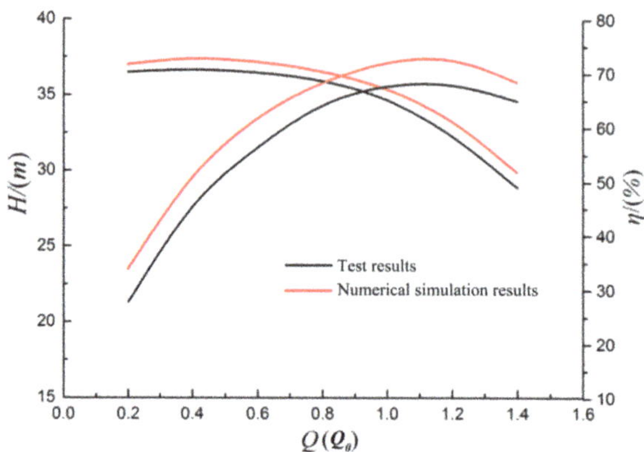

Figure 8. External characteristic curve of numerical simulation and test.

3. Analysis of Numerical Simulation Results

3.1. Analysis of Fluid Excitation Calculation Results

3.1.1. Force Analysis of the Volute Wall

The reason for the vibration of the marine centrifugal pump caused by the fluid excitation is the result of the pulsation force of the fluid acting on the structure. Setting up monitoring points at different locations to monitor changes in pressure pulsation in real time can more clearly reflect the characteristics of fluid excitation. In this paper, the pressure pulsation monitoring points were set at each volute cross-section and the position of the tongue. A schematic diagram of the monitoring points is shown in Figure 9.

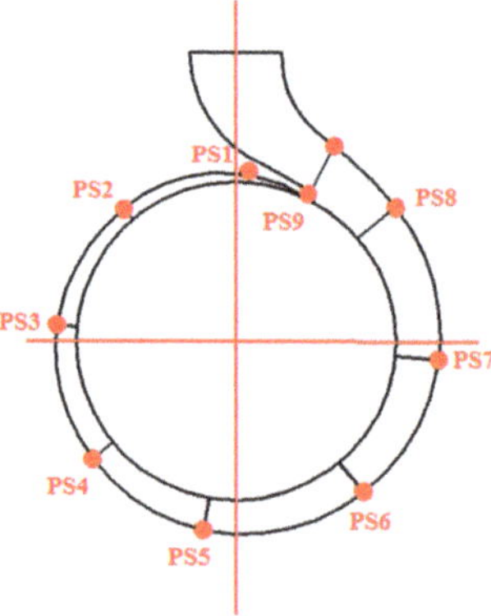

Figure 9. Location of pressure pulsation monitoring point on the volute wall.

Figure 10 shows the time-domain and frequency-domain amplitude changes of the pressure pulsation in one cycle at each monitoring point of the volute after the pressure pulsation has stabilized. It can be seen from Figure 10a that the pressure pulsation of each measuring point had obvious periodicity, and there were six peaks in one cycle. The number of peaks corresponded to the number of blades. Comparing the amplitude of pressure pulsation at each measuring point, it can be seen that the amplitude of the pressure pulsation gradually increased from the first cross-section to the eighth cross-section. At the volute tongue position, the pulsation amplitude increased sharply from 450 kPa at the eighth cross-section to 470 kPa. Figure 10a clearly reflects the variation law of internal pressure pulsation with time.

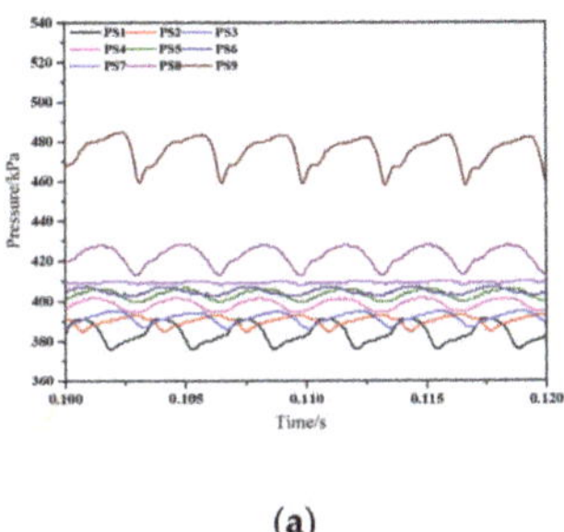 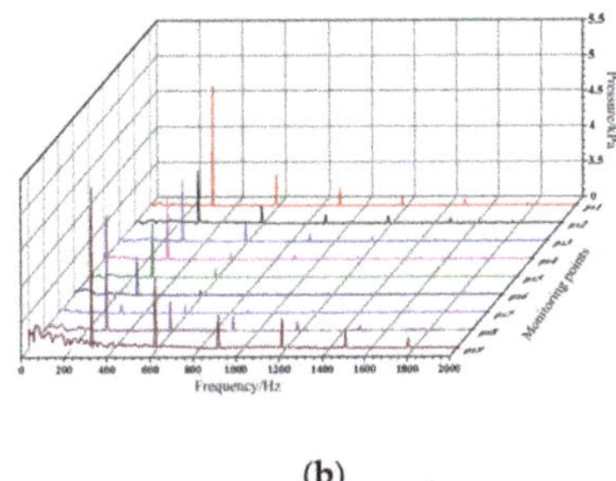

(**a**)   (**b**)

Figure 10. Time-domain and frequency-domain changes of pressure pulsation at each monitoring point: (**a**) time-domain characteristics; (**b**) frequency-domain characteristics.

It can be seen from Figure 10b that the pressure pulsation at each monitoring point had a clear peak at the 1× blade passing frequency (295Hz). Obvious peaks also appeared in multiples of the 1× blade passing frequency, indicating that the main reason for the pressure pulsation is the interaction between the impeller and the diaphragm. It can be seen from the variation of the main frequency amplitude at each monitoring point that the pressure pulsation amplitude at the tongue position was the highest. The main frequency amplitude of the pressure pulsation from the first to the seventh cross-section gradually decreased, and the main frequency amplitude suddenly increased at the eighth cross-section. In conclusion, because the fluid was greatly affected by the tongue, the pressure pulsation near the tongue had a large amplitude. With the increase in distance from the tongue, the main frequency amplitude of the pressure pulsation at each measuring point showed a downward trend.

3.1.2. Force Analysis of the Impeller

The fluid excitation force acting on the impeller surface causes rotor vibration and transfer to the pump body, which is one of the main factors leading to vibration of the marine pump unit. In order to explore the law of pump vibration caused by fluid excitation, the characteristics of fluid excitation force on the impeller were analyzed. Figure 11 shows the time-domain and frequency-domain amplitude changes of the horizontal radial X, horizontal radial Y, and vertical axial Z of the impeller under the rated condition, and these three directions were perpendicular to each other. It can be seen from the time-domain diagram of the three directions in Figure 11a,c,e that the force of the impeller had obvious periodicity, and the length of each period was equal to the time required for the impeller to rotate once. In addition, because the motor and the marine pump used the same rotor, eccentricity failure would not occur, and the radial forces on the impeller in the X and Y directions were very similar. Radial force is one of the main causes of vibration of the impeller. It can be seen from Figure 11a, and Figure 10c that the blades were most affected by the radial force in each part of the impeller, followed by the front cover plate, whereas the back cover plate was the least affected. Although the force of each part of the impeller was large, the force in the Z direction of the impeller was small due to the superposition of radial forces.

Fourier transform was performed on the forces in the three directions, and the frequency-domain characteristics of the forces in the three directions of the impeller were obtained as shown in Figure 11b,d,f. It can be seen from Figure 11b,d that the main frequency of the radial force of the impeller was at the 1× shaft frequency, and other characteristic frequencies were distributed at multiples of the 1× shaft frequency. The main frequency of the axial force of the impeller was at the 2× blade passing frequency, and there were higher peaks at the 1× and 3× blade passing frequencies.

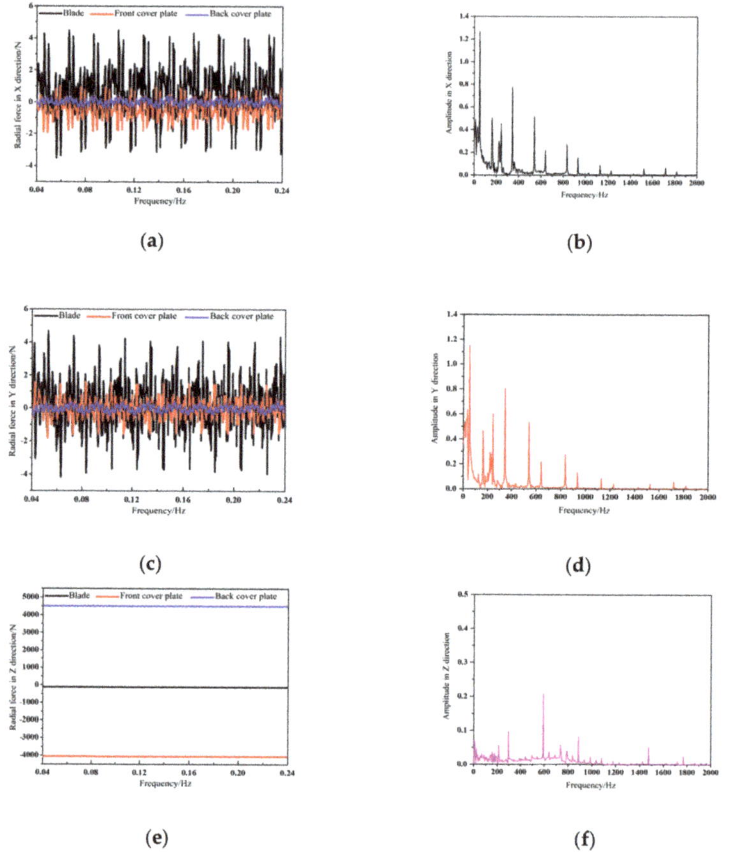

Figure 11. The radial force on each part of the impeller in the time domain and frequency domain: (**a**) time domain (*X*); (**b**) frequency domain (*X*); (**c**) time domain (*Y*); (**d**) frequency domain (*Y*); (**e**) time domain (*Z*); (**f**) frequency domain (*Z*).

3.2. Analysis of Electromagnetic Excitation Calculation Results

Because the air gap magnetic field of the motor is affected by many factors, the distribution of the internal magnetic field is complex. The finite element method can reflect the distribution of the internal magnetic field more comprehensively and clearly. Figures 12 and 13 show the magnetic line distribution and magnetic flux density distribution at a certain moment after the motor was stable.

It can be seen from Figure 12 that the high-energy region was distributed on the surface of the stator and rotor and the stator slot. It can be seen from Figure 13 that the high flux density was concentrated in the stator slot, the rotor slot, and the gap. The maximum value of the flux density was located in the stator and rotor slot, up to 2.6 T. When the stator is subjected to radial electromagnetic force, it vibrates and causes the motor to vibrate. In order to better explore the influence of radial electromagnetic force on the motor, the electromagnetic radial force on the surface of the stator slot was extracted for further analysis in this paper. Figure 14 shows the distribution nephogram of the electromagnetic force on the surface of the stator slot at different rotation angles. It can be seen from Figure 14 that the electromagnetic excitation force was distributed in the air gap of the fixed rotor and at the contact between the stator and the winding. The region excited by electromagnetic force was symmetrical and varied with the rotation angle of the rotor. In addition, the maximum electromagnetic excitation force did not exceed 3 N.

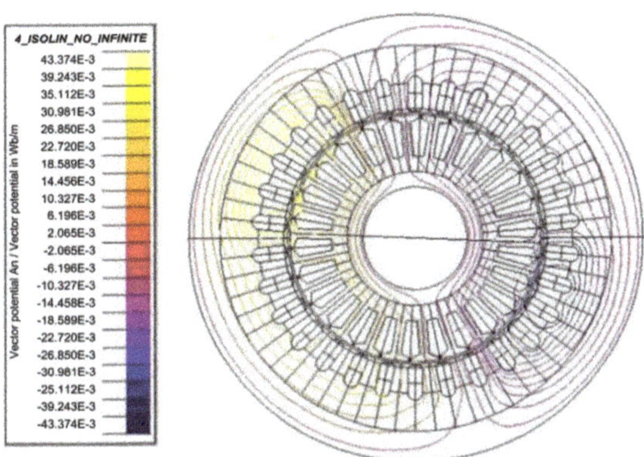

Figure 12. The distribution nephogram of magnetic lines during normal operation of the motor.

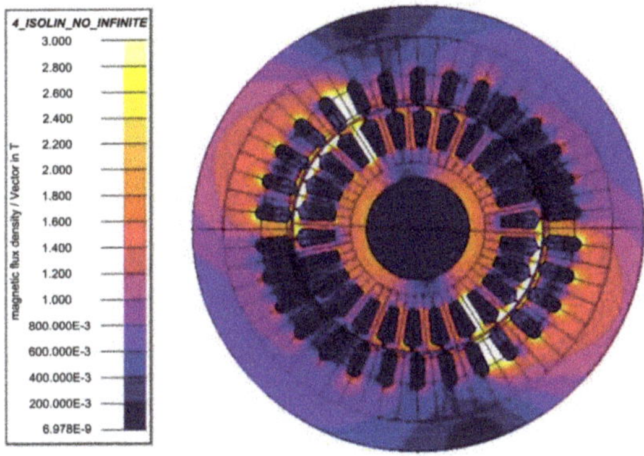

Figure 13. The distribution nephogram of flux density during normal operation of the motor.

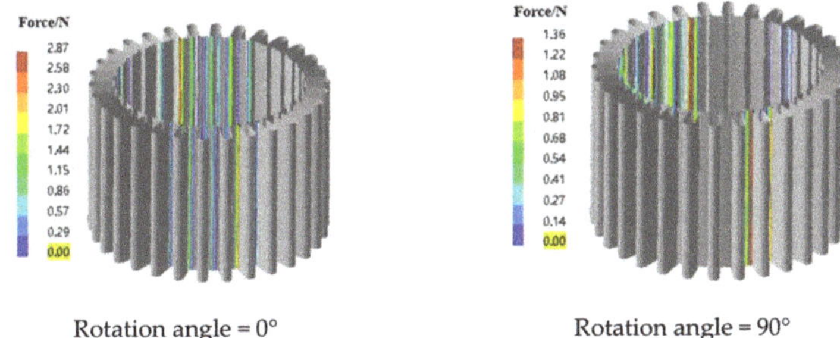

Rotation angle = 0°   Rotation angle = 90°

Figure 14. *Cont.*

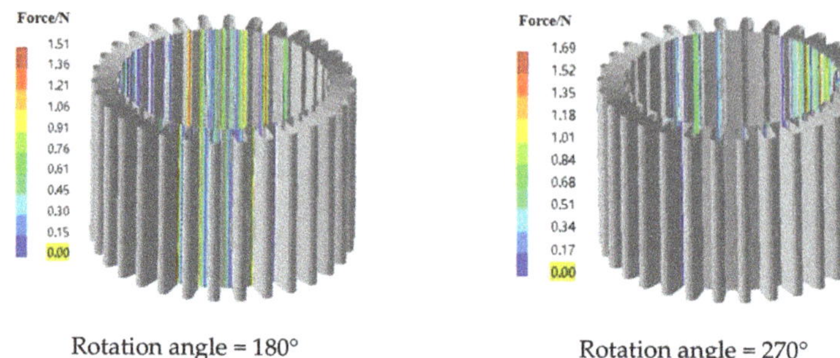

Figure 14. The distribution nephogram of electromagnetic force on the surface of the stator slot at different rotation angles.

3.3. Analysis of Vibration Calculation Results

3.3.1. Fluid Excitation on the Inner Surface of the Pump-Induced Vibration Analysis

The fluid can gain a lot of kinetic energy when the impeller works on the fluid. The fluid acting on the volute and the pump chamber produces high pulse force on the wall, which is the fundamental reason for the pump unit vibration caused by the fluid excitation on the inner surface of the pump. Figure 15 shows the location of the vibration measuring point of the pump unit. The vibration measuring points were located at the connecting plate, inlet flange, outlet flange, and base.

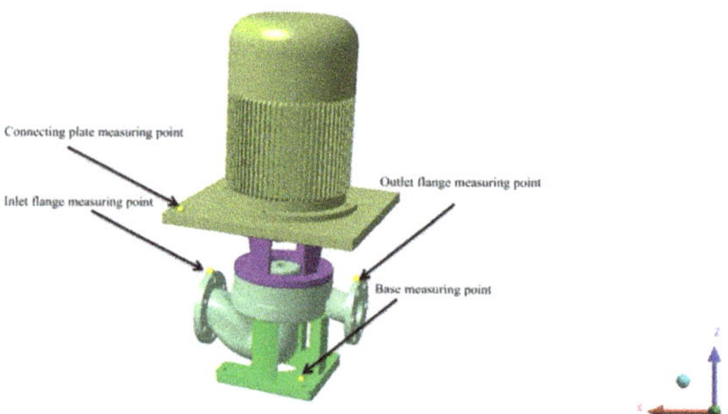

Figure 15. The position of each vibration measuring point of the pump unit.

The frequency characteristics of vibration velocity at each measuring point were obtained by calculating the vibration of pump unit caused by fluid excitation on the inner wall of volute and pump cavity, as shown in Figure 16. It can be seen from Figure 16a–d that the distribution of the characteristic frequency of vibration at each measuring point was similar. The main frequency of vibration at each measuring point was at the $1\times$ blade passing frequency (295 Hz), and obvious peaks also appeared in multiples of the $1\times$ blade passing frequency. In addition, it can be seen from Figure 16d that there were also obvious peaks in the low-frequency region such as the $1\times$ shaft frequency (49 Hz). Since the pump unit may have been resonated when passing through the natural frequency, peaks also appeared at multiple natural frequencies, for example, at 36 Hz in Figure 16a–d and at 117 Hz in Figure 16b,c. In addition, the vibration velocity levels of the radial direction at the measuring points of import and export flanges were the highest in the whole frequency domain, indicating that the import and export flanges were mainly affected by the radial force. At the measuring points of the connecting plate and the base, the axial vibration velocity level was the highest, which indicates that the main vibration directions of the measuring points were different.

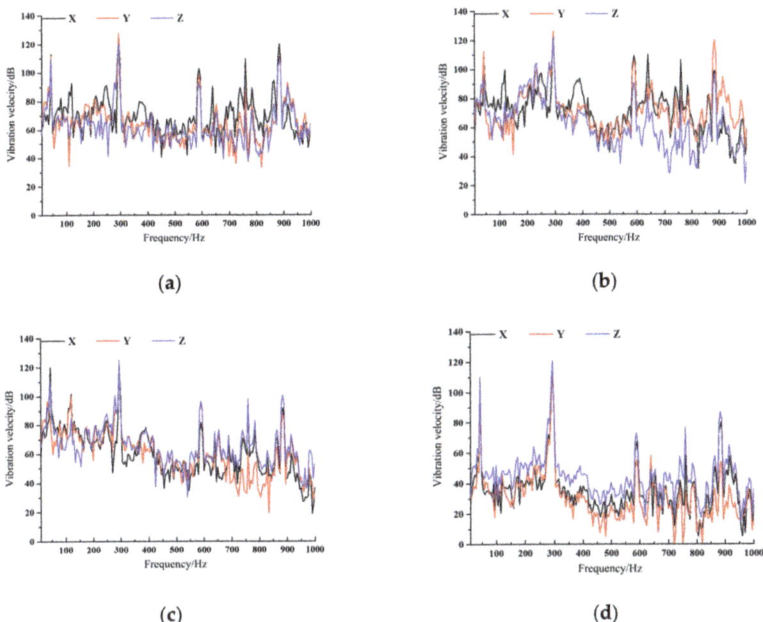

Figure 16. The frequency spectrum of the vibration of the pump unit caused by fluid excitation on the inner surface of the pump: (**a**) export flange; (**b**) import flange; (**c**) connecting plate; (**d**) base.

3.3.2. Fluid Excitation in Impeller-Induced Vibration Analysis

When the impeller works on the fluid, it is also subjected to the reaction force of the fluid on it, which causes the vibration of the rotor, and the vibration of the rotor is transmitted to the pump unit through the bearing housing. Figure 17 shows the frequency spectrum of the pump unit vibration caused by the fluid excitation at each measuring point through the rotor when the fluid passed through the impeller. It can be seen from Figure 17 that the vibration peak of the marine pump unit caused by fluid excitation in the impeller was near the 1× blade passing frequency at different measuring points, It can be seen from Figure 17a–d that the vibration peak of the marine pump unit caused by fluid excitation in the impeller was near the 1× blade passing frequency at different measuring points, which indicates that the 1× blade passing frequency was the main frequency of the vibration of the marine pump unit induced by the fluid. In addition, due to the influence of the rotor itself, an obvious peak appeared at the 1× shaft frequency. Compared with the vibration velocity level in different directions of each measuring point, the vibration velocity level of horizontal radial Y at the inlet and outlet flange was the highest, and the vibration velocity level of axial Z at the motor base and the base was the highest. The characteristic frequency of the pump unit vibration at each measuring point caused by the fluid excitation in the impeller was basically consistent with the fluid excitation on the inner surface of the pump. However, the vibration of the pump unit shown in Figure 17 was relatively stable in the whole frequency domain, and the vibration velocity level showed a downward trend with the frequency increasing. In the whole frequency domain, the variation law of the vibration of the pump unit caused by the fluid excitation in the impeller was quite different from the fluid excitation on the inner surface of the pump.

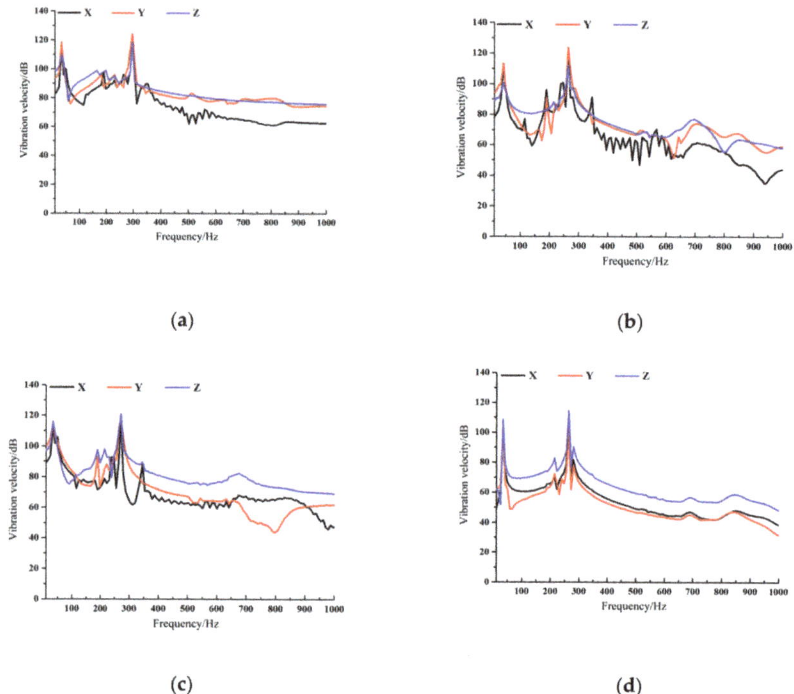

Figure 17. The frequency spectrum of the vibration of the pump unit caused by fluid excitation in impeller: (**a**) export flange; (**b**) import flange; (**c**) connecting plate; (**d**) base.

3.3.3. Electromagnetic Excitation-Induced Vibration Analysis

Radial electromagnetic force produced by the interaction between the stator and rotor in a motor can cause vibration of the marine pump unit. Figure 18 shows the vibration spectrum of each measuring point of the pump unit under radial electromagnetic force. It can be seen from Figure 18a–d that the vibration law of the pump unit caused by electromagnetic excitation was completely different from that caused by fluid excitation. Under the action of electromagnetic force, the main frequency of the vibration of each measuring point was at the 2× utility frequency (100 Hz), which conformed to the law of rotor vibration caused by electromagnetic force in general. There were many obvious peaks in the frequency domain of the vibration of the pump unit caused by electromagnetic excitation. The frequencies corresponding to these peaks were the natural frequencies of the marine pump structure, indicating that the structure of the marine pump unit was greatly affected by electromagnetic excitation, and there would be obvious peaks under multiple natural frequencies. It can be seen from Figure 18 that the vibration velocity level at the seventh natural frequency (295 Hz) had the largest amplitude compared with the other natural frequencies. Furthermore, the amplitude of the vibration at the seventh natural frequency is the largest at the base position, which is basically the same as the vibration amplitude at the 2× utility frequency.

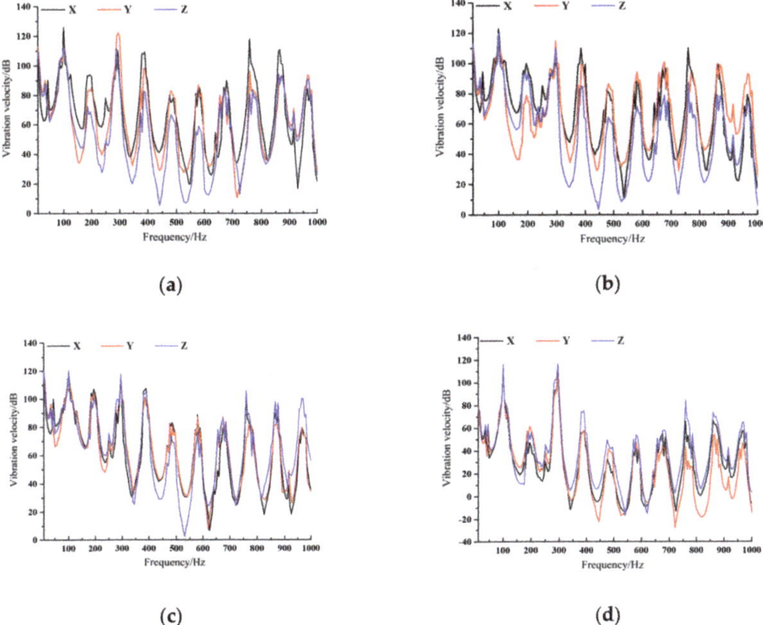

Figure 18. The frequency spectrum of the vibration of the pump unit caused by electromagnetic excitation: (**a**) export flange; (**b**) import flange; (**c**) connecting plate; (**d**) base.

In addition, it can be seen from Figure 18 that the vibration of the pump unit caused by electromagnetic excitation inside the motor showed obvious periodicity. Moreover, the amplitude change was very large; the minimum vibration velocity level at the base position was about −30 dB, which is already lower than the reference vibration level, and the maximum was about 117 dB.

3.4. Comparison of Numerical Simulation and Test Results

In order to intuitively compare the contribution of pump unit vibration caused by different excitation sources, the total vibration velocity level was used to evaluate it in this paper. Chen [27] used the total vibration level evaluation method to analyze the vibration of ships with different structures through experiments and simulations, and he effectively obtained the optimal hull structure model. Liu [28] studied the vibration characteristics of a centrifugal pump. The vibration of centrifugal pump unit caused by the change of speed and flow rate was evaluated by the total vibration level evaluation method. The results showed that the total vibration level evaluation method can better reflect the vibration variation law of centrifugal pump unit.

The vibration acceleration level (Lva) is widely used in the vibration analysis of cars, tracks, bridges, etc. According to the principle of energy superposition, Jiang [29] carried out a superposition calculation of vibration acceleration and obtained the calculation method of total vibration acceleration level (Lvat). By referring to Jiang's method, the vibration velocity was calculated by superposition in this paper. The total vibration velocity level evaluation method can evaluate the overall impact of multiple vibration sources on a measuring point. The vibration caused by different excitation sources may affect the marine centrifugal pump in different directions or in different frequency bands. The total vibration velocity level evaluation method was used to superposition the measured vibration energy in three directions of each vibration measuring point of the marine centrifugal pump unit

blade passing frequency. In the numerical simulation results, the vibration spectrum at the inlet and outlet flanges showed a high peak at the $3\times$ blade passing frequency. However, the test results were quite different from the numerical simulation results for frequencies greater than 900 Hz, which shows that fluid excitation in the pump mainly affects the low-frequency vibration of the marine pump unit. In addition, since the finite element method has a better ability to express the characteristics of the low-frequency band, the numerical simulation values of vibration in the high-frequency band were different from the test values.

4. Conclusions

In order to explore the contribution of different excitation sources to the vibration of the marine centrifugal pump unit, a marine centrifugal pump with a specific speed of 66.7 was studied in this paper. The numerical calculation method was used to analyze the vibration characteristics of the marine centrifugal pump caused by fluid excitation and electromagnetic excitation, and the numerical calculation results were compared with the test results. The conclusions were as follows:

(1) The pressure pulsation in the volute of marine centrifugal pump showed obvious periodicity. The main frequency of pump unit vibration caused by pressure pulsation was at the $1\times$ blade passing frequency, and obvious peaks also appeared in multiples of the $1\times$ blade passing frequency. As the distance between the monitoring point and the tongue increased, the main frequency amplitude of the pressure pulsation at each measuring point showed a downward trend. In addition, the force of the fluid acting on the impeller also showed obvious periodicity. The main frequency of the radial vibration of the impeller was at the $1\times$ shaft frequency, and other characteristic frequencies were distributed at multiples of the $1\times$ shaft frequency. The main frequency of axial vibration was at the $2\times$ shaft frequency.

(2) The main frequency of vibration caused by fluid excitation on the inner surface of the pump was at the $1\times$ blade passing frequency. Moreover, other characteristic frequencies appeared at the $1\times$ shaft frequency and at multiples of the $1\times$ blade frequency. The main frequency of vibration caused by electromagnetic excitation was at the $2\times$ utility frequency. In addition, because the marine pump unit was greatly affected by the structural natural frequency, there were obvious peaks at each natural frequency.

(3) The numerical calculation results were compared with the test results in this paper. The contribution of pump unit vibration caused by different excitation sources to the total vibration of the marine centrifugal pump unit was in the following order: fluid excitation on the inner surface of the pump > electromagnetic excitation > fluid excitation in the impeller. The maximum difference between the total vibration velocity level caused by the fluid excitation on the inner surface of the pump at each measuring point and the test value was less than 1.5%. The maximum difference between the test and the electromagnetic excitation was about 2.5%, and the maximum difference between the test and the fluid excitation in the impeller was greater than 4.5%. The fluid excitation on the inner surface of the pump could more accurately describe the actual operating characteristics of the marine centrifugal pump, especially at the base position. In addition, the fluid excitation on the inner surface of the pump had a better ability to express the low-frequency vibration characteristics of the marine pump unit. The impact of all excitation sources on different positions of the pump unit was in the following order: outlet flange > import flange > connecting plate > base.

Due to the huge amount of calculation and the lack of calculation methods, it is difficult to calculate the overall vibration of the pump unit combined with fluid excitation and electromagnetic excitation. How to reduce the computational complexity and how to establish a more accurate calculation method will be the future research direction of vibration analysis of marine centrifugal pumps.

Author Contributions: Conceptualization, C.D. and H.L.; writing—original draft preparation, Y.Z.; writing—review and editing, C.D. and L.D.; numerical calculation and experiment, C.D. and Q.P. All authors have read and agreed to the published version of the manuscript.

Funding: This research was funded by the National Natural Science Foundation of China (51879122, 51579117, 51779106), the National Key Research and Development Program of China (2016YFB0200901, 2017YFC0804107), the Zhenjiang Key Research and Development Plan (GY2017001, GY2018025), the Open Research Subject of Key Laboratory of Fluid and Power Machinery, Ministry of Education, Xihua University (szjj2017-094, szjj2016-068), the Sichuan Provincial Key Lab of Process Equipment and Control (GK201614, GK201816), the Jiangsu University Young Talent Training Program—Outstanding Young Backbone Teacher, Program Development of Jiangsu Higher Education Institutions (PAPD), and the Jiangsu Top Six Talent Summit Project (GDZB-017).

Conflicts of Interest: The authors declare no conflict of interest.

References

1. Liu, H.-M.; Liu, Y. Research Status and Developing Tendency of Malfunction Diagnosis in Centrifugal Pumps. *Agric. Sci. Technol. Equip.* **2019**, *1*, 70–74, 77.
2. Zhao, Y.-Q. Research on Cavitation Diagnosis and Its Characteristics of Flow Field and Acoustic Field in Centrifugal Pump. Master's Thesis, Jiangsu University, Zhen Jiang, China, 2018.
3. Duan, X.-H.; Tang, F.-P.; Duan, W.-Y.; Zhou, W.; Shi, L.-J. Experimental investigation on the correlation of pressure pulsation and vibration of axial flow pump. *Adv. Mech. Eng.* **2019**, *11*, 168781401988947. [CrossRef]
4. Zhao, W.-Y.; Bai, S.-B.; Ma, P.-F. Vibration of Rotor in Centrifugal Pump Status and Prospects. *Fluid Mach.* **2011**, *3*, 37–39.
5. Ren, Y.-X.; Chen, H.-X. Introduction. In *Fundamentals of Computational Fluid Dynamics*; Liu, J.-L., Song, Y.-Q., Eds.; Tsinghua University Press: Beijing, China, 2006; pp. 5–6.
6. Zhou, Y.-L. Analysis on Pressure Fluctuation and Vibration of a Centrifugal Pump for Off-design Conditions. *Fluid Mach.* **2015**, *2*, 52–55.
7. Lucius, A.; Brenner, G. Unsteady CFD simulations of a pump in part load conditions using scale-adaptive simulation. *Int. J. Heat Fluid Flow* **2010**, *31*, 1113–1118. [CrossRef]
8. Park, S.H.; Morrison, G.L. Centrifugal pump pressure pulsation prediction accuracy dependence upon CFD models and boundary conditions. In Proceedings of the ASME 2009 Fluids Engineering Division Summer Meeting, Vail, CO, USA, 2–6 August 2009; pp. 207–220.
9. Wang, Y. Research on Cavitation and Its Induced Vibration and Noise in Centrifugal Pimps. Ph.D. Thesis, Jiangsu University, Zhen Jiang, China, 2011.
10. Wang, Y.; Dai, C. Analysis on Pressure Fluctuation of Unsteady Flow in a Centrifugal Pump. *Trans. Chin. Soc. Agric. Mach.* **2010**, *41*, 91–95.
11. Jin, Y.-B.; Dong, K.-Y.; Yu, J.; Wu, X.-R. Research progress of centrifugal pump fluid-induced vibration. *Pump Technol.* **2015**, *3*, 1–5.
12. Ye, J.-P. Research on Optimization of Vibration and Structural Noise of Centrifugal Pump. Master's Thesis, Wuhan University of Technology, Wuhan, China, 2006.
13. Jiang, Y.-Y.; Yoshimura, S.; Imai, R.; Katsura, H.; Yoshida, T.; Kato, C. Quantitative evaluation of flow-induced structural vibration and noise in turbomachinery by full-scale weakly coupled simulation. *J. Fluids Struct.* **2007**, *23*, 531–544. [CrossRef]
14. Wang, Y.; Luo, K.-K.; Wang, K.; Liu, H.-L.; Li, Y.; He, X.-H. Research on pressure fluctuation characteristics of a centrifugal pump with guide vane. *J. Vibroeng.* **2017**, *19*, 5482–5497. [CrossRef]
15. He, T.; Yi, Z.-Y.; Sun, Y.-D. Numerical analysis for flow induced vibration of a centrifugal pump. *J. Vib. Shock.* **2012**, *31*, 96–102.
16. Jiang, A.-H.; Li, G.-P.; Zhou, P.; Zhang, Y. Vibration incited by fluid forces on centrifugal pump from volute path and impeller path. *J. Vib. Shock.* **2014**, *33*, 1–7.
17. Luo, B.; Wang, C.-L.; Xia, Y.; Ye, J.; Yang, X.-Y. Numerical simulation of flow-induced vibration of double-suction centrifugal pump as turbine. *J. Drain. Irrig. Mach. Eng.* **2019**, *37*, 313–318.
18. Yao, T.-T.; Zheng, Y. Finite element analysis of stress, deformation and modal of head cover in axial-flow hydro-turbine. *J. Drain. Irrig. Mach. Eng.* **2020**, *38*, 39–44.
19. Pei, J. Investigations on Fluid-Structure Interaction of Unsteady Flow-Induced Vibration and Flow Unsteadiness Intensity of Centrifugal Pumps. Ph.D. Thesis, Jiangsu University, Zhen Jiang, China, 2013.
20. Zhang, D.-S.; Zhang, L.; Shi, W.-D.; Chen, B.; Zhang, H. Optimization of Vibration Characteristics for Centrifugal Pump Volute Based on Fluid-structure Interaction. *Trans. Chin. Soc. Agric. Mach.* **2013**, *44*, 40–45.
21. Guo, W.-J. Analysis of Unsteady Flow and Vibration Characteristics of Low Specific Speed Centrifugal Pump Based on Two-way Fluid-Structure Interaction. Master's Thesis, Zhejiang Sci-Tech University, Hang Zhou, China, 2017.
22. El-Gazzar, D.M. Finite element analysis for structural modification and control resonance of a vertical pump. *Alex. Eng. J.* **2017**, *56*, 695–707. [CrossRef]
23. Bae, D.-M.; QI, D.L.; Cao, B.; Cuo, W. A study on the method vibration analysis of marine pump. *J. Korean Soc. Fish. Ocean. Technol.* **2015**, *51*, 279–284. [CrossRef]

24. Wu, J.-H.; He, T.; Yi, Z.-Y. FEM/BEM analysis for flow induced noise and vibration of a centrifugal pump. *Ship Sci. Technol.* **2016**, *38*, 49–55.
25. Jiang, Y.; Zhao, J.-T. Reduce vibration measures for ship centrifugal pump based on modal analysis and CFD simulation. *Ship Sci. Technol.* **2012**, *34*, 109–114.
26. Choi, B.K. Abnormal Vibration Diagnosis of High Pressure LNG Pump. *J. Power Syst. Eng.* **2005**, *2*, 45–49.
27. Chen, W. Numerical Simulation an Experimental Study on Damping Vibration of Ships. Master's Thesis, Shanghai Jiao Tong University, Shanghai, China, 2019.
28. Liu, Z.; Li, B.; Ma, Q.-N.; Zhu, D.-P. Experiment on vibration characteristics of centrifugal pump with high head. *J. Drain. Irrig. Mach. Eng.* **2013**, *31*, 938–942.
29. Jiang, T. Vibration Level Evaluation of the Environmental Vibration Affected by Multi-vibration Sources. *Urban Mass Transit* **2010**, *13*, 26–29.

Journal of
Marine Science and Engineering

Article

The One-Way FSI Method Based on RANS-FEM for the Open Water Test of a Marine Propeller at the Different Loading Conditions

Mobin Masoomi [1] and Amir Mosavi [2,3,*

[1] Department of Mechanical Engineering, Babol Noshirvani University of Technology, Babol, Iran; m.mo.masoomi@gmail.com
[2] Faculty of Civil Engineering, Technische Universität Dresden, 01069 Dresden, Germany
[3] John von Neumann Faculty of Informatics, Obuda University, 1034 Budapest, Hungary
* Correspondence: amir.mosavi@mailbox.tu-dresden.de

Abstract: This paper aims to assess a new fluid–structure interaction (FSI) coupling approach for the vp1304 propeller to predict pressure and stress distributions with a low-cost and high-precision approach with the ability of repeatability for the number of different structural sets involved, other materials, or layup methods. An outline of the present coupling approach is based on an open-access software (OpenFOAM) as a fluid solver, and Abaqus used to evaluate and predict the blade's deformation and strength in dry condition mode, which means the added mass effects due to propeller blades vibration is neglected. Wherein the imposed pressures on the blade surfaces are extracted for all time-steps. Then, these pressures are transferred to the structural solver as a load condition. Although this coupling approach was verified formerly (wedge impact), for the case in-hand, a further verification case, open water test, was performed to evaluate the hydrodynamic part of the solution with an e = 7.5% average error. A key factor for the current coupling approach is the rotational rate interrelated between two solution domains, which should be carefully applied in each time-step. Finally, the propeller strength assessment was performed by considering the blades' stress and strain for different load conditions.

Keywords: fluid–structure interaction; OpenFOAM; one-way approach; structural analysis; open water test; computational fluid dynamics; numerical analysis; fluid mechanics; blade design; propeller

Citation: Masoomi, M.; Mosavi, A. The One-Way FSI Method Based on RANS-FEM for the Open Water Test of a Marine Propeller at the Different Loading Conditions. *J. Mar. Sci. Eng.* **2021**, *9*, 351. https://doi.org/10.3390/jmse9040351

Academic Editor: Yuriy Semenov

Received: 26 January 2021
Accepted: 16 March 2021
Published: 24 March 2021

Publisher's Note: MDPI stays neutral with regard to jurisdictional claims in published maps and institutional affiliations.

1. Introduction

The propeller is the main part of a propulsion system by which engine power can move the marine vessels. Marine propellers work in an intense and complicated flow field and high-risk work conditions; those two aspects must be considered to design a new propeller. First, evaluating the hydrodynamic coefficients like efficiency and thrust and torque coefficient. Second, strength due to loads and manufactured material. The propeller blade strength role is essential in the cavitation phenomenon and propellers' efficiency. In essence, the blades' structural behavior has fully interacted with the hydrodynamic propulsion qualification, particularly propeller efficiency. There are two main approaches for hydrodynamic calculations of marine propellers. The first, the inviscid numerical methodologies, involve the lifting line method, the boundary element method (BEM) [1], and the vortex lattice method (VLM) [2]. The second is computational fluid dynamic (CFD) involving large eddy simulation (LES) [3], or Reynolds-averaged Navier–Stokes (RANS) [4–6]. Because marine propeller operates in a viscous flow and complex current of the wake, CFD methods are more suitable and efficient than the inviscid methods (BEM, VLM), albeit some researchers use the inviscid method for propeller simulation due to the low-cost and lower simulation power needed.

Many researchers have used the RANS method to overcome the rotating blades' solution complexity, especially for marine propellers. Maksoud et al. [7] carried out how the

171

propeller hub could change the propeller efficiency by using CFX. Another main factor in operational marine propeller condition is the propeller and rudder interaction, investigated by Simenson et al. [8]. Valentine [9] used the RANS equations to predict the propeller blades' flow characteristics by considering turbulence inflow characteristics. In this paper, two main issues should be evaluated; first hydrodynamic calculations based on the RANS method. Second, structural analysis, that fluid–structure interaction must be engaged in the computational prediction approach. The nonlinear hydrodynamic load exerted on propeller blades due to the propeller's rotational motion inducing a centrifugal force.

The majority of the fluid-structure interaction studies on marine propeller focus on the composite propeller; Das et al. [10] used a reverse-rotation propeller in a CFD analysis. Mulcahy [11] investigated a comprehensive study on the composite propellers' hydro-elastic tailoring. In 2011, Blasques et al. [12] investigated the propeller's hydrodynamic improvements using different laminate layups for optimal speed and fuel consumption. In 2008, Young [13] investigated marine propeller fluid-structure interaction analysis to assess the composite blade's behavior. Also, the Tsai-Wu strength criterion is considered to evaluate the blade strength. Lee et al. [14] investigated a two-way coupling approach consisting of added mass based on the coupling between boundary element and finite element method. He et al. [15] used a hydro-elastic approach to evaluate composite propeller's performance, especially vibration due to loads on propeller hub and the composite layup scheme shaped based on coupling CFD and FEM methods. Finally, a comprehensive study for four propellers with different concerning materials was published in 2018 by Maljaars et al. [16], consisting of RANS–FEM and BEM–FEM results versus experimental results. Hong et al. [17] developed a pre-twist approach to gentrify the propeller's hydrodynamic characteristics using FEM/CFD-based software, ANSYS/CFX. Han et al. [18] used Star-CCM+ and Abaqus coupling approach to study the marine composite propellers; the results were reasonably close to experimental outputs. Paik et al. [19] investigated different composite propellers numerically and experimentally. In 2020, Shayanpoor et al. [20] performed an analysis by considering the CFD–FEM-based approach under the two-way coupling method on the KP458 propeller.

A simple FSI surrogate modeling is introduced in the present study by considering two distinct solvers, Linux-based/open access solvers (OpenFOAM) and windows-based/commercial software (Abaqus); thus, the major challenge is the coupling approach by considering the propeller's dynamic motion. Accordingly, the pressures extracted from OpenFOAM transfer to Abaqus in each predetermined time-step, to use as structural load, the rotation rate and the number of time-steps should be first evaluated. In the following sections, for the first step, hydrodynamic solution verified with experimental tests. For the structural solution, the current FSI approach compared with the wedge impact case was verified later; the justification of using the wedge impact verification is the similarity of the two cases. For the second step, the verified numerical model performs to analyze the advance coefficients' effect on the forces and stresses imposed on the propeller blades.

2. Materials and Methods

Fluid–structure interaction approaches are divided into monolithic and partitioned methods. In addition, partitioned methods are divided into one-way and two-way approaches. Moreover, two-way coupled is divided into strong and weak approaches [21]. From the accuracy aspect, the two-way coupling approach is more accurate than the one-way approach, especially for cases with more significant deformations and deflections. On the other hand, the one-way coupling requires less data for a single iteration per time-step. In addition, the mesh advocated for the fluid domain needless to be recalculated at each time-step. This leads the numerical solution to remain stable with unchanged mesh quality. Thus, the needed time related to the numerical solution is lower than two-way coupling, updated only after each time-step for a new iteration. Therefore, an overriding advantage of the one-way coupling approach is decreasing in the numerical solution time. Piro's [22] compared the one-way and two-way coupling approaches (RQS-RDyn-TC) for different plate thicknesses, the discrepancies between the methods were small for thick plates; whatever the plate's thickness becomes small, the accuracy of one-way coupling decrease. Due to the small deflection that occurred for the propeller blades in the present study, the one-way coupling method could be accurate enough.

2.1. Governing Equations of the Flow Around the Propeller

The most applicable and usable approach for simulating turbulent regimes is based on solving the Reynolds-averaged Navier–Stokes (RANS) equations. In the present study, InterDymFoam, a multiphase solver of OpenFOAM libraries, is used for hydrodynamic simulation. InterDymFoam is a proper solver based on the RANS equation by considering multi turbulence models [23] for dynamic mesh cases. The fluid is regarded as an incompressible Newtonian fluid that should inherently satisfy the mass conservation and momentum equations. The RANS equations are based on time-averaged variables decomposing the velocity, pressure fields into:

$$
\begin{aligned}
u &= \bar{u} + u' \\
p &= \bar{p} + p' \\
\bar{u} &= \bar{u}i + \bar{v}j + \bar{w}k \\
u' &= u'i + u'j + w'k
\end{aligned}
\tag{1}
$$

$$
\frac{\partial u_i}{\partial x_i} = 0
\tag{2}
$$

$$
\frac{\partial(u_i)}{\partial t} + \frac{\partial(u_i u_j)}{\partial x_j} = f_i - \frac{1}{\rho}\frac{\partial p}{\partial x_i} + \frac{\partial}{\partial x_j}\left[v\left(\frac{\partial u_i}{\partial x_j} + \frac{\partial u_j}{\partial x_i}\right) + \tau_{ij} \right]
\tag{3}
$$

$x_i = (x, y, z)$ represents coordinates, $u_i = (u, v, w)$ are the component of Reynolds-averaged velocity. f_i denotes the body forces presented as forces per unit volume and in the present study assumed that $f_i = 0$. Moreover, u, ρ, and P are fluid velocity vectors, density, and pressure, respectively. The Boussinesq assumption is considered to represent the Reynolds stress for incompressible flows, which is commented below:

$$
\tau_{ij} = v_t\left(\frac{\partial u_i}{\partial x_j} + \frac{\partial u_j}{\partial x_i}\right) - \frac{2}{3}\delta_{ij}k
\tag{4}
$$

$$
k = \frac{1}{2}\left((\overline{u'})^2 + (\overline{v'})^2 + (\overline{w'})^2\right), \quad u' = u - \bar{u}, \quad \bar{u'} = \frac{1}{T}\int_0^T (u(t) - \bar{u})dt
\tag{5}
$$

where v_t represent the turbulence eddy viscosity, k denotes turbulent kinetic energy (TKE) per mass. In addition, δ_{ij} surrogate as the Kronecker delta. $\bar{u}$ is time-averaged velocity, in which $\bar{u'}$ and $\bar{u'}^2$ are the mean and variance velocity, respectively. A two-equation turbulence model (k-ε) is used for the present study. ε denotes the dissipation rate of energy per mass, which determines the amount of energy lost by the viscous forces in the turbulent flow that should be introduced. (μ_t) is turbulent viscosity:

$$
\mu_t = \rho C_\mu \frac{k^2}{\varepsilon}
\tag{6}
$$

$$
\varepsilon = \frac{1}{2}\frac{\mu}{\rho}\overline{\left\{\nabla u' + (\nabla u')^T\right\} : \left\{\nabla u' + (\nabla u')^T\right\}}
\tag{7}
$$

To track the particles and capture the interface for the multiphase solution easily, the volume of fluid (VOF) could be a practical approach. The VOF method uses a volume fraction variable α to represent the air and water portion in each finite volume cell [24], where ρ_1 and μ_1 represent the physical properties of water, and ρ_2 and μ_2 also mean the physical properties of air, which introduces new conservation equations:

$$\rho = \alpha\rho_1 + (1-\alpha)\rho_2, \ \mu = \alpha\mu_1 + (1-\alpha)\mu_2, \quad \begin{array}{l} \alpha = 0 : air \\ \alpha = 1 : water \end{array} \tag{8}$$

The two-phase dynamic solution similar to the present case (propeller in water) is transient with a high turbulent regime that caused the solution to be inherently unstable. There were some algorithms to couple the mass conservation and momentum equations, the semi-implicit method for pressure-linked equations (SIMPLE), and the pressure implicit with the splitting of operators (PISO) that in the present study, the high-fidelity algorithm based on merging the PISO and SIMPLE called PIMPLE used.There are two important parameters, inner and outer correctors; inner corrector is the number of times the pressure is corrected, and the outer corrector is the number of times the equations are solved in each time-step. Outer corrector puts an obligation to stop the solution for each time-step apart from the solution be converged or not.

2.2. Governing the Structural Equations

Structural analysis is an essential step for one-way coupling approaches that should be performed correctly. The cantilever beam model is the initial theory to calculate the propeller blade strength introduced by Taylor [25]. This method was implemented and developed by some researchers. The method's drawback was poor results for the points with a low thickness on the propeller's blade compared to thicker blade's sections near the propeller's root. This problem continued until the introduction of shell theories developed by Cohen [26] and Conolly [27], the limitation of this method was the propellers' geometry complexity. For instance, wide-blade or high-skew propellers could not be assessed accurately, but in recent years, the finite element method used widely by dividing into solid or shell element approaches. Many investigations are based on both approaches, but Young [13] and Blasques et al. [12] performed a study to evaluate the output results' differences. Their investigation indicates that, although both methods are sufficient, the Shell element model needs lower computational power than the solid element method. Moreover, the solid element method has some prominency rather than a shell element model, Young [13]; this is why most FSI problems used a solid element model for the structural solver.

The deflection due to the structure's imposing loads is the main issue [28] performed by the finite element method. This technique broadly consists of discretizing a structure into several elements that should be assembled at the end. In addition, internal stresses are in equilibrium due to the continuity of stress for interface elements. The present finite element uses the explicit method, in which a time-based approach (central difference method) is used to integrate the equations of motion. In this method, the period is considered small enough to prevent divergence [29]. The equation of motion for the structural deformation corresponding to the propeller blade fixed coordinate is introduced by Equation (9):

$$M_s \ddot{d} + C_s \dot{d} + K_s d = F_{ST} \tag{9}$$

where M_s is the mass matrix, C_s belongs to damping matrix and K_s represent the matrix for the structure stiffness. the variables $\ddot{d}, \dot{d}, d$ are the acceleration, velocity, and displacement, respectively. F_{ST} is the summation of all loads imposed on the structure. Importantly, for the cases like a propeller, this load comprises force due to rotation, centrifugal force and moments, Coriolis force, and external load on the structure. For the case in hand, due to static analysis of the propeller and motionless blades in each time-step, the only pressure used for calculation is the fluid pressure extracted from the CFD solution. The calculation algorithm's first step is to solve the dynamic equilibrium relation, Equation (11). The kinematic conditions solve the next iteration's kinematic constraint in each distinct increment.

$$\ddot{U} = (M)^{-1}(P - I)_t, \; M\ddot{U} = P - I \tag{10}$$

$$\dot{U}_{(t+\frac{\Delta t}{2})} = \dot{U}_{(t-\frac{\Delta t}{2})} + \frac{(\Delta t_{(t+\Delta t)} + \Delta t_{(t)})}{2}\ddot{U}_t \tag{11}$$

$$U_{(t+\Delta t)} = U_{(t)} + \Delta t_{(t+\Delta t)}\dot{U}_{(t+\frac{\Delta t}{2})} \tag{12}$$

All of these parameters belong to nodal points, where M is the nodal mass matrix, U is nodal displacement and $\ddot{U}$ is nodal acceleration. To govern net forces act on nodal points (P–I) is used, that p is the external loads imposed on the structure. This parameter is considered nodal forces. The integrated accelerations are used to calculate velocity variations; this new added velocity value from the previous middle increment determines the middle of the current increment Equation (11). then The time-integrated velocities are added to the beginning displacements' increment to determine the final displacements' increment, Equation (12); after estimation of the nodal displacement in time(t), the element strain increments are calculated from the strain rate, The stress components can be calculated from constitutive equations and the solution process repeated for time ($t + \Delta t$).

2.3. Modeling and Computational Setup
2.3.1. Open Water Test Characteristic

The present study's framework is a numerical solution related to the Potsdam propeller test case (PPTC). For this purpose, the same propellers' geometry with the same material is accepted from the experiment cases, represented in Tables 1 and 2, respectively [29]. The International Towing Tank Conference (ITTC) recommended that the propeller rotational speed is considered constant, but propellers' advanced speed varies for different advance coefficients. The incident flow into the propeller is the opposite of real working conditions; propellers must be rotated in the opposite direction. We will hereafter comply with this rule to perform open water tests. The solution domain is modeled cylindrically with the following dimensions; 3.5 D forward, 10 D rearward, and 5 D in diameter, D is propeller diameter [30] (Figure 1).

Table 1. Geometrical specification of the propeller.

Propeller Model	Vp1304
Diameter	0.25 m
Hub coefficient	0.3
Number of blades	5
pitch coefficient (r/R = 0.7)	1.635
A_E/A_0	0.779

Table 2. Structural specification of the propeller.

Material	Al-Alloy
Elasticity	120 Gpa
Poisson's ratio	0.34
Mass density	7400 kg/m^3

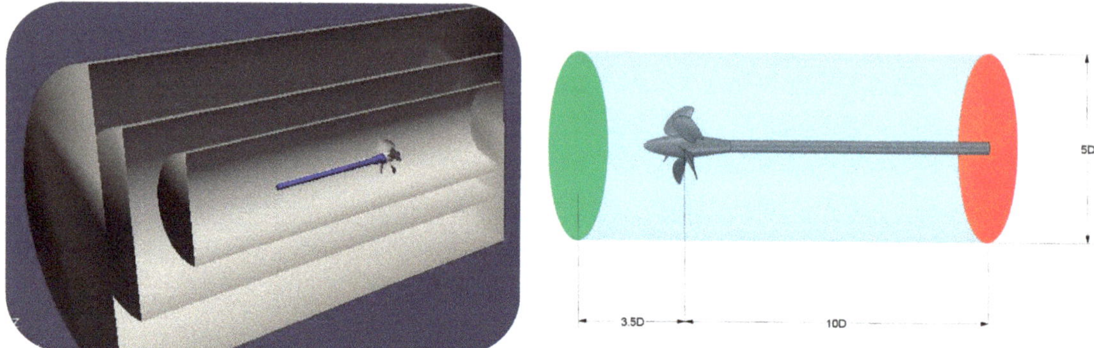

Figure 1. Numerical solution domain used for computational fluid dynamics (CFD) part of the present study.

For the open water tests, The main parameters are the thrust coefficient k_t, and torque coefficient k_Q, represented by the dimensionless values mentioned in equation (13). The coefficients are directly related to rotational speed, n, diameter, D of the propeller, and water density, ρ. Furthermore, T represents thrust [N], Q is equal to torque [Nm], V_a denote the advance speed [m/s], and η is proportional to the efficiency [-].

$$J = \frac{V_a}{nD}, k_t = \frac{thrust}{\rho n^2 D^4}, Kq = \frac{Torque}{\rho n^2 D^5}, \eta = \frac{K_T J}{2\pi K_Q} \tag{13}$$

2.3.2. Applied Boundry Condition and Dynamic Motions Method

Although various boundary conditions can be devoted to, accurate results without divergence need a proper allocation of boundary conditions In the present case, the inlet and outlet boundary condition was set based on the downstream of the outlet boundary domain. For the inlet boundary condition, free-flow velocity is considered constant, dependent on advance velocity for each advance coefficient. Moreover, the inlets' turbulence intensity is considered, I = 5%. Two main approaches could simulate the propeller's dynamic motions, multi-reference frame (MRF) and arbitrary mesh interface (AMI), since the AMI is more practical for propeller case studies used in the present numerical solution. This method is based on the interpolation between two distinct but adjacent domains connected with an interface [30]. Two similar cylindrical domains encompass the propeller, one is static, and another one is dynamic, moves with the propeller's rotation. Although there was no physical relationship between the two zones, the fluid and numerical calculations were transported through the interface.

An appropriate mesh quality and structural-based domain around the propeller needs some consecutive cylinders for dividing the domain into some sub-domain. The smallest cylinder is a small grid size, and the subsequent cylinder is larger than the former cylinder. In this regard, snappyHexMesh is used as the main tool and rhinoceros' role as an assistant tool to generate the numerical solution mesh. As shown in Figure 2, a high-quality structured mesh can be obtained by considering these techniques. A mesh independence study was established for the accuracy of the CFD solutions besides keep the computational cost.

The mesh generation framework is explained, but the mesh independence study must be performed simultaneously. The mechanism whereby the performance of griding qualified is highly dependant on the main propeller characteristics shown in Equation (12); that is how the mesh could be alleviated the computational cost without losing the accuracy. The propeller rotational speed, ω = 15 rps, and advance speed, v_a = 2 m/s, the advance coefficient, J = 0.53 [-] is considered to perform the numerical solution, Table 3. The calculation for this advanced coefficient involved five different initial grid sizes, Table 4, dynamic multiphase solutions, InterDyMFoam, which comes from OpenFOAM libraries with the same underlying physics relative to InterFoam. The thrust and torque are extracted from the postprocessing tool as the initial value; then these values substitute in Equation (13); finally, the results for each simulation are gathered in Table 4; all these cases provided acceptable results. Consequently, fine (IIII) resolution leads to a reasonable prediction of thrust and torque coefficient with optimum computational cost compared with other cases.

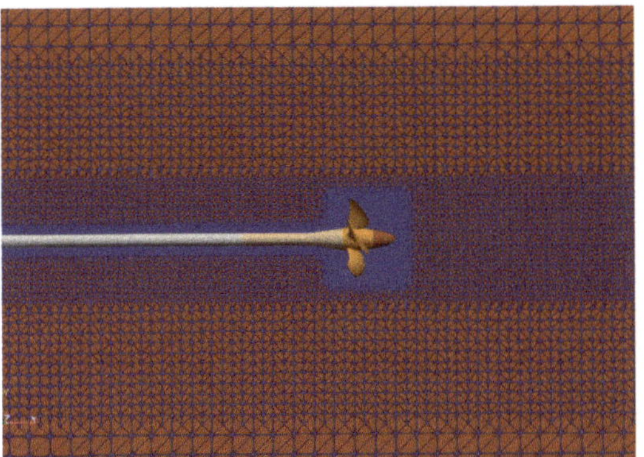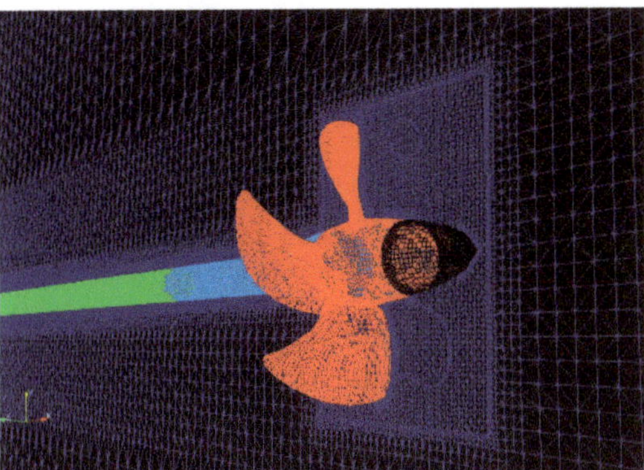

Figure 2. Mesh generated structure around the propeller.

Table 3. Primary data assumed in calculations.

Parameter	Unit	Model	Real
Density (water)	kg m^{-3}	999.0	1025
Kinematic viscosity (water)	$\text{m}^2\,\text{s}^{-1}$	1.139×10^{-6}	1.188×10^{-6}
Revolution (propeller)	s^{-1}	15	4.33

Table 4. Mesh independency study for vp1304 propeller.

Quality	Base Grid	Cell.NUM	$\dfrac{k_t}{(k_t)_{excellent}}$	$\dfrac{k_q}{(k_q)_{excellent}}$	NUM
Coarse	0.11	245,210	1.1	1.11	(I)
Mid	0.09	315,402	1.05	1.055	(II)
Mid-fine	0.08	335,183	1.025	1.024	(III)
Fine	0.064	425,060	1.015	1.013	(IIII)
Excellent	0.0325	835,205	≈ 1	≈ 1	(V)

3. Results and Discussion

3.1. CFD Validation

The numerical model tests are performed base on the advanced coefficient shown in Table 5, the same as the experimental tests [31]. As before said, the numerical investigations performed based on InterDymFoam to derive the thrusts and torques forces, then these values substitute in Equation (13) to obtain the trust coefficients, torque coefficients, and efficiency. These coefficients are calculated and compared with the experimental values; the resemblance between the present study results and the experiments is sufficient, Figure 3. Neither this numerical method nor any other methods could not achieve accurate results for low advance coefficients. That is why the error is an intrinsic part of the simulation, especially for low advance coefficients. In fact, the J = 0.266 error percentage does not account for the average error calculation. The reason is, due to severe turbulence flows around the propeller, the maximum error percentage belongs to the highest turbulent rate, J = 0.266, which caused unavoidable discrepancies. Finally, e = 7.5% is considered as the average error percentage for the present method's efficiency.

Table 5. Different advance coefficient specifications.

J [-]	ω [rps]	V_a [m/s]	Number
0.266	15	1	I
0.533	15	2	II
0.8	15	3	III
1.06	15	4	IIII
1.23	15	5	V
1.6	15	6	VI

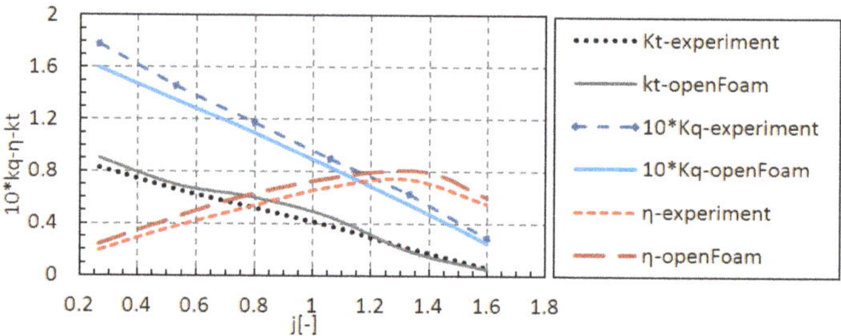

Figure 3. Hydrodynamic comparison between experimental and numerical results.

Hydrodynamic Analysis of the vp1304 Propeller

Four advanced coefficient-related contours are illustrated in Figure 4 to show the motion of the flow's particles around the propeller, particularly around the faces and backside. In essence, for the bigger advance coefficient, the blade's pressure gradient decrease; this leads the propeller's thrust and torque to be lower than the smaller advance coefficient. That is why the maximum propeller's thrust occurred at bollard state (advance velocity = 0), and after this, it gradually decreased until it reached near zero for j = 1.6 onwards. Because the pressure and velocity are interrelated, the flow field's evaluation for the velocity should be considered. The discrepancies between j = 0 and j = 1.23, minimum and maximum advanced coefficient, for the velocity contour are affected by the direction and disparity of the propeller's flow. As the advanced velocity increases, the propellers' backflow becomes more parallel with low dispersion, and the velocity is smoother around the propeller; accordingly, the hydrodynamic gradient pressure for propeller blades decreases.

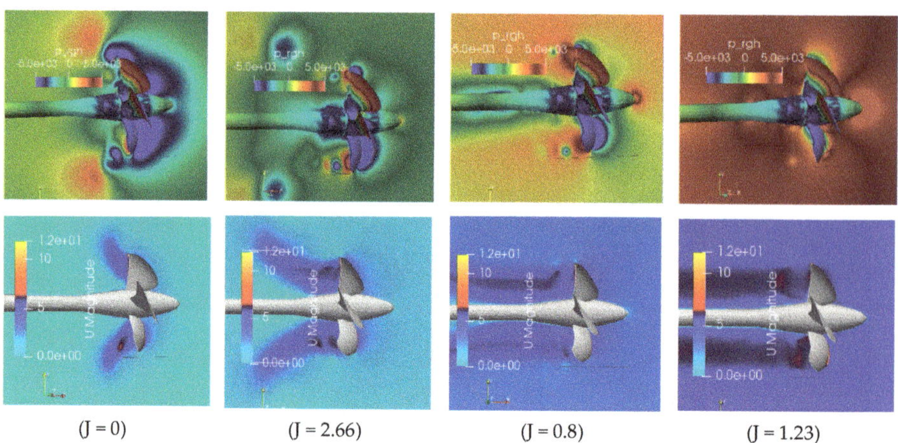

(J = 0)   (J = 2.66)   (J = 0.8)   (J = 1.23)

Figure 4. Pressure (**top**) and velocity (**bottom**) contours for different advance coefficients.

Table 6. Three gauge positions on the propeller blade.

Gauge	Distance from the Center	
p-1	0.04 m	
p-2	0.08 m	
p-3	0.12 m	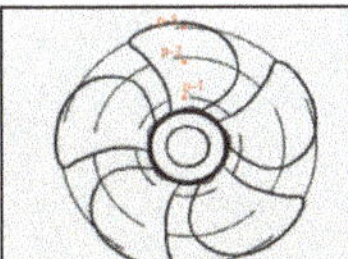

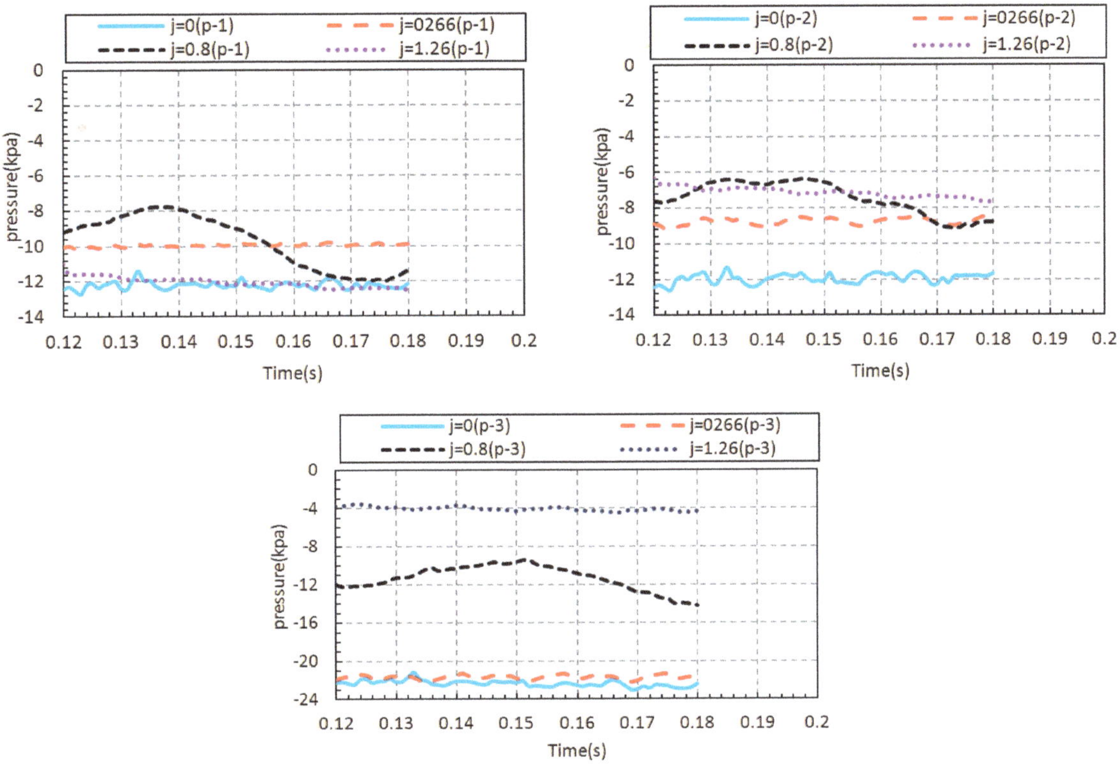

Figure 5. Gauge pressure variations of one rotation (t = 0.12 s–t = 0.18 s) for different advanced coefficients.

The valid question is, to what extent has the force dispersion on the propellers' blade changed? This question is answered by following the same approach of the pressure evaluation. Upon that, ParaView, a visualization and postprocessing tool for OpenFOAM, is accomplished for force evaluation. A framework for the force solution is generated and captured each propeller blade surface and adopted normal vectors on these surfaces. After this, by multiplying the pressure with these normal vectors, the force could be extracted from the hydrodynamic solution. There is another method to capture the imposed force by adding force-Library to the OpenFOAM solver. This method's drawback is that these codes must be added to the solver before the numerical solution started; thus, it is useless when a solved case wants to be evaluated. As shown in Figure 6, the maximum force imposed on the propeller occurred at J = 0; the smallest advance coefficient, J = 0.266, experienced the force (F = 600 N) similar to the bollard pull value (J = 0).

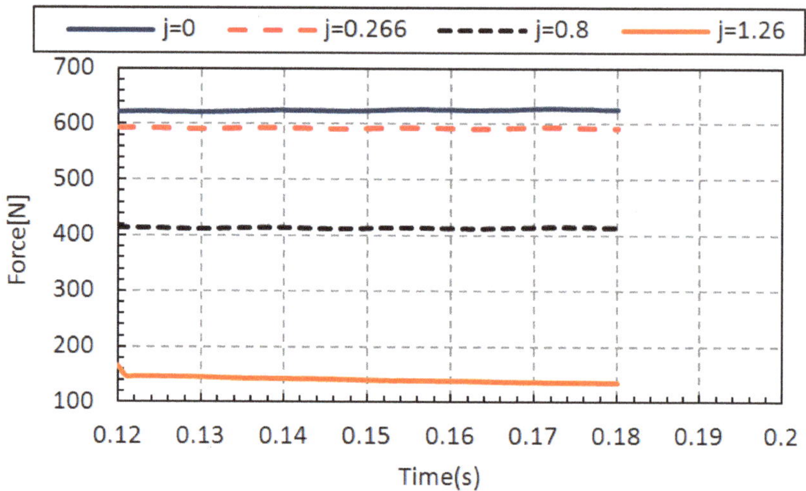

Figure 6. Integral forces act on the propeller at the different advanced coefficients.

3.2. Fluid–Structure Interaction Validation

3.2.1. Finite-Element Method

Von Mises stress calculation is at the scope coverage of the present method by implementing the static/general solution of Abaqus solvers. Substantially the propeller's dynamic motions are only considered in the CFD approach, and it is reasonable to suppose that the propeller is fixed for each time-step in the structural solver. As before said, the FEM solver used solid elements with three translation degrees of freedom, the displacements, and rotations (u, v, w, ϕ_1, ϕ_2). Moreover, weight functions (δu, δv, δW_1, δw_2, δw_3) are approximated:

$$u = \sum_{j=1}^{n} u_j.\psi_j, \ \delta u = \psi_i \quad \varphi_1 = \sum_{j=1}^{n} w^2{}_j.\psi_j, \ \delta w_2 = \psi_i$$
$$v = \sum_{j=1}^{n} v_j.\psi_j, \ \delta v = \psi_i \quad \varphi_2 = \sum_{j=1}^{n} w^3{}_j.\psi_j, \ \delta w_3 = \psi_i \tag{14}$$
$$w = \sum_{j=1}^{n} w_j.\psi_j, \ \delta w_1 = \psi_i$$

These Lagrange interpolation functions (ψ_i) are substituted in the differential equations' weak form [32]. These functions are nodal parameters (x and y) in which x and y are nodal displacements. At the finite element methods based on displacement, the displacement's manner in the element boundaries is not separated; unlike the strains, that the manner of strain is continuous only within one element. The point here is, choose between the linear or quadratic elements. Indeed, the strains have a constant value in linear elements, but in quadratic elements, the strains are nonlinear with more accurate strain or stress results than linear elements. According to Barlow [33], strains and stresses can be solved without limitation in the element, just for points, including defined nodes.

3.2.2. One-Way Coupling Approach

There are two main modes, dry and wet modes, with the critical factor of "added mass-generated effects" due to structural deformations. Dry condition considers only material/structural damping and wet condition consider added mass due to the blades' vibration. Wet modes are computed by finite element embedding the structure in a fluid domain modeled by acoustic elements. When the propeller reaches maximum load at real state condition, blades begin to vibrate on their natural frequencies that, analysis under "wet" condition can give more accurate and reliable results, especially for large deformations. Due to the Investigation of Lee et al. [14], the difference between the results of the dry and wet conditions is not significant, especially for the case with low deformations.

Thus, the one-way coupling with the dry condition used in the present study could perform accurate results. Nevertheless, the underlying physics dominant on the one-way coupling caused fluctuating trends due to large deformation and membrane forces rather than two-way approaches.

An aluminum wedge's results are considered and verified with Agard and Pancirolis' investigation [34] to evaluate the model's applicability. The wedge characteristic used for the numerical solution is shown in Tables 7 and 8; two variables were considered for the verification purpose, von Mises stress and strain. Since the maximum value for these variables occurred at the midpoint of the wedge wing, 150 mm from the wedge apex at the interior side (1 [mm] above the neutral axis), and Whatever the deformations larger, the accuracy challenge for the represented one-way approach is more dependable: thus, the method is verified under the most complicated state. In the present method by considering the wedge impact verification case, Agard [24], which was previously verified, Figure 7, the method adoption performed well for the propeller rotation. The similarity of steps and methods for the present study rather wedge impact can be reliable enough to use as a base approach of the vp1304 propeller. With this justification, this verified process can be utilized to analyze the propeller with a small but essential amended step; this involved the propeller's rotation at each time-step for the structural solver, which is discussed further in the next sections.

Table 7. Wedge characteristic used for one-way coupling verification solution.

Characters	Wedge Length	Wedge Thickness	Deadrise Angle
value	0.3 m	0.002 m	20°

Table 8. Material properties for the wedge used for one-way coupling verification solution.

Characters	Material	E [Gpa]	ρ [kg/m^3]	ν [-]
value	Aluminum	68	2700	0.3

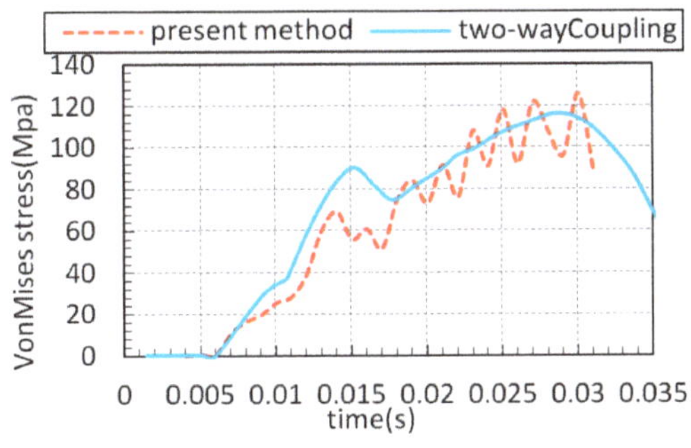

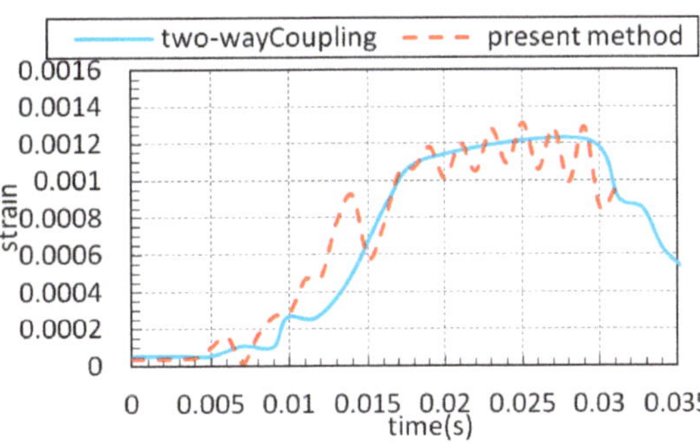

Figure 7. Present method (by considering the wedge impact) verification versus two-way coupling [24].

In the present study, the structure is considered as a rigid body for fluid simulations that the flexural mass is neglected in rigid/quasi-static (RQS) approximation. Thus, the hydrodynamic forces and pressures are independent of the structure deformations. In Equation (15), $f_R(t)$ considered as the fluid force and the deformations δ_{RQS} extracted from this method are always smaller than the real deformations. As said in Heller and Jasper [35], a "dynamic amplification factor" must be applied due to neglecting the flexural mass to correct error predicting.

The coupling process's first step is gathering the hydrodynamic data by performing the numerical solution for the marine propeller accomplished with InterDyMFoam akin to OpenFOAM libraries. The hydrodynamic data for a rigid body has mainly involved the point-by-point exerted pressure on the propeller's surface, collected as an Excell file for the ease of transferring. Now the question is how the pressures should impose on the propeller in the structural solver. Two steps were needed to perform the Abaqus analysis. The first, the time-step assessment, involved the number of separate cases in the Abaqus solver must be determined based on Equation (16). The extracted pressures from the CFD solver were considered as a load condition at the FEM solver for the related designated time-steps. Each case at Abaqus was solved distinctly, and the results were recorded for the final datasheet. All these procedures are represented in the chart shown in Figure 8.

$$k.\delta_{RQS} = f_R(t) \tag{15}$$

$$(a),\ \omega = \frac{2\pi}{T}\ (b),\ n = \frac{\nabla t}{T}\ (c),\ m = \frac{360}{n} \tag{16}$$

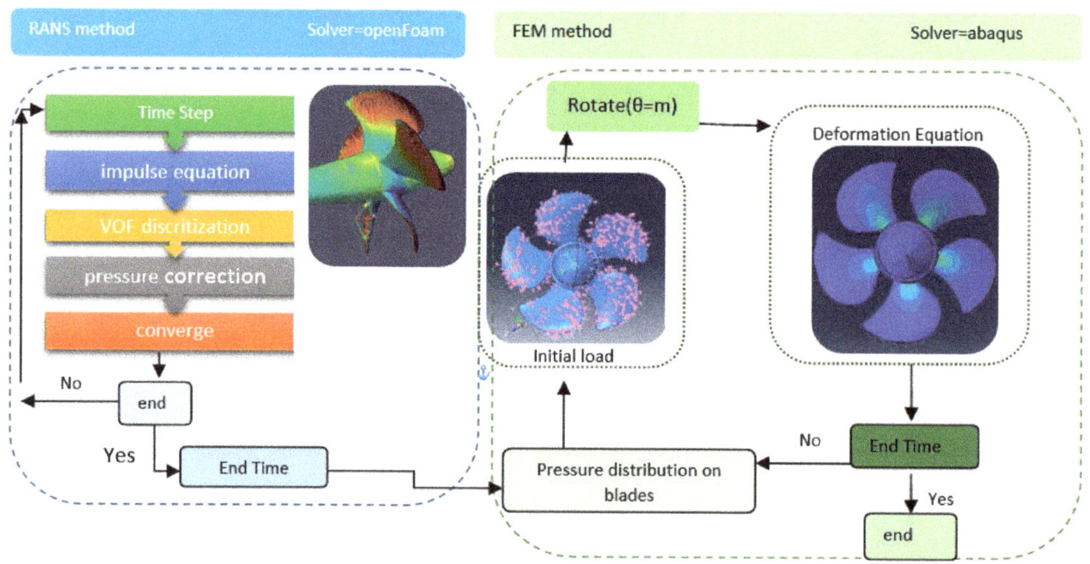

Figure 8. The comprehensive chart of the present method by considering the solvers engaged.

In more detail, the FEM part of the method could involve a set of data for different cases in each time step; the rotation rate is constant for all advance coefficients due to constant rotational speed (ω = 15 rps). The rotational rate is obtained at each time step by substituting the related values in Equation (15), $\theta = 27°$. The number of time-steps for the two complete revolutions is C = 23 distinct cases. Although there were five tables for each advance coefficient, Table 9 is represented as a sample to show how the time-steps were selected and to what extent the propeller must rotate at a specific time to match the propeller's position with the CFD solver. The number of cases, number of time-steps in Abaqus is selectable and can be increased to achieve smoother diagrams with lower fluctuations; also targeted reduction in the number of time-steps to achieve lower cost and time of the simulation. As shown in Table 9, the maximum stress and strain for each time-step were evaluated to engage in the final outputs.

Table 9. The datasheet for the coupling procedure of v = 5 m/s test case.

Case (FEM)	1	2	3	4	5	6	7
Time step(s)	0.07	0.075	0.08	0.085	0.09	0.095	0.1
Rotation Angle	$18°$	$45°$	$72°$	$99°$	$126°$	$153°$	$180°$
Stress (pa)	6.01×10^6	5.92×10^6	5.47×10^6	5.59×10^6	5.91×10^6	9.16×10^6	5.90×10^6
Strain (m)	4.7×10^{-5}	4.67×10^{-5}	4.34×10^{-5}	4.4×10^{-5}	4.6×10^{-5}	8.4×10^{-5}	4.6×10^{-5}
8	9	10	11	12	13	14	15
0.105	0.11	0.115	0.12	0.125	0.13	0.135	0.14
$207°$	$234°$	$261°$	$288°$	$315°$	$342°$	$369°$	$396°$
8.87×10^5	8.40×10^5	5.20×10^6	5.70×10^6	5.80×10^6	5.80×10^6	5.70×10^6	5.70×10^6
8.2×10^{-6}	7.8×10^{-6}	4.11×10^{-5}	4.5×10^{-5}	4.6×10^{-5}	4.59×10^{-5}	4.56×10^{-5}	4.56×10^{-5}
16	17	18	19	20	21	22	23
0.145	0.15	0.155	0.16	0.165	0.17	0.175	0.18
$423°$	$450°$	$477°$	$504°$	$531°$	$558°$	$585°$	$612°$
5.70×10^6	5.60×10^6	5.68×10^6	5.63×10^6	5.60×10^6	5.55×10^6	5.40×10^6	5.30×10^6
4.51×10^{-5}	4.44×10^{-5}	4.45×10^{-5}	4.44×10^{-5}	4.41×10^{-5}	4.36×10^{-5}	4.28×10^{-5}	4.17×10^{-5}

The comprehensive question is, how do we diagnose the reliability of the one-way rather than two-way coupling or the advantages of using the one-way method? First, for the small deflections, significantly smaller than the body's thickness, classified in the nonlinear deflections, the one-way is similar to two-way coupling results, and whatever the deflections became larger, the accuracy of the one-way approach decrease. For the case at hand, propeller deflections were classified in the range of low-deflection cases; thus, this approach could be reliable enough. Agard [24] used a parameter, the wetting equation (WQ), to classified the cases upon eigenfrequency and Young's modulus for wedge impact studies to estimate the intrinsic error belong to one-way coupling. The one-way coupling named as an industrial approach due to:

- Decrease the complexity of the numerical solution by dividing it into two parts;
- Create two distinct mesh generation schemes depending on the grid dimension needed;
- The ability to use one hydrodynamic solution for many structural sets, using different materials, thickness and different structural design;
- Lower numerical solution cost and time rather than a two-way approach, the solution time is evaluated in the present study illustrated in Table 10;
- High-fidelity results for the cases with low deflection;
- One-way coupling is more useful for cases with large domain and multiphase systems like investigations on marine vessels, propellers, etc.

Table 10. The time needed to analyze different cases.

Advance Velocity	V = (1, 2, 3, 4, 5)	m/s
FEM solution time + gathering datasheet	2 + 1	hour
CFD solution time	24	hour
Cumulative time (present method)	27	hour

3.3. Structural Behavior of Propellers' Blade

The propellers' work conditions indicate that the blades should sufficiently withstand long work cycles without failure or permanent distortion. The initial research on the propeller's structure and the analysis method was introduced by Taylor [36]. Since then, research on propellers' hydroelastic started to include the deformations of the (high-skew) bronze propeller in the 1980s. The marine propeller's design with the systematic propeller series was performed by Ekinci [37], who investigated B-series propellers using some empirical methods with different load conditions. The superiority of the present method is related to perform a quick structural calculation. Indeed, CFD and FEM are not correlating with each other, and the emphasis is put on using one hydrodynamic calculation for the several structural solvers.

Different cases could be performed for different materials involved Alloy or composite materials. A reasonable method to judge how the materials affect the propellers' structures depends on two main structural variables, stress and strain. The discrepancy in value and the maximum or minimum occurrence positions are the most important factors in evaluating the propeller's strength. To cast light on the mechanism whereby how the efficient case selected, the maximum von Mises stress imposed on the propeller used as a key factor to illustrate the stress distribution for each advance coefficient.

Apart from some exceptions, maximum stress occurred near the blades' root, and whatever far away from the root, the value for stress decreased by a gentle slope. As predicted, when the propellers' thrust reaches the maximum value, the propellers' blade has deformed in the load vector's direction. The highest value for strain and stress occurs at bollard pull (j = 0) due to thrust and torque values. As shown in Figure 9, in the bollard states' (j = 0) maximum stress is about, $s = 3 \times 10^6$ pa, but for another advanced coefficient, this value is oscillating about, $s = 1 \times 10^6$ pa, and $s = 1 \times 10^6$ pa.

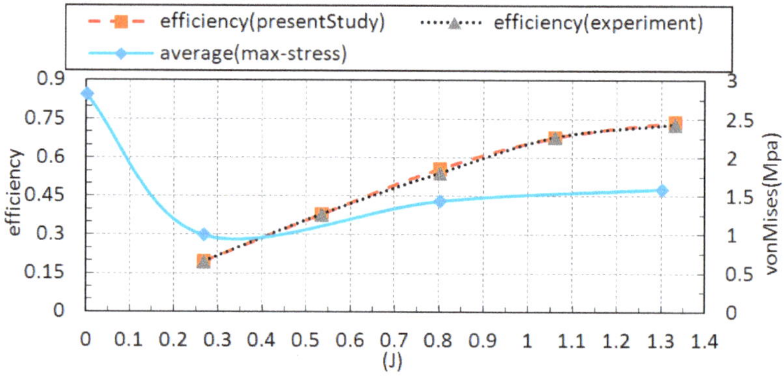

Figure 9. The comparison between a maximum stress and hydrodynamic efficiency.

For further explanation, Figure 10 used to evaluate the maximum strain and stress trends; maximum stress locations are different for each advance coefficient, at J = 1.23, the maximum value for stress is about, S = 5.2 Mpa, occurred at t = 0.08 s ((rev-1/(t = 0.66 s to t = 0.12 s)), this value is different for J = 0.8, maximum von Mises stress appears at t = 0.12 s (end of the rev-1). Such a different trend is valid for other advanced coefficients. The oscillation occurs for von Mises stress due to neglecting the damping and added mass effects for structural behavior. A contour-based figure, Figure 11, was constructed to show the stress for one revolution versus the advance coefficient The propeller's design is such that the blade's thickness near the hub is greater than the tip; as a result, the stress distribution indicates that stress concentration at the blade-hub intersection. The von Mises range is between S = 1.8 × 10⁶ pa and 2.4 × 10⁶ pa, except for t = 0.12 s, that is, S = 8.2 ×10⁶ pa. The main assessment comprises how stress and strain distribute on the propeller's blades and which blade absorbs the maximum load.

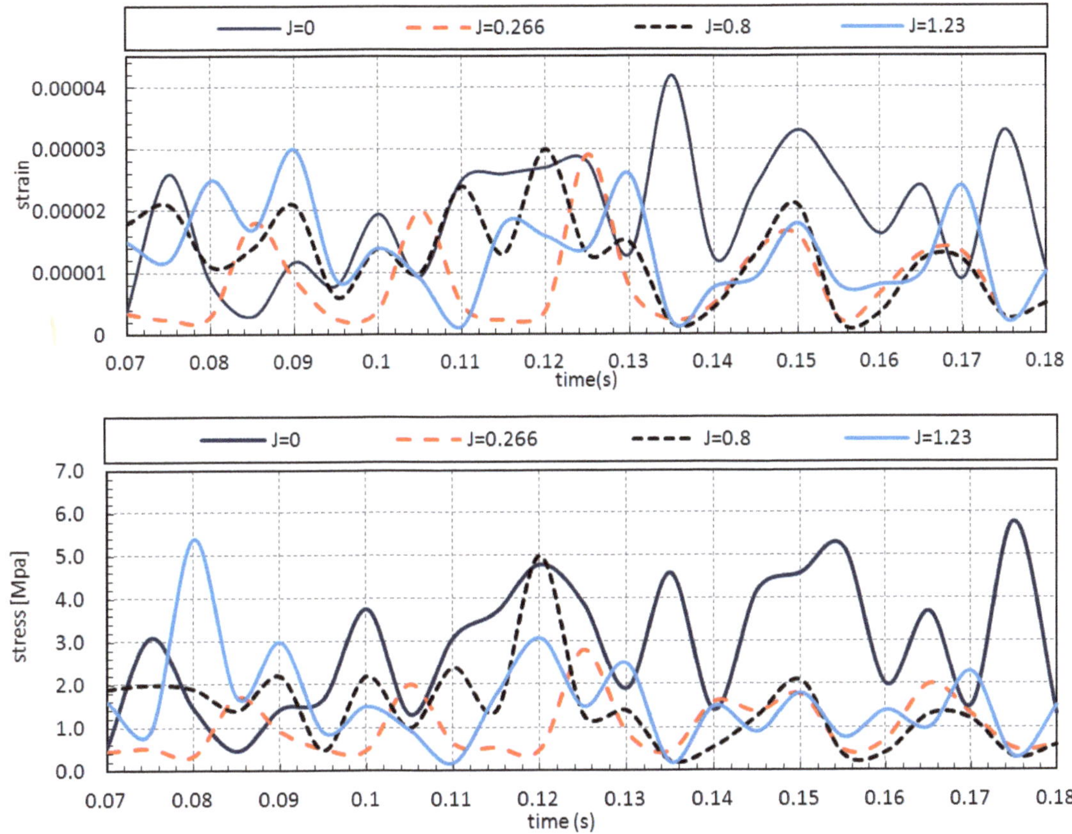

Figure 10. Maximum stress and strain values, for two revolutions, at J = 0, 0.266, 0.8, 1.23.

3.4. Propellers' Structural Behavior in Rotation

Three different points were selected based on Table 6 to evaluate the blade rotation angle's effects on stress distribution. The initial impression from Figure 12 is that the point near the blade root (P1), apart from the advance coefficient, has the greatest von Mises stress value, and the minimum value belongs to the point at the top of the blade (P3). Albeit, the diagrams' harmony for three points is similar to each other. Consequently, a set of different graphs for each advance coefficient are used to illustrate how the stresses on the propeller blade's surface are changed. Following the same approach, the von Mises stress versus the rotation angle is shown in Figure 13 for four different advance coefficients. Although the von Mises diagrams for all advance coefficients (J) are not harmonic (rev-1:0–360 and rev-2:360–720), the trend for the J = 0.266 distribution is more harmonic than other advance coefficients. j = 0.266 is a minimum point for the stress diagram; thus, the propeller's structural behavior is more stable and has a minimum value because of maximum k_t and k_q, which are occurred at j = 0.266. Therefore, the main portion of exerted pressure on the propeller uses to generate thrust.

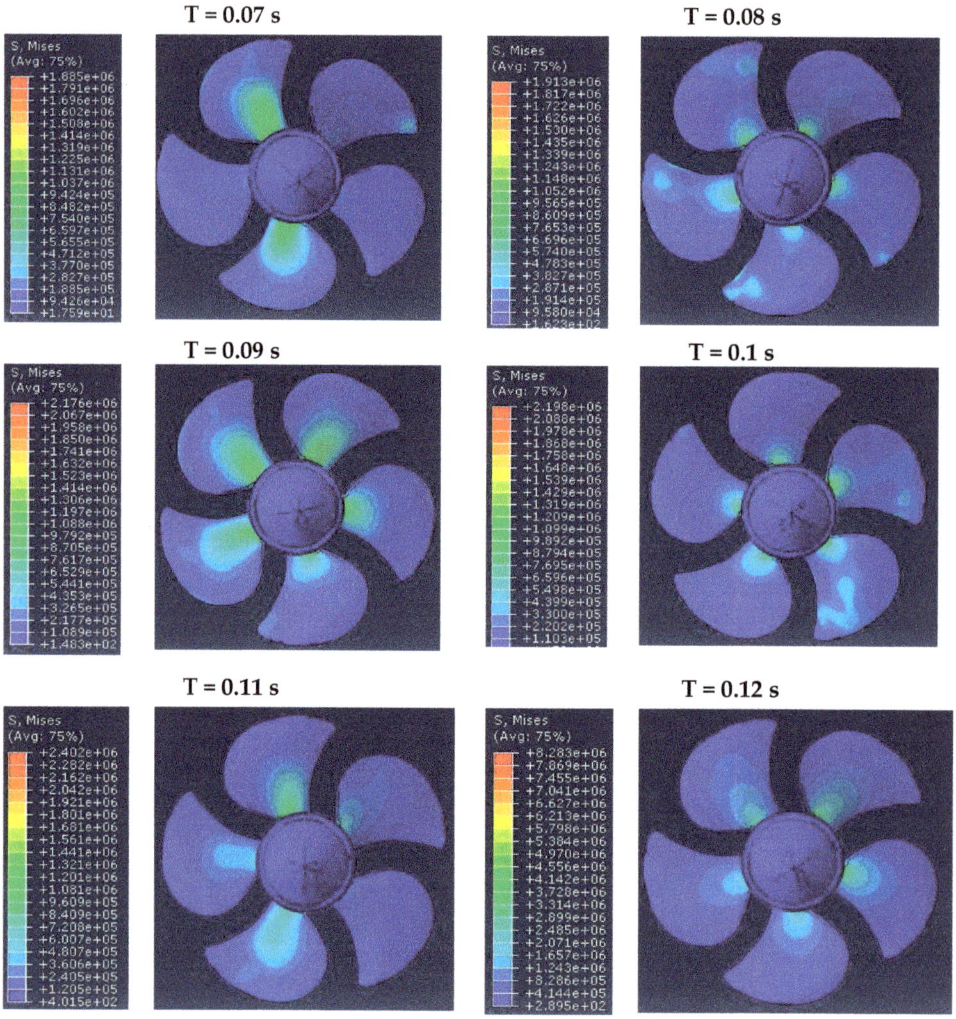

T = 0.07 s   T = 0.08 s

T = 0.09 s   T = 0.1 s

T = 0.11 s   T = 0.12 s

Figure 11. Von Mises stress evaluation on propeller surface at different time-steps.

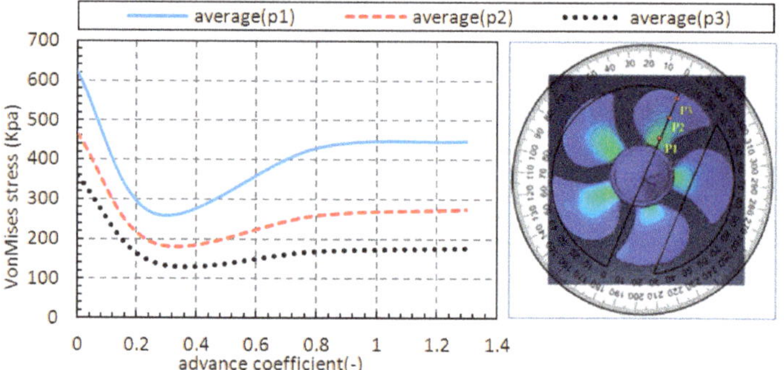

Figure 12. Maximum von Mises stress on a blade versus advance coefficient.

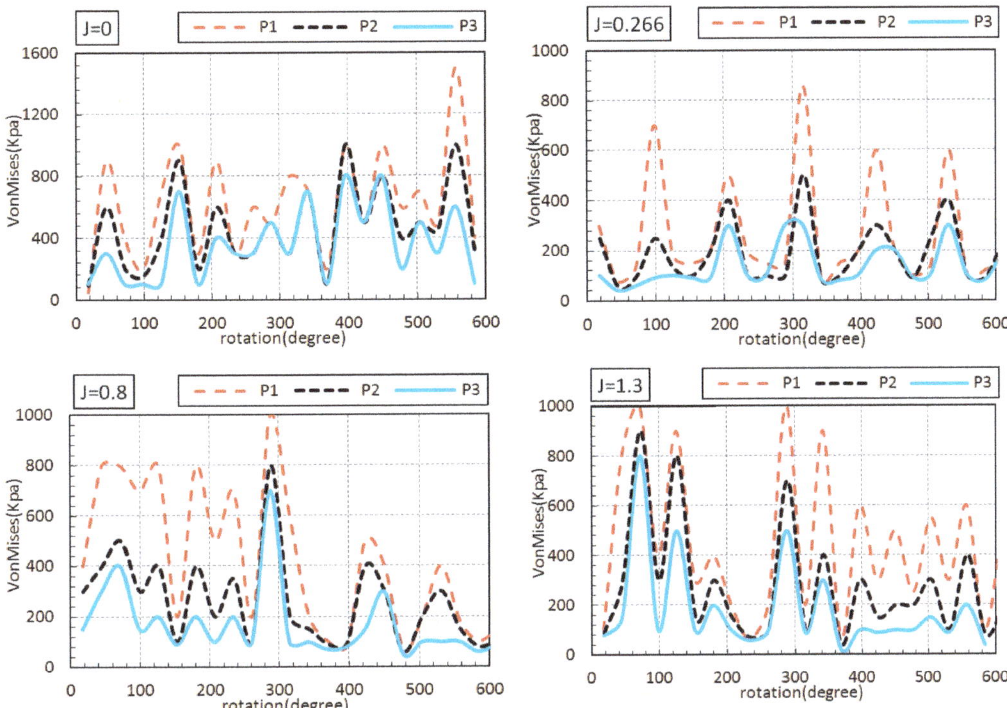

Figure 13. Maximum von Mises stress versus rotation angle for j = 0, 0.266, 0.8, 1.23.

4. Conclusions

In the present study, simple surrogate modeling for rigid/quasi-static approach is used to investigate hydroelastic simulation for open water propeller test cases (PPTC); the coupling approach comprises a CFD-FEM method solved separately. In the first step, the hydrodynamic solver, InterDyMFoam, is verified with an experimental study that the average efficiency error for different advanced coefficients was about e = 7.5% for efficiency on average. For further evaluation in the hydrodynamic section, a force analysis was performed using the ParaView postprocessing toolkit for each advanced coefficient; also, pressure and velocity contours were demonstrated versus different advance coefficients.

For the second step, the pressure distributions were obtained from OpenFOAM visualization software, ParaView, and used as the initial structural loads in Abaqus software. The point is that the propeller must have an appropriate rotational motion in each time-step; this procedure continued until it reaches any complete revolutions needed. Emphasis is put on von Mises stress, which is vital for evaluating the propeller's structural strength. A fact that is borne out is that the maximum stress occurs at the bollard pull state, J = 0. In addition, j = 0.266 has the minimum stress value apart from the advance coefficient. Although von Mises stress's value remained stable without a notable change after J = 0.8, the value and maximum stress position could be changeable for each advance coefficient under different work conditions. Consequently, the propeller's structural behavior can effectively be analyzed by the one-way coupling approach, which is a simple, but efficient model by considering the OpenFOAM and Abaqus solvers as the CFD and FEM solutions, respectively. The present method aims to assess the appropriate material or design framework for a wide range of propellers and working conditions with an accurate and low-cost method.

Author Contributions: Conceptualization, methodology, M.M.; software, original draft preparation, M.M.; validation, A.M.; writing—review and editing, visualization, project administration, M.M. and A.M.; All authors have read and agreed to the published version of the manuscript.

Funding: This research received no funding.

Institutional Review Board Statement: Not applicable.

Informed Consent Statement: Not applicable.

Data Availability Statement: Not applicable.

Conflicts of Interest: Authors declare no conflict of interests.

Abbreviations

Arbitrary mesh interface	AMI
Boundary element method	BEM
Computational fluid dynamic	CFD
Fluid–structure interaction	FSI
Finite element method	FEM
International Towing Tank Conference	ITTC
Large eddy simulation	LES
Multi-reference frame	MRF
Pressure implicit with splitting of operators	PISO
Potsdam propeller test case	PPTC
Vortex lattice method	VLM
Volume of fluid	VOF
Reynolds-averaged Navier–Stokes	RANS
Rigid/quasi-static	RQS
Semi-implicit method for pressure-linked equations	SIMPLE
Turbulent kinetic energy	TKE
Wetting time equation	WQ

Abbreviations

K_t	Thrust coefficient
K_q	Torque coefficient
V_a	Advance velocity
η	Efficiency
J	Advance coefficient
E	Elasticity
ρ	Density
υ	Poissons' ratio
ω	Angular frequency
S	von Mises stress
P_i	Pressure gauge (i = 1–2–3)
ε	Strain
F	Force
T,t	Time
e	Error percentage
rev	Propeller revolution

References

1. Young, Y.L. Hydroelastic behavior of flexible composite propellers in wake inflow. In *Proceedings of the 16th International Conference on Composite Materials*, Kyoto, Japan, 8–13 July 2007.
2. Olsen, A.S. Optimisation of propellers using the vortex-lattice method. In *Mechanical Engineering, Maritime Engineering*; Technical Univercity Of Denmark: Lyngby, Denmark, 2002.
3. Cokljat, D.; Caridi, D. Embedded LES methodology for general-purpose CFD solvers. In *Sixth International Symposium on Turbulence and Shear Flow Phenomena*; Begel House Inc.: Danbury, CT, USA, 2009.

4.   Yao, J. Investigation on hydrodynamic performance of a marine propeller in oblique flow by RANS computations. *Int. J. Nav. Archit. Ocean Eng.* **2015**, *7*, 56–69. [CrossRef]

5.   Krasilnikov, V.; Zhang, Z.; Hong, F. Analysis of unsteady propeller blade forces by RANS. In Proceedings of the First International Symposium on Marine Propulsors smp, Trondheim, Norway, 6 June 2009.

6.   Dubbioso, G.; Muscari, R.; Di Mascio, A. Analysis of a marine propeller operating in oblique flow. Part 2: Very high incidence angles. *Comput. Fluids* **2014**, *92*, 56–81. [CrossRef]

7.   Abdel-Maksoud, M.; Hellwig, K.; Blaurock, J. Numerical and experimental investigation of the hub vortex flow of a marine propeller. In Proceedings of the 25th Symposium on Naval Hydrodynamics, St. John's, NL, Canada, 12 June 2004.

8.   Simonsen, C.D.; Stern, F. RANS maneuvering simulation of Esso Osaka with rudder and a body-force propeller. *J. Ship Res.* **2005**, *49*, 98–120. [CrossRef]

9.   Valentine, D.T. *Reynolds-Averaged Navier-Stokes Codes and Marine Propulsor Analysis*; Carderock Division, Naval Surface Warfare Center: Bethesda, MD, USA, 1993.

10. Das, H. CFD Analysis for Cavitation of a Marine Propeller. In Proceedings of the 8th Symposium on High Speed Marine Vehicles, Naples, Italy, 22–23 May 2008.

11. Mulcahy, N.; Prusty, B.; Gardiner, C. Hydroelastic tailoring of flexible composite propellers. *Ships Offshore Struct.* **2010**, *5*, 359–370. [CrossRef]

12. Blasques, J.P.; Berggreen, C.; Andersen, P. *Hydro*-elastic analysis and optimization of a composite marine propeller. *Mar. Struct.* **2010**, *23*, 22–38. [CrossRef]

13. Young, Y.L. Fluid–structure interaction analysis of flexible composite marine propellers. *J. Fluids Struct.* **2008**, *24*, 799–818. [CrossRef]

14. Lee, H.; Song, M.-C.; Suh, J.-C.; Chang, B.-J. Hydro-elastic analysis of marine propellers based on a BEM-FEM coupled FSI algorithm. *Int. J. Nav. Archit. Ocean Eng.* **2014**, *6*, 562–577. [CrossRef]

15. He, X.; Hong, Y.; Wang, R. Hydroelastic optimisation of a composite marine propeller in a non-uniform wake. *Ocean Eng.* **2012**, *39*, 14–23. [CrossRef]

16. Maljaars, P.; Bronswijk, L.; Windt, J.; Grasso, N.; Kaminski, M. Experimental validation of fluid–structure interaction computations of flexible composite propellers in open water conditions using BEM-FEM and RANS-FEM methods. *J. Mar. Sci. Eng.* **2018**, *6*, 51. [CrossRef]

17. Hong, Y.; Hao, L.F.; Wang, P.C.; Liu, W.B.; Zhang, H.M.; Wang, R.G. Structural design and multi-objective evaluation of composite bladed propeller. *Polym. Polym. Compos.* **2014**, *22*, 275–282. [CrossRef]

18. Han, S.; Lee, H.; Song, M.C.; Chang, B.J. Investigation of Hydro-Elastic Performance of Marine Propellers Using Fluid-Structure Interaction Analysis. In Proceedings of the ASME International Mechanical Engineering Congress and Exposition, Houston, TX, USA, 13–19 November 2015; American Society of Mechanical Engineers: New York, NY, USA, 2015.

19. Paik, B.-G.; Kim, G.-D.; Kim, K.-Y.; Seol, H.-S.; Hyun, B.-S.; Lee, S.-G.; Jung, Y.-R. Investigation on the performance characteristics of the flexible propellers. *Ocean Eng.* **2013**, *73*, 139–148. [CrossRef]

20. Shayanpoor, A.A.; Hajivand, A.; Moore, M. Hydroelastic analysis of composite marine propeller basis Fluid-Structure Interaction (FSI). *Int. J. Marit. Technol.* **2020**, *13*, 51–59.

21. Vassen, J.-M.; De Vincenzo, P.; Hirsch, C.; Leonard, B. Strong coupling algorithm to solve fluid-structure-interaction problems with a staggered approach. *ESASP* **2011**, *692*, 128.

22. Piro, D.J. A Hydroelastic Method for the Analysis of Global Ship Response Due to Slamming Events. In *Naval Architecture and Marine Engineering*; University of Michigan: Ann Arbor, MI, USA, 2013.

23. Moukalled, F.; Mangani, L.; Darwish, M. *The Finite Volume Method in Computational Fluid Dynamics*; Springer: Berlin/Heidelberg, Germany, 2016; Volume 6.

24. Aagaard, O. *Hydroelastic Analysis of Flexible Wedges*; Department of Marine Technology, Norwegian University of Science and Technology: Trondheim, Norway, 2013.

25. Taylor, D. *The Speed and Power of Ships*; Ransdell Inc.: Washington, DC, USA, 1933.

26. Cohen, J.W. On stress calculations in helicoidal shells and propeller blades. In *Mechanical, Maritime and Materials Engineering*; Delft University of Technology: Delft, The Netherlands, 1955.

27. Conolly, J. Strength of propellers. *Trans. RINA* **1974**, *103*, 139–204.

28. Greening, P.D. *Dynamic Finite Element Modelling and Updating of Loaded Structures*; University of Bristol: Bristol, UK, 1999.

29. Taghipour, R. Efficient prediction of dynamic response for flexible and multi-body marine structures. In *Marine Technology*; Norwegian University of Science and Technology: Trondheim, Norway, 2008.

30. Cosden, I.A.; Lukes, J.R. A hybrid atomistic–continuum model for fluid flow using LAMMPS and OpenFOAM. *Comput. Phys. Commun.* **2013**, *184*, 1958–1965. [CrossRef]

31. Klasson, O.K.; Huuva, T. Potsdam propeller test case (PPTC). In Proceedings of the Second International Symposium on Marine Propulsors, SMP, Hamburg, Germany, 15–17 June 2011.

32. Khan, A.M. Flexible composite propeller design using constrained optimization techniques. In *Aerospace Engineering*; Iowa State University: Ames, IA, USA, 1997.

33. Barlow, J. More on optimal stress points—reduced integration, element distortions and error estimation. *Int. J. Numer. Methods Eng.* **1989**, *28*, 1487–1504. [CrossRef]

34. Panciroli, R. *Dynamic Failure of Composite and Sandwich Structures*; Springer: Dordrecht, The Netherlands, 2013; Volume 192.
35. Heller, S.; Jasper, N. Strength on the Structural Design of Planing Craft. Available online: https://repository.tudelft.nl/islandora/object/uuid%3Adc7516f5-6d26-4419-8955-37af96d86878 (accessed on 22 March 2021).
36. Taylor, J.L. Natural vibration frequencies of flexible rotor blades. *Aircr. Eng. Aerosp. Technol.* **1958**, *30*, 331. [CrossRef]
37. Ekinci, S. A Practical Approach for Design of Marine Propellers with Systematic Propeller Series. *Brodogr. Teor. Praksa Brodogr. Pomor. Teh.* **2011**, *62*, 123–129.

Journal of
Marine Science and Engineering

MDPI

Article

Effect of Boundary Conditions on Fluid–Structure Coupled Modal Analysis of Runners

Dianhai Liu [1], Xiang Xia [2], Jing Yang [1] and Zhengwei Wang [3,*]

[1] Technology Center State Grid XinYuan Company Ltd., Beijing 100761, China; dianhai-liu@sgxy.sgcc.com.cn (D.L.); yangjingshirley@163.com (J.Y.)
[2] College of Water Resources and Civil Engineering, China Agricultural University, Beijing 100083, China; xiaxiangcau@163.com
[3] Department of Energy and Power Engineering, Tsinghua University, Beijing 100084, China
* Correspondence: wzw@mail.tsinghua.edu.cn; Tel.: +86-13601363209

Abstract: To predict the resonance characteristics of hydraulic machinery, it is necessary to accurately calculate the natural modes of the runners in the operating environment. However, in the existing research, the boundary conditions of the numerical modal analysis of the runner were not unified. In this paper, numerical modal analysis of a prototype Francis pump turbine runner was carried out using the acoustic–structure coupling method. The results of three different constraints were compared. The influence of the energy loss on the chamber wall on the natural modes of the runner was studied by the absorption boundary. The results show that the constraint condition (especially the rotating shaft) has significant impacts on the torsional mode, the radial mode, the 1 nodal-diameter mode, and the 0 nodal-circle mode, and the maximum differences in the natural frequencies under different conditions are 69.3%, 56.4%, 35.1%, and 9.4%, respectively. The change of the natural frequencies is closely related to the modal shapes. On the other hand, the energy loss on the wall mainly affects the nodal-circle modes, and the influence on other modes is negligible. The results can provide references for the design and resonance characteristics analysis of hydraulic machinery runners.

Keywords: hydraulic machinery runner; wet modal analysis; acoustic–structure coupling; boundary condition

Citation: Liu, D.; Xia, X.; Yang, J.; Wang, Z. Effect of Boundary Conditions on Fluid–Structure Coupled Modal Analysis of Runners. *J. Mar. Sci. Eng.* **2021**, *9*, 434. https://doi.org/10.3390/jmse9040434

Academic Editor: Eva LOUKOGEORGAKI

Received: 1 April 2021
Accepted: 14 April 2021
Published: 17 April 2021

1. Introduction

The natural mode is an inherent vibration characteristic of the structural systems. For hydraulic machinery runners, the natural mode is an important technical index. The accurate calculation or measurement of the natural mode is of great significance to the safety of the unit, and even the whole power station.

The goal of modal analysis of the runners has evolved from the dry mode to the wet mode. It is well known that the natural modes of structures in water are different from those in air (or a vacuum) [1]. Therefore, the wet mode and its prediction method have been widely studied. At first, the test method is the most direct and effective. For example, Rodriguez et al. [2] and Han et al. [3] used the experimental method to analyze the wet modes of a scale model of a Francis turbine runner and a current turbine blade, respectively. Presas et al. [4] carried out modal tests on a scale model of a pump turbine runner outside and inside the casing. The effect of the added mass on the dynamic response was proposed and discussed. Østby et al. [5] studied the added mass effect of water on a simplified low specific speed Francis turbine runner using an experimental method. The measurements revealed a frequency reduction of about 40% when the runner was hanging in water. Egusquiza et al. [6] measured the natural modes of a prototype pump turbine runner in air and water. Considering the modal shapes, the mechanism of the effect of the added mass on the natural modes was analyzed. The main defect of the test method was that it costed too much. Thus, the numerical simulation has become the popular research method because it is easy to implement. Hübner et al. [7] took the modal shapes of the structure in air as the initial displacement condition of the two-way fluid–structure coupling simulation, analyzed the frequency reduction coefficient of each mode of the structure in water, and then obtained the wet modes of the structure. The defect of this method is that only one mode can be obtained in each calculation. Liu et al. [8] proposed a two-way fluid–structure coupling method based on a PolyMax modal identification algorithm. This method can obtain multiple modes at once. However, because of the complexity of the two-way coupling, the efficiency of this method is still low. Egusquiza et al. [9,10] measured the natural frequencies of a runner in a tank by experiment and acoustic theory-based fluid–structure coupling method. The results showed that the natural frequencies obtained by the above two methods were in good agreement, which proved that the acoustic–structure coupling method is accurate. Compared with the two-way fluid–structure coupling method, the acoustic–structure coupling method simplifies the Navier–Stokes equation to a certain extent, and is highly efficient; thus, it is widely used. Huang et al. studied the effect of cavitation on the modes of a Francis turbine runner using the acoustic–structure coupling method [11] and calculated the static and dynamic stresses of the runner during operation [12]. Rodriguez et al. [13] measured the modes of an underwater circular plate using the acoustic–structure coupling method and compared it with the experimental data, which further proved that this method can accurately predict the natural frequencies of underwater structures considering the influence of rigid walls.

Due to the reflection of the pressure wave on the wall, the additional mass of the water around the structure increases, which will reduce the natural frequencies. Askari et al. [14] studied the influence of the axial and radial clearances on the modes of a circular plate in water tanks, and the results showed that the natural frequencies changed with the thickness of the clearance. Additionally, the smaller the thickness, the more significant the influence. Similarly, He et al. [15] studied the effect of the clearance on the dynamic characteristics of a pump turbine runner using the acoustic–structure coupling method. The results showed that the thickness of the clearance between the runner and the wall has a significant impact on the resonance condition. In current numerical research, the flow passage vessel is usually simplified as a rigid wall to analyze the wet mode of the internal structure, i.e., the system composed of the structure and water is assumed to have no energy loss at the boundary [9]. However, flow passage vessels (such as runner chambers) are not completely rigid. Therefore, the outer wall will vibrate along with the structure–water system and absorb some of the energy of the system. The effect of the energy loss on the modes of the runners needs to be further discussed.

In practical engineering, the runner of hydraulic machinery is fixed on the shaft by bolts. However, in the research of the dynamic characteristics of the runners, the setting of the constraints is not uniform. Without considering the rotating shaft, line constraint (i.e., fixed constraint on bolt center line) [11,12], surface constraint (i.e., fixed constraint on connecting surface) [15] and rope suspension [2,5,6] are often adopted. Among them, the rope suspension is usually used in the experimental modal analysis to eliminate the influence of constraints or other structures on the dynamic characteristics. However, some researchers have carried out modal analysis of the runner, the shaft, and even the rotor as a whole [16,17], to make the numerical model closer to the actual operating environment.

In this paper, the modal analysis of a prototype pump turbine runner is carried out by the acoustic–structure coupling method. The purpose is to investigate the influence of the constraints and the boundary conditions of the wall on the prediction of the natural modes of the runner.

2. Numerical Model

2.1. Governing Equation

The governing equation of the acoustic–structure coupling method has been derived in detail [18,19], and its discrete form can be written as follows:

$$\begin{bmatrix} [M_S] & 0 \\ \bar{\rho}_0[R]^T & [M_F] \end{bmatrix} \left\{ \begin{array}{c} \{\ddot{u}\} \\ \{\ddot{p}_e\} \end{array} \right\} + \begin{bmatrix} [C_S] & 0 \\ 0 & [C_F] \end{bmatrix} \left\{ \begin{array}{c} \{\dot{u}\} \\ \{\dot{p}_e\} \end{array} \right\} + \begin{bmatrix} [K_S] & -[R] \\ 0 & [K_F] \end{bmatrix} \left\{ \begin{array}{c} \{u\} \\ \{p_e\} \end{array} \right\} = \left\{ \begin{array}{c} \{F_S\} \\ \{0\} \end{array} \right\} \tag{1}$$

where $[M_S]$, $[C_S]$, and $[K_S]$ are the mass matrix, the damping matrix, and the stiffness matrix of the structure, respectively; $[M_F]$, $[C_F]$, and $[K_F]$ are the mass matrix, the damping matrix, and the stiffness matrix of the fluid; $\{u\}$ is the node displacement vector; $\{p_e\}$ is the node pressure; $\bar{\rho}_0$ is the density of the acoustic fluid; $[R]^T$ is the boundary matrix of the fluid; and $\bar{\rho}_0[R]^T$ and $-[R]$ represent the coupling mass matrix and the coupling stiffness matrix of the system, respectively.

In this paper, the modes of a runner under different constraints are compared, and the wall of the runner chamber is set as the absorption boundary to consider the energy loss. Among them, the change of the constraint condition mainly affects the stiffness matrix of the structure, and the change of the absorption coefficient α on the wall affects the damping matrix of the fluid. When the energy dissipation on the fluid boundary is considered, the damping matrix can be expressed as [19]:

$$[C_F] = \frac{\alpha}{c} \int_S \{N_p\}\{N_p\}^T dS \tag{2}$$

where c is the sound velocity in the fluid; $\{N_p\}$ is the shape function of pressure element; and S is the fluid boundary. It can be seen that when all fluid boundaries are set as total reflection boundaries (i.e., $\alpha = 0$), the fluid damping is 0.

In fact, the absorption boundary mentioned here is an acoustic impedance boundary condition. The relationship between the acoustic impedance Z and absorption coefficient can be described as [20]:

$$Z = Z_0 \frac{1 + \sqrt{1 - \alpha}}{1 - \sqrt{1 - \alpha}} \tag{3}$$

where $Z_0 = \bar{\rho}_0 \cdot c$ is the dielectric characteristic impedance of the fluid.

2.2. Finite Element Model

A prototype Francis pump turbine runner with nine blades was taken as the object of this article. The runner was made of structural steel: the density was 7850 kg/m^3, the Poisson's ratio was 0.3, and the elastic modulus was 200 GPa. The cut-away view and 3D modeling of the runner and the surrounding fluid are shown in Figure 1. Several geometric parameters of the runner are given here in dimensionless form.

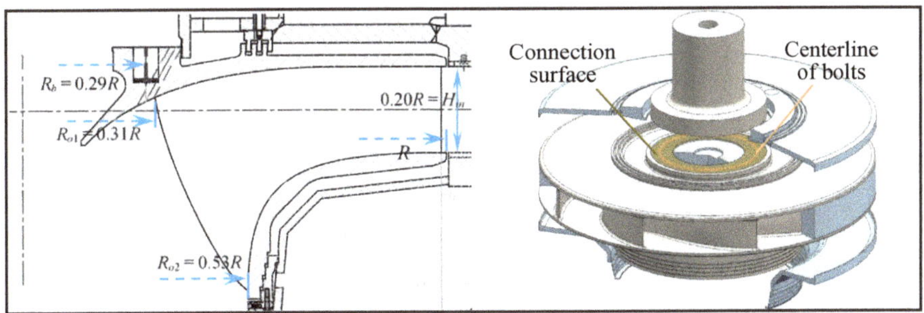

Figure 1. Schematic diagram of the runner and surrounding fluid.

First, the independence of the grid was checked. The fixed constraint was set at the bolt centerline, the interface between the fluid domain and the structure was set as the fluid–structure interaction surface, the interface between the fluid domain and the runner chamber was set as the rigid wall, and the interface between the fluid domain and the external fluid domain was set as the full absorption boundary.

The geometry of the clearance was very narrow; therefore, the prediction accuracy of the pressure wave could be greatly improved by using the hexahedral mesh to discretize it. However, the runner and its internal fluid were divided by tetrahedral mesh. In addition, the nodes on the fluid–structure interaction surface were kept at one-to-one correspondence. Four sets of grids were divided according to a certain proportion, named A, B, C, and D, and their parameters are shown in Table 1. For the first five modes of the runner, the frequency ratio f_i/f_{iD} was used to analyze the grid independence, and the results were shown in Figure 2. f_i is the ith natural frequency of the runner, and the subscript D represents the result corresponding to the grid D.

Table 1. Grid parameters of fluid structure coupling modal analysis of runner.

Grid	A	B	C	D
Unit number of structure (-)	271,271	638,713	1,209,795	2,033,087
Unit number of water (-)	366,915	834,053	1,551,437	2,414,909
Total unit number (-)	638,186	1,472,766	2,761,232	4,447,996

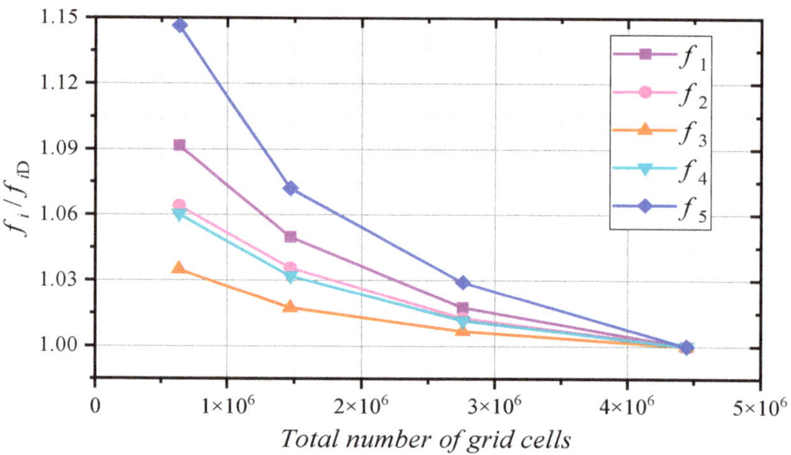

Figure 2. The first five natural frequencies of the runner in channels with different grid sizes.

It can be seen that the frequencies of each order tended to a stable value with the increasing number of elements. Considering the accuracy and efficiency of the solution, grid C was selected for the simulations. The finite element model is shown in Figure 3.

Figure 3. Finite element model of fluid–structure coupling modal analysis of runner.

2.3. Definition of Vibration Mode

For hydraulic machinery runners, the vibration mode is similar to that of a disk. To describe and analyze the modes of the runner conveniently, the concepts of nodal diameter (ND) and nodal circle (NC) in the vibration mode of the disks should be introduced. ND and NC refer to the straight or circumferential line where the displacement of the disc structure remains at zero during vibration. In fact, ND and NC modes refer specifically to axial vibration. However, there are also circumferential and radial vibrations in the modes of the disk-like structure, as shown in Figure 4.

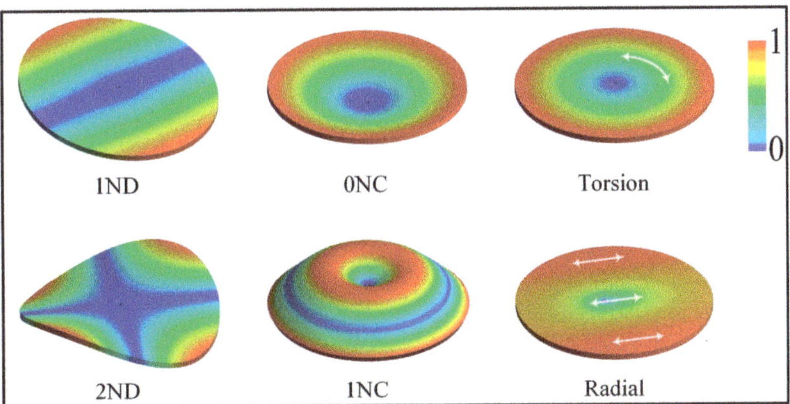

Figure 4. Schematic of vibration modes of a disk.

The geometry of the runner is much more complicated than that of the disk. Therefore, there are some special vibration patterns in ND and NC modes, such as crown dominant (CD), band dominant (BD), and counter phase (CP) [15]. In addition, the modes of the runner also include the vibration modes dominated by blades.

3. Results and Discussion

3.1. Modes under Different Constraints

The fixed constraints were set on the shaft-connecting surface, bolt centerline, and the top of the short shaft, respectively (as shown in Figure 1), and then the modes of the runner in channels were calculated by the acoustic–structure coupling method. The modal shapes and the natural frequencies under different constraints were compared.

The first 10 modes of the runner under surface constraint are shown in Figure 5. Among them, the first five modes are typical ND and NC modes; the sixth is the torsional mode; the seventh is the radial vibration mode; the eighth and tenth order modes are dominated by blade vibration; and the ninth is an 0NC-CP mode. The deformation of the 0NC-CP mode was concentrated at the position where the inlet edge was far away from the blade due to restriction of the blade to the upper crown and lower band.

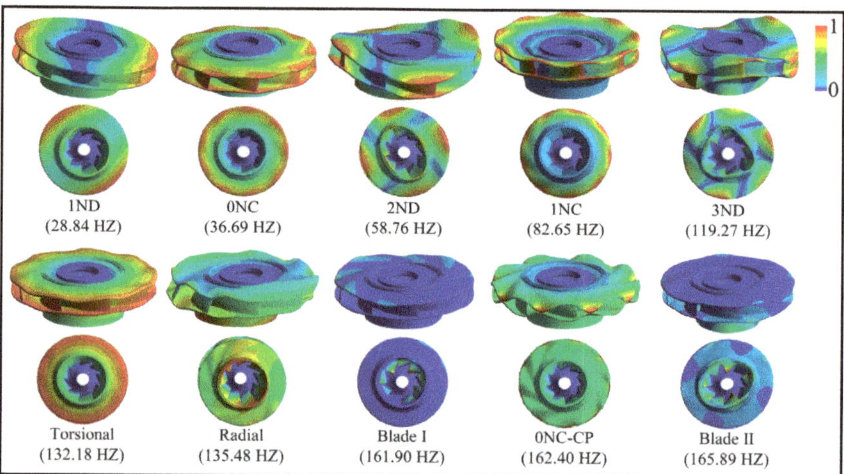

Figure 5. The first 10 modes of the runner with surface constraint.

Comparing the modes under different constraint conditions, it was found that the change of the modal shapes was generally small. The shapes of the 1ND mode, the 0NC mode, the Torsional mode, and the Radial mode changed obviously, as shown in Table 2. For the radial mode with the most obvious change, the maximum vibration point moved upward and even showed a certain axial swing when the shaft was considered.

Table 2. Modal shapes under different constraints.

	1ND	0NC	Torsional	Radial
Surface constraint				
line constraint				
shaft constraint				

To analyze the influence of the constraint condition on the natural frequencies, the frequency f_S under surface constraint was defined as the reference value. Then, the frequency ratios f_L/f_S and f_{Sh}/f_S were obtained by dividing the natural frequencies f_L (under line constraint) and f_{Sh} (under shaft constraint) with the corresponding reference frequencies. The frequency ratio of each mode is shown in Figure 6. It can be seen that the influence of the constraints on the frequencies of different modes was quite different. Among them, the

Torsional and the Radial modes were the most variable modes, especially affected by the rotating shaft. For example, the frequency of the Torsional mode was reduced to $0.851\,f_S$ under the line constraint, and the frequency could be further reduced to $0.307\,f_S$ when the shaft was considered. For the ND modes, the 1ND mode was most affected by the constraints. Additionally, with the increase in the order, the influence decreased rapidly. A similar rule also appeared in the NC modes, but it changed slowly. For the modes dominated by blade vibration, the change of the frequency with the constraint condition could be ignored.

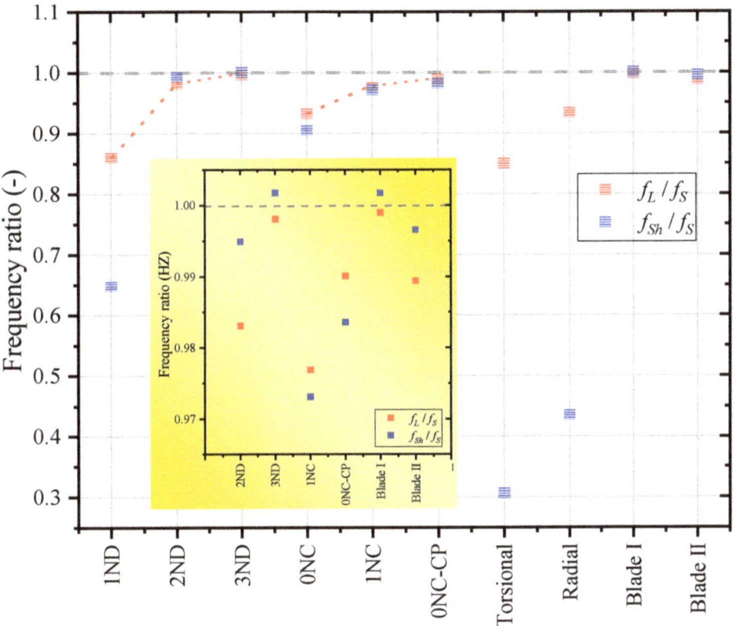

Figure 6. Influence of the constraint condition on the natural frequencies.

If only the maximum difference of natural frequencies under different constraints is concerned, it can be described by $\Delta m = (f_{max} - f_{min})\,f_S$. f_{max} and f_{min} are the maximum and minimum natural frequencies of a mode under different constraints. The Δm of each mode is shown in Table 3. It can be seen that the Torsional mode, the Radial mode, the 1ND mode, and the 0NC mode varied most significantly with the constraint condition. The maximum differences in the natural frequencies of these four modes under different constraints were 69.3%, 56.4%, 35.1%, and 9.4%, respectively. When these modes are involved in practical engineering problems, it is necessary to carefully analyze whether the setting of constraint condition is reasonable, especially considering the influence of the shaft or even the rotor system. For example, if the constraint condition is not accurate, it is likely to misjudge the resonance characteristics of the runner when the 1ND exciting force appears in the unit.

Table 3. Maximum differences of natural frequencies under different constraints.

Modal	1ND	2ND	3ND	0NC	1NC
Δm	0.351	0.017	0.004	0.094	0.027
Modal	0NC-CP	Torsional	Radial	Blade I	Blade II
Δm	0.016	0.693	0.564	0.003	0.011

Theoretically, the above results are reasonable. When the surface constraint was adopted, the local stiffness was the largest, and the natural frequencies of the runner were the highest. When the line constraint or shaft constraint was used instead, the local stiffness was reduced, and the natural frequencies were reduced. It can be seen from the modal shapes that the stiffness here strongly restricted the movement of the Torsion mode, the Radial mode, the 1ND mode, and the ONC mode, and the restriction on the movement of other modes was weak. Therefore, the change of the natural frequencies presented such a rule.

3.2. Modes Considering Energy Loss on the Wall

The rigid wall boundary on the runner chamber was changed to the absorption boundary, and the absorption coefficient α was set to 0.2, 0.4, and 0.6, respectively. Firstly, the results showed that the absorption coefficient had no impact on the shapes of most modes except the 1NC mode. When α increased to 0.2, the position of the nodal circle of the 1NC mode obviously moved outward, and the position of maximum displacement shifted from the outer edge to the center; when α increased to 0.4, the 1NC mode disappeared. The change of the modal shape was shown in Figure 7.

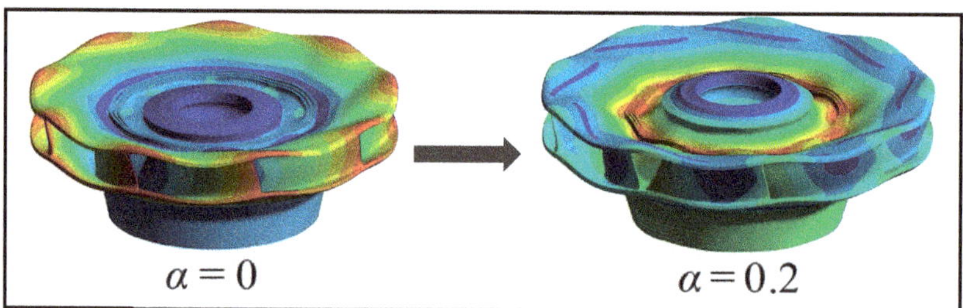

Figure 7. Variation of the 1NC mode with absorption coefficients.

Figure 8 shows the variation of the natural frequency of each mode with the absorption coefficient. Here, the frequency ratio f_α/f_0 was used for demonstration. f_α is the natural frequency under different absorption coefficients, and f_0 is the frequency when the absorption coefficient is 0.

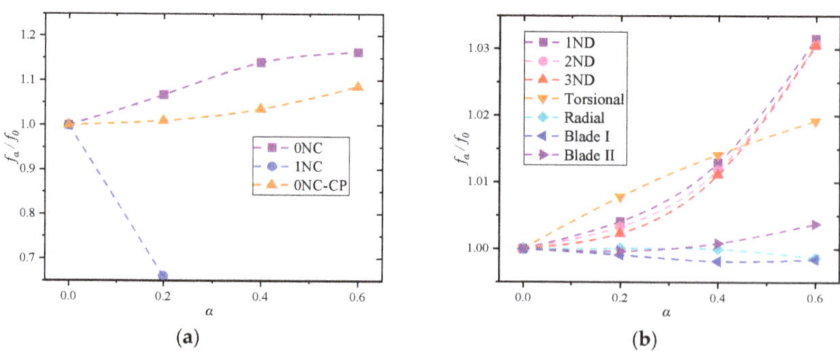

Figure 8. Variation of the natural frequencies with absorption boundary: (**a**) NC modes; (**b**) other modes.

The results showed that, with the increase in the absorption coefficient on the runner chamber, the natural frequencies of the runner increased except for a few modes. It is well known that the proximity of the outer wall increases the added mass of water, thereby reducing the natural frequencies of the underwater structure. When the outer wall was not rigid, it vibrated along with the structure–water system and absorbed some of the energy in the pressure wave. At this time, the influence of the wall on the added mass was weakened, so the natural frequencies of the structure increased. Among them, the frequency change of the NC modes was the most significant, followed by the ND modes and the Torsional mode. The variation of the Radial mode and the blade vibration modes could be ignored. Compared with other modes, the frequencies of the NC modes changed too much; thus, they are shown separately in Figure 8a. For the NC modes, the change rule of the frequency of each mode was different. In particular, the frequency of the 1NC mode was abnormal. When α increased to 0.2, the frequency of the 1NC mode decreased by approximately 35%. Meanwhile, the frequencies of 0NC and 0NC-CP modes increased by 7% and 1%, respectively. For the ND modes, the change of the frequency of each mode with the absorption coefficient was almost the same. Additionally, with the increase in α, the frequency gradient of the ND modes increased.

In fact, the proximity of the wall has very different effects on the added mass of different modes. For example, during the movement of the nodal circle modes, the runner moves upward or downward as a whole to squeeze the water in the side chamber. At this time, the wall has a strong reaction force against the runner–water system, which causes a great added mass. For the blade vibration modes, the Radial mode, etc., the influence of the wall on the added mass is very small. The above results can be found in the study of Li et al. [21]. In the same way, when the absorption coefficient on the wall increased, the change in the added mass of the nodal circle modes was the most obvious, and the change in the blade vibration modes and the Radial mode was small. As for the abnormal change of the 1NC modal frequency, we think that it was caused by the change of the modal shape.

In general, the energy loss on the runner chamber had little impact on the runner's modes. When α was increased from 0 to 0.2, the frequency variation of each mode was within 0.5%, except for the NC modes. Therefore, in the numerical modal analysis of the runner, only the influence of the energy loss on the NC modes needs to be considered.

4. Conclusions

In this paper, the influence of the constraint condition and the energy loss on the natural modes of hydraulic machinery runner was analyzed. The conclusions are as follows.

The constraint condition determines the local stiffness of the structure; therefore, it has a significant impact on the modes of the runner. The influence of constraint condition on the frequencies of the different modes is quite different, and the degree of the effect is closely related to the modal shapes. Among them, the Torsional mode, the Radial mode, the 1ND mode, and the 0NC mode varied most significantly with the constraint condition. The maximum differences in the natural frequencies of these four modes under different constraints were 69.3%, 56.4%, 35.1%, and 9.4%, respectively. When the above modes are involved in practical engineering problems, the rationality of the setting of the constraint condition must be carefully analyzed. In particular, the correct modelling of the shaft region is essential for the correct determination of natural frequencies and avoiding exciting frequencies.

When the chamber wall was not rigid, it vibrated along with the structure–water system and absorbed some of the energy in the pressure wave. At this time, the influence of the wall on the added mass was weakened, so the natural frequencies of the structure increased. The frequency variations of the Radial mode and the blade vibration modes can be ignored, and the variations of the ND modes and the Torsional mode are also small, but the variations of the NC modes are significant. Except for several specific modes, the modal shapes of the runner were almost unchanged. The results showed that it is necessary to pay attention to the influence of the energy loss on the chamber on the NC modes in the numerical modal analysis of the runner.

Author Contributions: Conceptualization, Z.W.; methodology, Z.W.; software, X.X.; validation, D.L. and J.Y.; formal analysis, D.L.; investigation, J.Y.; resources, Z.W.; writing—original draft preparation, X.X.; writing—review and editing, D.L.; funding acquisition, Z.W.; All authors have read and agreed to the published version of the manuscript.

Funding: This research was funded by the National Natural Science Foundation of China, grant number 51876099.

Institutional Review Board Statement: Not applicable.

Informed Consent Statement: Not applicable.

Data Availability Statement: Not applicable.

Conflicts of Interest: The authors declare no conflict of interest. The funders had no role in the design of the study; in the collection, analyses, or interpretation of data; in the writing of the manuscript, or in the decision to publish the results.

References

1. Liang, Q.W.; Wang, Z.W.; Fang, Y. Modal analysis of Francis turbine with considering FSI. *J. Hydroelectr. Eng.* **2004**, *23*, 116–120. (In Chinese)
2. Rodriguez, C.G.; Egusquiza, E.; Escaler, X.; Liang, Q.; Avellan, F. Experimental investigation of added mass effects on a Francis turbine runner in still water. *J. Fluids Struct.* **2006**, *22*, 699–712. [CrossRef]
3. Han, Q.L.; Xing, W.T.; Li, W.H.; Zhang, Z.; Guo, S.S. Experimental modal analysis of blades in different media. *Acta Energ. Sol. Sin.* **2019**, *40*, 285–290. (In Chinese)
4. Presas, A.; Valero, C.; Huang, X.X.; Egusquiza, E.; Farhat, M.; Avellan, F. Analysis of the dynamic response of pump-turbine runners—Part I: Experiment. In Proceedings of the IOP Conference Series: Earth and Environmental Science, Beijing, China, 19–23 August 2012; IOP Publishing: Bristol, UK, 2012; Volume 15.
5. Østby, P.T.K.; Sivertsen, K.; Billdal, J.T.; Haugen, B. Experimental investigation on the effect off near walls on the eigen frequency of a low specific speed francis runner. *Mech. Syst. Signal Process.* **2019**, *118*, 757–766. [CrossRef]
6. Egusquiza, E.; Valero, C.; Liang, Q.W.; Coussirat, M.; Seidel, U. Fluid added mass effect in the modal response of a pump-turbine impeller. In Proceedings of the 2009 International Conference on Mechatronic and Embedded Systems and Applications, San Diego, CA, USA, 30 August–2 September 2009; Volume 48982, pp. 715–724.
7. Hübner, B.; Seidel, U.; Roth, S. Application of fluid-structure coupling to predict the dynamic behavior of turbine components. In Proceedings of the 25th IAHR Symposium on Hydraulic Machinery and Systems, Timisoara, Romania, 20–24 September 2010.
8. Liu, X.; Luo, Y.Y.; Karney, B.W.; Wang, Z.; Zhai, L. Virtual testing for modal and damping ratio identification of submerged structures using the PolyMAX algorithm with two-way fluid-structure Interactions. *J. Fluids Struct.* **2015**, *54*, 548–565. [CrossRef]
9. Liang, Q.W.; Rodriguez, C.G.; Egusquiza, E.; Escaler, X.; Farhat, M.; Avellan, F. Numerical simulation of fluid added mass effect on a francis turbine runner. *Comput. Fluids* **2007**, *36*, 1106–1118. [CrossRef]
10. Egusquiza, E.; Valero, C.; Huang, X.X.; ou, E.; Guardo, A.; Rodriguez, C. Failure investigation of a large pump-turbine runner. *Eng. Fail. Anal.* **2012**, *23*, 27–34. [CrossRef]
11. Huang, X.X.; Escaler, X. Added mass effects on a Francis turbine runner with attached blade cavitation. *Fluids* **2019**, *4*, 107. [CrossRef]
12. Huang, X.X.; Oram, C.; Sick, M. Static and dynamic stress analyses of the prototype high head Francis runner based on site measurement. In Proceedings of the IOP Conference Series: Earth and Environmental Science, Montreal, QC, Canada, 22–26 September 2014; IOP Publishing: Bristol, UK, 2014; Volume 22, p. 032052.
13. Rodriguez, C.G.; Flores, P.; Pierart, F.G.; Contzen, L.R.; Egusquiza, E. Capability of structural-acoustical FSI numerical model to predict natural frequencies of submerged structures with nearby rigid surfaces. *Comput. Fluids* **2012**, *64*, 117–126. [CrossRef]
14. Askari, E.; Jeong, K.; Amabili, M. Hydroelastic vibration of circular plates immersed in a liquid-filled container with free surface. *J. Sound Vib.* **2013**, *332*, 3064–3085. [CrossRef]
15. He, L.Y.; Zhou, L.J.; Ahn, S.H.; Wang, Z.; Nakahara, Y.; Kurosawa, S. Evaluation of gap influence on the dynamic response behavior of pump-turbine runner. *Eng. Comput.* **2019**, *36*, 491–508. [CrossRef]
16. Valentín, D.; Ramos, D.; Bossio, M.; Presas, A.; Egusquiza, E.; Valero, C. Influence of the boundary conditions on the natural frequencies of a Francis turbine. In Proceedings of the IOP Conference Series: Earth and Environmental Science, Grenoble, France, 4–8 July 2016; IOP Publishing: Bristol, UK, 2016; Volume 49, p. 072004.
17. Egusquiza, E.; Valero, C.; Presas, A.; Huang, X.; Guardo, A.; Seidel, U. Analysis of the dynamic response of pump-turbine impellers. Influence of the rotor. *Mech. Syst. Signal Process.* **2016**, *68*, 330–341. [CrossRef]
18. Woyjak, D.B. *Acoustic and Fluid Structure Interaction, A Revision 5.0 Tutorial*; Swanson Analysis Systems: Houston, TX, USA, 1992.
19. Rajakumar, C.; Ali, A. Acoustic boundary element eigenproblem with sound absorption and its solution using lanczos algorithm. *Int. J. Numer. Methods Eng.* **1993**, *36*, 3957–3972. [CrossRef]
20. ANSYS Inc. *Mechanical APDL Theory Guide*; ANSYS: Canonsburg, PA, USA, 2017.
21. Li, D.K.; Xia, X.; Zhou, L.J.; Gong, K.; Wang, Z. Effect of outer edge modification on dynamic characteristics of pump turbine runner. *Eng. Fail. Anal.* **2021**, *124*, 105379. [CrossRef]

Journal of
*Marine Science
and Engineering*

MDPI

Article

Effect of Roughness of Mussels on Cylinder Forces from a Realistic Shape Modelling

Antoine Marty [1], Franck Schoefs [2,*], Thomas Soulard [3], Christian Berhault [4], Jean-Valery Facq [1], Benoît Gaurier [1] and Gregory Germain [1]

[1] Ifremer, Marine Structure Laboratory, 150 Quai Gambetta, 62200 Boulogne sur Mer, France; antoine.marty@ifremer.fr (A.M.); jvfacq@ifremer.fr (J.-V.F.); bgaurier@ifremer.fr (B.G.); ggermain@ifremer.fr (G.G.)

[2] Research Institute of Civil Engineering and Mechanics, Sea and Litoral Research Institute, Ecole Centrale de Nantes, Université de Nantes, CNRS, 44322 Nantes, France

[3] Laboratoire de Recherche en Hydrodynamique, Énergétique et Environnement Atmosphérique, Sea and Littoral Research Institute, Ecole Centrale de Nantes, CNRS, 44322 Nantes, France; Thomas.Soulard@ec-nantes.fr

[4] Scientific Expert, Self-Employed, 83140 Six-Fours-les-Plages, France; christian.berhault22@gmail.com

* Correspondence: franck.schoefs@univ-nantes.fr

Abstract: After a few weeks, underwater components of offshore structures are colonized by marine species and after few years this marine growth can be significant. It has been shown that it affects the hydrodynamic loading of cylinder components such as legs and braces for jackets, risers and mooring lines for floating units. Over a decade, the development of Floating Offshore Wind Turbines highlighted specific effects due to the smaller size of their components. The effect of the roughness of hard marine growth on cylinders with smaller diameter increased and the shape should be representative of a real pattern. This paper first describes the two realistic shapes of a mature colonization by mussels and then presents the tests of these roughnesses in a hydrodynamic tank where three conditions are analyzed: current, wave and current with wave. Results are compared to the literature with a similar roughness and other shapes. The results highlight the fact that, for these realistic roughnesses, the behavior of the rough cylinders is mainly governed by the flow and not by their motions.

Keywords: marine growth; hydrodynamic loading; roughness; mussels; morison coefficients

Citation: Marty, A.; Schoefs, F.; Soulard, T.; Berhault, C.; Facq, J.-V.; Gaurier, B.; Germain, G. Effect of Roughness of Mussels on Cylinder Forces from a Realistic Shape Modelling. *J. Mar. Sci. Eng.* **2021**, *9*, 598. https://doi.org/10.3390/jmse9060598

Academic Editor: Apostolos Papanikolaou

Received: 15 April 2021
Accepted: 12 May 2021
Published: 31 May 2021

Publisher's Note: MDPI stays neutral with regard to jurisdictional claims in published maps and institutional affiliations.

1. Introduction

Since 2010, Floating Offshore Wind Turbines (FOWT) have been shown to be very promising for producing offshore wind energy in deeper water (>60 m) while reducing the need for spatial area nearshore where the sharing of space with other activities creates conflicts. Few prototypes and pilot farms have proven the maturity of floating concepts for wind turbines. It is now facing the reduction of cost that relies on the optimization of design, the development of new installation processes and new technologies for inspection and maintenance [1]. The need for new components in comparison with Oil and Gas floating platforms has also been shown. The size of the floater is smaller and that is the case for its underwater components: mooring lines and power cables. The order of magnitude of their diameter is 0.3–0.5 m [2,3]: these small diameters in comparison with components of Jackets offshore platforms lead to an increase of the relative roughness (i.e., ratio between the roughness and the smooth diameter) in comparison with previous tests carried out by the oil and gas industry. By increasing the role of the roughness, the effect of its shape should be reinvestigated.

Mooring lines and power cables are recognized to be the most critical components for which the feedback from the Oil and Gas industry cannot be transferred immediately [2,3]. Even if the Oil and Gas sector invested in research and development for mooring line design, it experienced unexpected failures: according to Ma et al.'s review on failures of permanent mooring systems between 2001 and 2011 [4], the annual probability of failure was estimated to be around 3×10^{-3} over an average sample of 300 permanent mooring systems from oil and gas industry. This assessment steered operators towards a strengthening of safety factors, which can involve solutions such as mooring lines redundancy or thicker mooring lines. However, reducing mooring system CAPEX leads to avoid redundancy, to lighten mooring lines components, to shorten their length and to use high-performance nonstandard materials such as synthetic ropes. The nascent floating offshore wind industry then faces a challenge: reducing mooring system CAPEX without increasing the risk of high consequences in case of failure. According to Fontaine et al.'s review of "past failures, pre-emptive replacements and reported degradations" (Figure 7 in [5]), over 74 analyzed failures, it became clear that fatigue is one of the main issues. Based on the same observations, the JIP led by Carbon Trust identified four major innovation needs for mooring systems ([6], p.48), among them the "Understanding of fatigue mechanisms in floating wind mooring systems". According to Braithwaite and McEvoy, offshore fish farms experienced failures due to the presence of biofouling [7]. The loading of these underwater beam components is usually modeled through the Morison quasi-static equation [8] where drag and inertia coefficients comprise as much as possible the complex hydrodynamic interactions between water and the cable. Macro-fouling, called marine growth in the following, has been shown to change drastically the value of these coefficients and thus the loading [9–11] and the structural reliability [12–14]. Three effects have been shown to drive the loading changes [10]: the change of the diameter, of the mass and of the roughness by both changing the quasi-static and the dynamic loading [15].

In 1990, Sarpkaya summarized more than 20 years of research on the effect of roughness [16]. It was shown that this changes both the bounds of hydrodynamic regimes (from laminar to turbulent) and the level of the loading. Ameryoun [17] simulated the effect of the growth of mussel's roughness through a flowchart of the load computation from the response surface model [18] and concluded it may lead to an increase of 50% of the drag force in a single year. Usually, the experimental hydrodynamic test over-simplifies the shape of the marine growth: it is usually modeled with sand or gravels. When the shape is more realistic by using a natural colonization by barnacles, anemones or seaweeds [18], the shape is not fully described as well as the surface density of specimens. It is usually summarized in a single value: relative roughness computed as the ratio between the roughness (surface to peak distance) and the smooth diameter of the component. This simplification of the real geometry explains part of the discrepancies of tests in basin reported in the standards [19,20]. Furthermore, there are only few published reports about on-site marine growth assessment from inspections and they usually register only a mean thickness and the type of species of a multi-layer marine growth [21–24]. It is thus not possible to depict a representative roughness. Recently, underwater image processing was improved [25–28] and the first quantitative data were extracted: among them, the roughness of mussels, a dominant species in Atlantic and North-Sea area [3,29]. This paper takes advantage of these data to provide a realistic reproduction of marine growth in terms of geometry and density.

With a view to simplifying further bench-marking studies, a dedicated test campaign is carried out with two homogeneous realistic shape roughnesses. The main objective of this study is to analyze how the taking into account of the real hard roughness geometry influences the loading estimation. This paper is split into four sections. Section 2 introduces the model of the two tested geometries, the experimental setup and the selected hydrodynamic conditions are chosen to be as representative as possible to those encountered by underwater mooring lines (submitted to wave and current effects). Section 3 gives the results in terms of drag and inertia Morison coefficients. Three types of conditions (current only, wave only, current plus wave) are imposed to the three studied geometries (1 smooth and 2 rough) and the results are compared with those of the literature in order to underline the benefit to take into account the real geometry of roughness for hydrodynamic load estimation. Section 4 summarizes the paper.

2. Hard Marine Growth Reproduction and Experimental Setup

The objective of the tests is to represent as close as possible the hydrodynamic conditions and bio-colonization by mussels encountered by underwater components of floating offshore wind turbines. In this section, we first propose a realistic reproduction of marine growth in terms of geometry and density. The experimental setup is presented a second time.

2.1. Realistic Shape of Colonization by Mussels

This paper focuses on the full coverage by adult mussels encountered on Atlantic and North Sea offshore structures (species Mytilus Edulis). Recently, underwater inspections and image processing were carried out on two test sites. A day and a depth were selected, with a view to get the best conditions according to [30], and three pictures on three separated parts of a chain were taken with the aksi3D® (Figure 1c) developed during the ULTIR project [25]. In these best conditions (luminosity, turbidity, distance to the target), the accuracy reaches 0.7 cm. This chain is the main anchoring material for the buoy equipment of the SEMREV site, operated by Ecole Centrale de Nantes, where adult mussels were observed. The same type of pictures was obtained 10 km away, on the test platform UN@SEA ee (called UN-SEA-SMS previously) [3] of Université de Nantes, two years after its installation in June 2017. The organization of each specimen and the roughness were measured. Figure 1 illustrates the organization of the specimen. On Figure 1a, the red frame represents a pattern of size 20 cm × 20 cm that was shown to be representative of an elementary representative organization of the specimens on the covered surface. Figure 1b represents the top view of mussels by an elliptical shape whose major axis inclination with respect to x axis is reported in Table 1. Note that for simplifying the presentation, mussels are aligned horizontally and vertically in Figure 1b that is not the case due to the important difference between the size of vertical and horizontal axes (see position X and Y in Table 1). For simplifying future modeling and bench-marking, the major axis is approximated by 8 values: $0°$, $+/-30°$, $+/-45°$, $+/-60°$ and $90°$. Similar absolute value of the inclination is plotted with the same color. Table 1 gives the position of the centers and the inclination for each of the 16 specimens in the patch. It is shown that the organization is not totally random and that similar angles are observed: it comes from the fact that an optimal organization of mussels should optimize the access to food, that is, phytoplankton obtained by filtering the sea water.

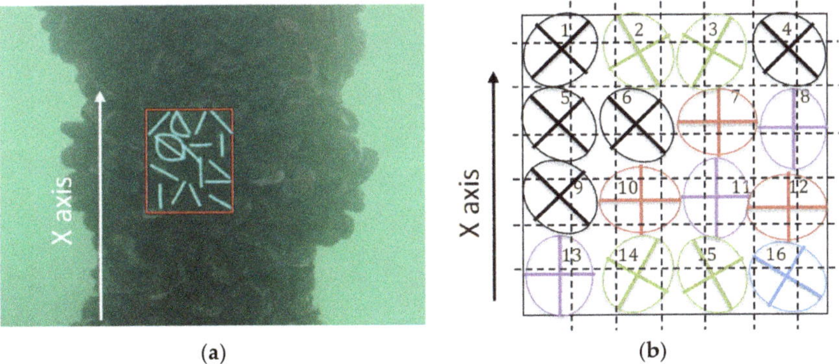

Figure 1. *Cont.*

(c)

Figure 1. (**a**) Typical underwater picture with the organization of the specimen; (**b**) corresponding elliptical shape with major and minor axis: purple (0°), green (+/− 30°), black (+/− 45°), blue (+/− 60°) and red (90°); (**c**) using the aksi3D® system (tested at IFREMER).

Table 1. Position of the center and inclination of each specimen.

N° Position/Angle	1	2	3	4	5	6	7	8	9	10	11	12	13	14	15	16
X	8	25	41	58	8	24	43	59	8	26	42	58	8	25	41	58
Y	59	58	58	59	42	42	42	41	26	25	26	23	9	8	9	8
Inclination of major axis/axis x	+45°	−30°	+30°	−45°	+45°	+45°	+90°	0°	+45°	+90°	0°	+90°	0°	−30°	+30°	+60°

Roughness was also measured and modeled according to the protocol described in [30]. Figure 2a shows the definition of a roughness k and typical numbers. Note that the ratio l/k varies between 1 and 1.15 for adult mussels. In this study, the value $l/k = 1.1$ is used. The relative roughness e is defined as the ratio k/D_e, where k is the dimension of the studied roughness and D_e the equivalent diameter. In the literature, several definitions of the roughness exist [10,31]. Decurey et al. [3] give a definition of D_e in line with on-site measurements. In Ameryoun et al. [17], they used a stochastic modeling of marine growth and hydrodynamic parameters to define the roughness as the ratio of the apparent height of the surface roughness (mussel length from the wider section to the external extremity, k) on the equivalent diameter of the studied configuration. Indeed, a mussel cover may be composed of several highly compact superimposed layers. As such, layers below the external one represent a thickness of closed surfaces where no fluid dynamics is permitted, with no entrapped water volume. This closed volume corresponds therefore to the difference between the whole thickness (from the internal diameter to the extremity, th) and the surface roughness (k). Figure 3 represents the different parameters for the calculation of the equivalent diameter.

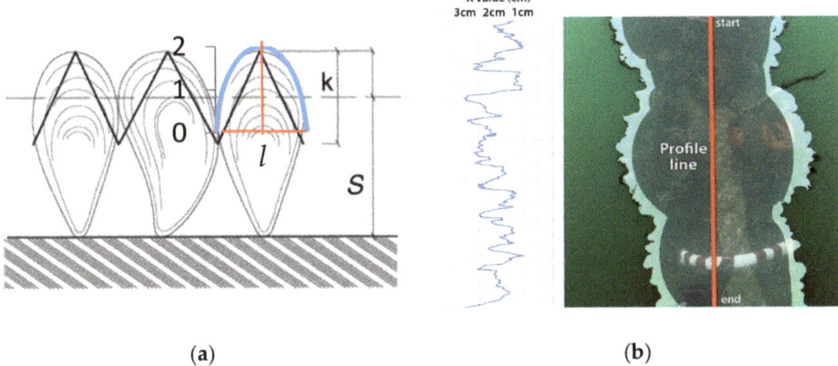

(**a**)   (**b**)

Figure 2. (**a**) Definition of the roughness for a given size of the shell S (numbers in cm); (**b**) typical extraction of the roughness from image processing.

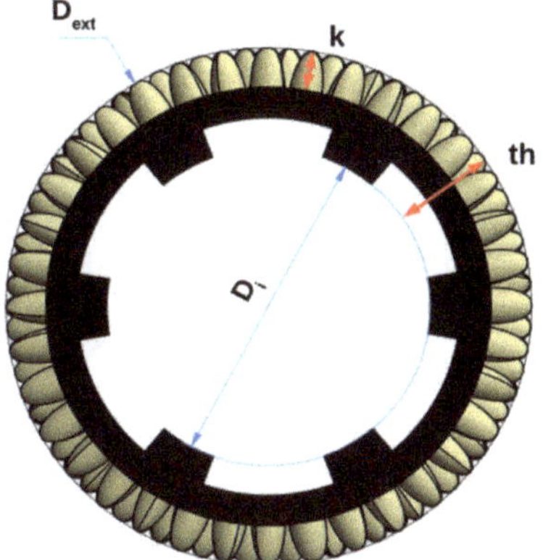

Figure 3. Definition of roughness parameters adapted to the experimental set-up.

Consequently, this thickness is assumed to be a diameter increase in a fluid dynamic point of view and thus the equivalent diameter is calculated as follow.

$$D_e = D_i + 2\,(th - k). \tag{1}$$

Then, the relative roughness e is defined only from the external layer, over a cylinder of equivalent diameter D_e. Applying the same principle on the external layer, the part below the wider section of the mussel is considered closed. Consequently, only the mussel height upon the wider section is considered to define the roughness k, representing the surface irregularities impacting the flow boundary layer. Several lines were inspected and Figure 4 provides the distribution of the roughness that were measured between 1.5 and 3 cm [30].

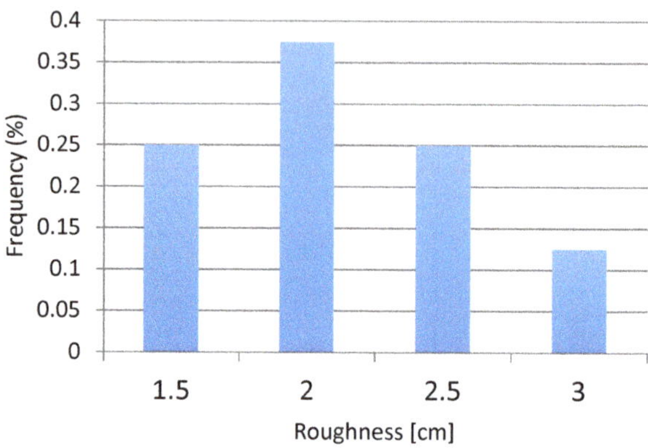

Figure 4. Distribution of roughness extracted from image processing.

The objectives of this paper are:

− to analyze the effect of a realistic roughness on the loading and to compare with other tests in the literature,
− to highlight whether the realistic size of mussels significantly impacts the loading.

For these reasons, two sizes are selected for the roughness: small size mussels (mussel roughness of 20 mm) and larger individual mussels (mussel roughness of 30 mm). This is to notice that when mussels cluster around the rope, they do not fill all the space and create interstices full of water. In the following, we consider that only the last level of mussels creates the surface roughness and the other levels create a close volume due to the high concentration of mussels. To this end and by means of 3D printing, two types of mussels shape and distributions have been considered. The first one is called *C1* with an outside diameter (Dext) equal to 260 mm with small size mussels (roughness of 20 mm) and the second one called *C2* with outside diameter equal to 280 mm composed of larger individual mussels (roughness of 30 mm). According to Figures 1 and 2, the design of the roughness due to mussel can be modelled as a semi-ellipsoid with the minor and major axes and its height. Both configurations have an ellipse base of 16 × 18 mm and 20 mm tall for *C1* and 24 × 27 mm and 30 mm tall for *C2*. Precise dimensions are given in the Figure 5.

For each mussel's shape, the distribution around the cylinder follows the same pattern. Mussels are arranged depending on their angle between the major axis of the ellipse and the cylinder axis according to Figure 1b in such a way as to generate a stochastic distribution network as shown on the Figure 5 Right. The eight angles pattern is repeated all along the circumference of the cylinder and then reproduced along the cylinder axis with a staggered positioning, represented with the arrows on the drawing. Note that, for printing reasons, the four specimens horizontally aligned on Figure 1b were arrayed in checkerboard; that agrees also the real organization of mussels for which there is an important difference between the size of vertical and horizontal axes. The experimental set-up is based on a smooth cylinder (called *S*) of diameter $D = 160$ mm, on which the roughness is superimposed in order to design a configuration with roughness (see Figure 6). The three cylinders' arrangement characteristics are summarized in Table 2.

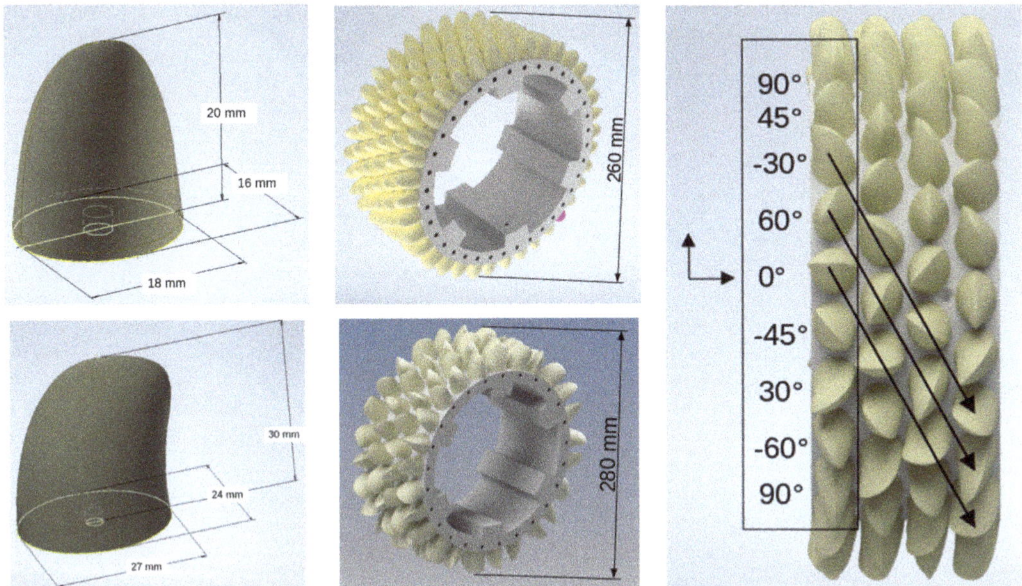

Figure 5. Mussels' roughness shape for C1 on top and C2 at the bottom. On the right, mussels distribution around the cylinder with the C2 shape.

Figure 6. On the left, from the top to the bottom, cases *S*, *C1* and *C2*. On the right, *C2* roughness mounted on the cylinder.

Table 2. Synthesis of the studied roughness parameters for all configurations.

Configurations	D_i [mm]	D_{ext} [mm]	k [mm]	th [mm]	D_e [mm]	$e = k/D_e$	Mass System [daN]	Areal Density for nb. Specimens/m^2
S	160	160	0	0	160	0	47	-
C1	160	260	20	50	220	0.091	105	2969
C2	160	280	30	60	220	0.136	110	1374.5

2.2. Realistic Hydrodynamic Configurations

Reynolds R_e and Keulegan-Carpenter KC numbers have been shown to drive the evolution of drag forces and inertia coefficient of Morison equations with the water particle velocity. Their definition in the presence of marine growth [11] is presented in Equations (2) and (3):

$$R_e = \frac{U\, D_e}{v} = \frac{A_x \omega\, D_e}{v} \tag{2}$$

with v the kinematic velocity, U, the flow velocity or the oscillation speed $A_x\omega$.

$$KC = 2\pi \frac{A_x}{D_e} \tag{3}$$

The reduced speed is defined as:

$$U_r = \frac{U}{f\, D_e} \tag{4}$$

with $f = \frac{\omega}{2\pi}$.

The objective is to cover common hydrodynamic conditions with $4.10^4 < R_e < 3.10^5$ and $4 < KC < 12$. The range of KC allows to detect the strong non-linearities of drag and inertia forces with particle velocity.

According to the potential of the equipment (see Section 3), values in Table 3 are reached for each configuration.

Table 3. Synthesis of the normalized numbers covered for all configurations.

Configurations	KC	U_r	Re/10^5
S	3.9–15.7	4.1–39.1	0.4–2.7
C1	2.5–11.4	3–56.8	0.55–3.8
C2	2.5–11.4	3–56.8	0.55–3.8

3. Experimental Setup

We seek to understand the hydrodynamic behavior of a submarine cable under waves and current conditions. In this way, we performed tests using a fixed cylinder (with or without roughness) under current conditions first, then the superimposition of an oscillating cylinder under these same current conditions with the aim of reproducing the effect of the wave and current interaction. The horizontal oscillating motions of the tested cylinder, which simulate the wave part, are made using a 6-axis hexapod.

3.1. Ifremer Flume Tank, Assembly and Instrumentation

The tests are carried out in the wave and current circulating flume tank of Ifremer located in Boulogne-sur-Mer (France) [32]. The test section is: 18 m long × 4 m wide × 2 m high. In this work, the three instantaneous velocity components are denoted (U; V; W) along the (X; Y; Z) directions respectively (Figure 7). The incoming flow ($\overline{U_\infty}$; $\overline{V_\infty}$; $\overline{W_\infty}$) is assumed to be steady and constant. By means of a grid and a honeycomb (that acts as a flow straightener) placed at the inlet of the working section, a turbulent intensity of I = 1.5% is achieved.

An overview of the global set-up is presented in Figure 7. The cylinder movements are generated using a 6-axis hexapod on which the structure and the instrumentation are fixed. As shown on Figure 7, the cylinder is horizontally freely mounted so that the cylinder is located in the middle of the test section (at one meter depth). The 2 m length cylinder is perpendicular to the direction of the upstream flow. To simulate wave conditions, the hexapode moves with an oscillating and periodic motion in parallel to the flow to represent the horizontal part of the waves' orbital velocity. The hexapode motions along the Ox axis are characterized by its amplitude A_x and its frequency f. The axis coordinate system (x, y, z) is chosen so that the Ox axis is in the same direction as the current. The Oz axis is across the width of the basin and the Oy axis is vertical and oriented upwards, see Figure 7 left.

Two 6 components load cells, with a maximal loads range of $F_{x;y;z}$ = 150 daN, fixed at each extremity of the cylinder, allow the measurement of the forces applied on the cylinder. The location of these load cells is identified by their own axis systems as shown in the Figure 7 (right). The two cylindrical load cells measure the forces applied on the cylinder only; half of the total load for each cell. The noise of the measurement is negligible. The data treatment from Morison equation requires a sinusoidal loading [33]. That explains the presence of residuals.

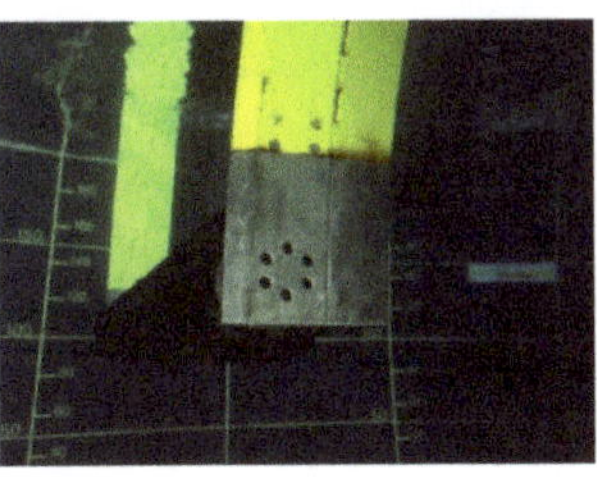
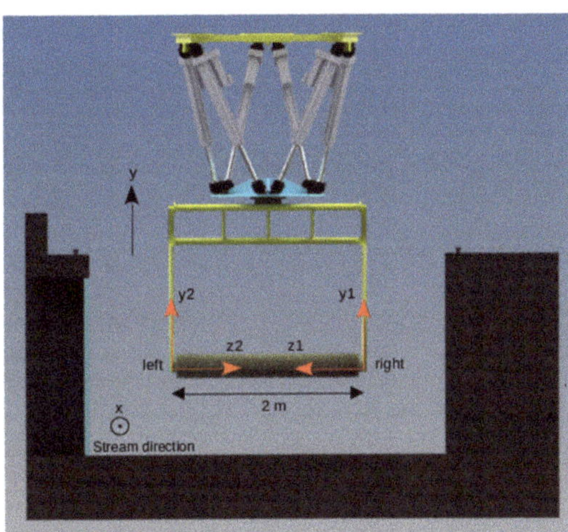

Figure 7. Presentation of the global set-up with the 6-axis hexapod (**left**), the smooth cylinder (**top center**) and one of the rough cylinder (**bottom center**) and axis coordinate system (x, y, z) used in tests (**left**). In black, the main system. The Ox axis is common to all systems and corresponds to the main flow direction. In red, the axes of the load cells (**right** and **left**).

3.2. Post-Processing of Results

The detailed procedure for computing forces and Morison coefficients is available in [34,35]. Inertia and drag coefficients are obtained. The following notations are used:

- C_D for the drag coefficient in steady flow (also written C_{DS} in standards)
- C_d for the drag coefficient in oscillating motion (also written C_D in standards)
- C_m for the inertia coefficient in oscillating motion (also written C_M in standards).

These three parameters are plotted as a function of the dimensionless numbers previously cited (R_e, KC and U_r). All the raw data can be found on the data share platform SEANOE [34,35].

Three tests are considered and their results are commented on in the next three subsections:

− Current only
− Oscillating motion
− Current and oscillating motion

4. Results and Discussion

4.1. Current Only Tests

Figure 8 shows the evolution of the drag forces in the function of the flow velocity. In the studied flow range, there are no drag force differences between the two roughnesses. The curves highlight the classical evolution according to a square power law for the two rough cylinders. This response is however different for the smooth cylinder with a linear evolution until the transition obtained at a flow speed of 1.25 m/s. For $U > 1.25$ m/s, the drag is quite constant around $F_D \approx 200$N.

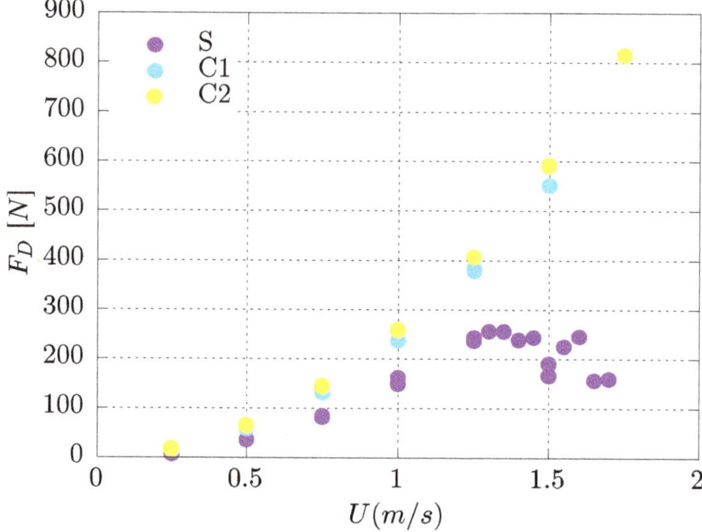

Figure 8. Drag force evolution for the three tested cases.

By comparing the results of smooth and shape C2, an increase of 58, 211 and 413% of the drag force for velocities of 1.25, 1.5 and 1.75, respectively, is observed for an increase of only 38% of diameter D_e. Moreover, by comparing C1 and C2, the small change of the roughness increases the drag force of around 8% for velocities between 0.5 and 1.5 m/s.

Let us now focus on the variation of the overall mean drag coefficient C_D, denoted C_{DS} in some standards, the Strouhal number $S_t = \frac{f_v D_e}{U}$ (with f_v the vortex shedding frequency) and the r.m.s. values of the lift with Reynolds number (Figure 9).

For the smooth case, the overall shape of the $C_D(R_e)$ curve clearly coincides with the results presented in the literature. In the subcritical Reynolds number regime, a nearly constant value for C_D of about 0.9 is found. For increasing Reynolds numbers, hence by approaching the critical flow state or lower transition that starts at $R_e \approx 2.1 \times 10^5$, this value gradually decreases. The minimum value of the drag coefficient of $C_D \approx 0.28$ at $R_e \approx 2 \times 10^5$ marks the transition from the critical Reynolds number regime to the upper transition. This phenomenon is well known [36,37] and confirms the accuracy of the experimental set-up and of the measurements.

For roughness cases (C1 and C2), the transition does not occur in the flow velocity range and the relative roughness (10^{-1}) studied. It was observed for the smallest relative roughness between 5×10^{-4} and 2×10^{-2} [38]. The results show that C_D increases with the size of the roughness, reaching a nearly constant value of about 1.05 for C1 and 1.15 for

C2. Note that API and DNV standards gathered studies from 1971 to 1986 and recommend values of 1.11 for the relative roughness of *C1* ($e = 0.09$). However, standards do not report the results of the PhD of Theophanatos [18] (p. 96), where a discussion about similar values of e is available. In this study, cylinders fully covered by a relative roughness close to *C1* were tested ($e = 0.085$) from a single layer of mussels of size 0.27 mm with a value of C_D of 1.2, close to the value obtained by pyramids and gravels. However, the areal density of the peaks was not given. For *C1* and *C2*, they are the following (Table 2):

– Areal density for *C1* = 2969 specimens/m^2.
– Areal density for *C2* = 1374.5 specimens/m^2.

It was shown that the percentage of cover (another estimate of the areal density) plays a significant role: C_D varies from 1.15 to 1.2 for percentages of cover of 75% and 100%, respectively. The results of the present study suggest 1.05 instead 1.11 (standards) or 1.2 (Theophanatos), which leads to a reduction of respectively 5% and 13% of the drag force.

Moreover, standards indicate no results among 56 experiences reported in the $C_D = f(e)$ curves with a drag coefficient larger than 1.14 for the range of relative roughness 2×10^{-6}–4.5×10^{-2}. The results of this paper show that standards could suggest a value of 1.15 for larger relative roughness up to $e = 0.14$.

It is thus crucial to report not only the relative roughness and the shape, but also the organization and areal density of the species in future studies.

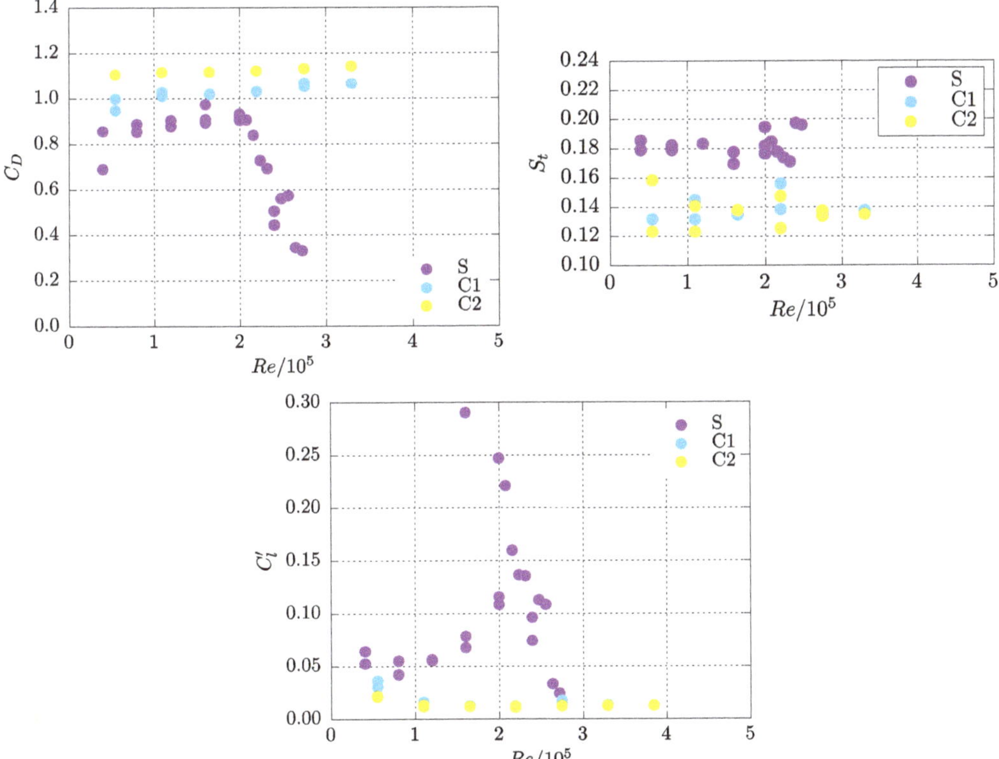

Figure 9. Distribution of the three main hydrodynamic parameters as function of the Reynolds number from direct force measurements for the three test cases: *S*, *C1* and *C2*.

Figure 9 presents the dependency of the Strouhal number on the Reynolds number. A constant value of $St = 0.18$ is observed in the subcritical regime for the smooth cylinder. This value is lower than the Strouhal number commonly used, which is generally equal to 0.21, see [39]. For both rough cylinders, the Strouhal number presents a nearly constant value of about 0.14.

The variation of the r.m.s values of the lift fluctuations with the Reynolds number is also shown in Figure 9. A maximum value of approximately 0.3 is obtained for $R_e \approx 2 \times 10^5$ in the subcritical state. For larger Reynolds numbers inside this flow regime, a steep decrease of the r.m.s. values is observed. For both rough cases, the fluctuations are very low with: $Cl' \ll 0.05$.

These results show that the surface roughness has an influence on the drag coefficient, the r.m.s. values of the lift fluctuations and the Strouhal number. The r.m.s. values are always lower for the rough circular cylinders. A similar trend is observed on the drag coefficient, where a difference of about 10% between cases is observed in this range ($R_e < 2 \times 10^5$). The vortices are shed into the wake with different frequencies. The Fourier transform of the lift forces shows (Figure 10) that the amplitude peaks of the vortex shedding frequencies are much higher for the smooth configuration with values of 25 N for $2 \leq R_e/10^5 \leq 2.5$ when it reaches only 2 N for the two rough configurations.

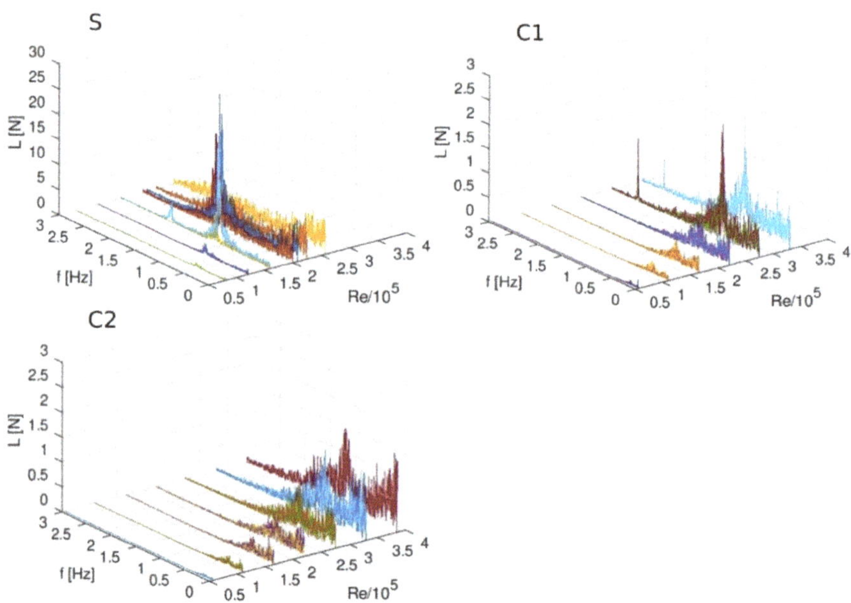

Figure 10. Lift forces Fourier transform as function of the Reynolds number for the three test cases: *S*, *C1* and *C2*.

The peak of the spectrum for *C1* is closer to that of *S* and is easily identified when it is more spread for *C2*. This can be explained by a two times larger areal density of mussels for *C1* in comparison with *C2* (Table 2). Configuration *C1* behaves dynamically as a smooth cylinder when turbulences appear with *C2*.

4.2. Oscillating Motions

For the oscillating motions test cases, the current velocity is equal to zero. Figure 11 presents the evolution of the oscillating drag coefficient C_d (left), denoted by C_D in some standards, and the inertia coefficient C_m, denoted by C_M in some standards, as a function of the Keulegan-Carpenter number *KC*. Several points are plotted per *KC* because several tests have been carried out at the same motion amplitude *Ax* but with different frequencies.

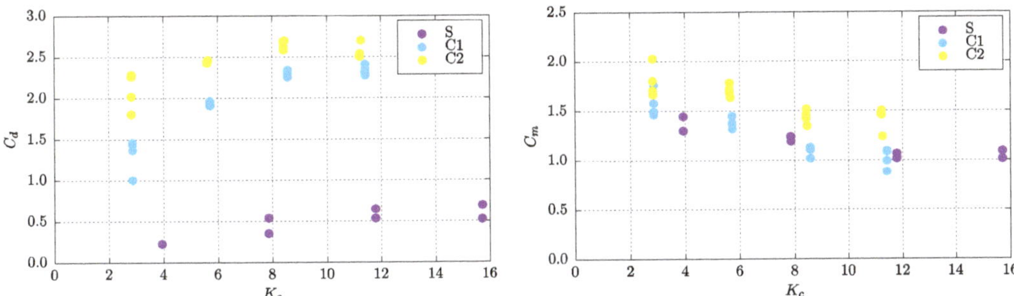

Figure 11. Evolution of C_d (**left**) and C_m (**right**) vs. *KC*.

Figure 11 (right) shows the usual trend of decreasing evolution of C_m with *KC* for *KC* < 15 [16,40,41]. The results show a constant difference between the inertia coefficient of 0.3 between the two rough cylinders with a higher value for the higher relative roughness (*C2*). The C_m of the smooth cylinder is slightly lower than the rough cases. To our knowledge, only the studies of Nath [42] reported values of C_m for *e* = 0.1 with artificial roughness represented by cones. The values reported are significantly higher: 2.8 and 2.5 for *KC* = 5 and 12, respectively, where our test results give 1.4 and 1 for *C1*. However, again, the areal density of peaks is not reported in the paper. A realistic shape for mussels appears to drastically change the inertia coefficient.

Concerning oscillating drag coefficients between smooth and rough cases, results show a significant difference. The calculated coefficients are more than three times higher for cases *C1* and *C2* compared to the smooth case, with $C_d \approx 2.5$ for *KC* > 6 for the rough cases and $C_d \approx 0.5$ for *KC* $\leq$ 16 for the smooth cylinder. The behavior of the rough cylinders is mainly governed by the flow and not by their motions, contrary to the smooth cylinder for which its behavior is mainly governed by its motions. Again, only the studies of Nath were carried out with a relative roughness close to ours (*C1*): they are compared with other studies in [18] (Figure 9.9). Again, the values reported are significantly higher: 3.2 and 2.7 for *KC* = 5 and 12, respectively, with large scatters where our tests give 1.7 and 2.3 for *C1*. The effect of roughness is shown to be significant, especially for low *KC* ($\approx$3) where $C_d \approx 1.3$ for *C1* and 2.1 for *C2*, leading to a 62% increase of drag forces.

4.3. Current and Oscillating Motions

This section presents results concerning current and oscillating motions tested cases. The coefficients introduced in the previous sections are calculated: the mean drag coefficient C_D, the oscillating drag coefficient C_d and the inertia coefficient C_m. These coefficients are at first presented configuration by configuration as a function of U_r in Figure 12. It is first observed that mean and oscillating drag are very close for both roughness cases. The mean drag coefficients are two times higher for the rough cases than for the smooth one. These results confirm that the behavior of the rough cylinders is mainly governed by the flow and not by their motions, contrary to the smooth cylinder for which its behavior is mainly governed by its motions. Inertia coefficients for the rough cases present less dispersion than for the smooth cylinder and show a value for *C2* 25% higher than for *C1* for U_r < 10.

Figure 13 presents each coefficient for the three studied configurations. In order to compare the behavior of each configuration, the current velocity is fixed at 1 m/s. These coefficients are represented as a function of the reduced speed for all the motion amplitudes in order to study the amplitude and the frequency parameters effects at the same time.

The results show several and opposite behaviors of the coefficients. First of all, the inertia coefficient C_m tends to be similar for each configuration. The higher the frequency (small U_r), the lower the coefficient. Moreover, the motion amplitude has no impact on the evolution of the inertia coefficient. Regarding drag coefficients C_d and C_D, their behaviors are totally the opposite. The value of C_d increases with the reduced velocity U_r. Moreover, for a fixed frequency (or U_r fixed) the amplitude parameter has a high impact and the value of the coefficient increases when the amplitude A_m decreases. The exact opposite phenomenon occurs concerning the mean drag coefficient C_D, with the value of the coefficient decreasing when the amplitude Am increases.

Finally, as for the previous case, there is an important difference concerning oscillating drag coefficients and mean drag coefficients between smooth and rough cylinders. The calculated coefficients are much higher for cases *C1* and *C2* compared to the smooth case for which there is no dependency on the motion amplitude and frequency. A strong dependency on the amplitude of the drag coefficients at a fixed frequency for the rough cases is here clearly highlighted.

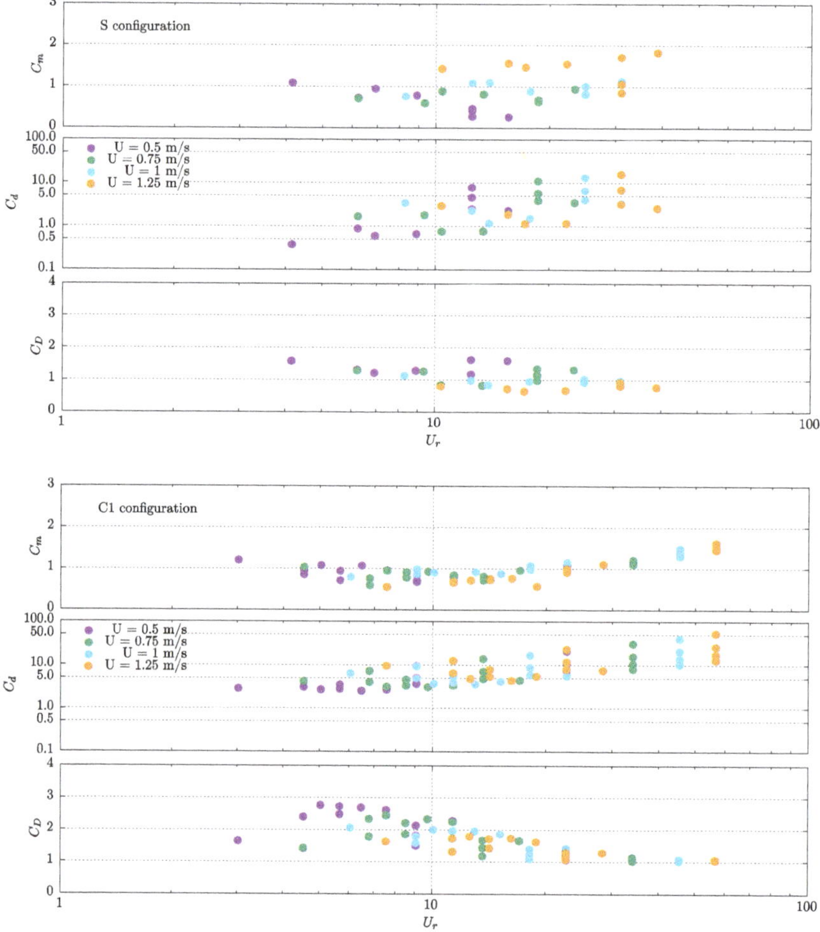

Figure 12. *Cont.*

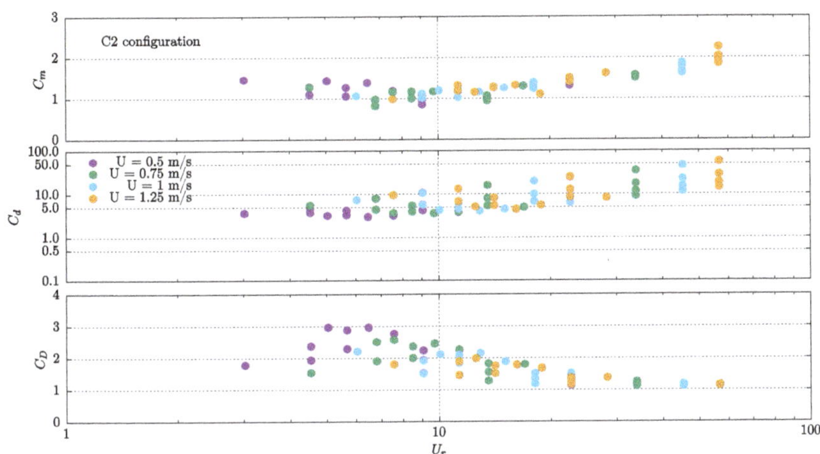

Figure 12. Evolution of C_m, C_d and C_D vs. KC for the S (**top**), *C1* (**middle**) and *C2* (**bottom**) cases.

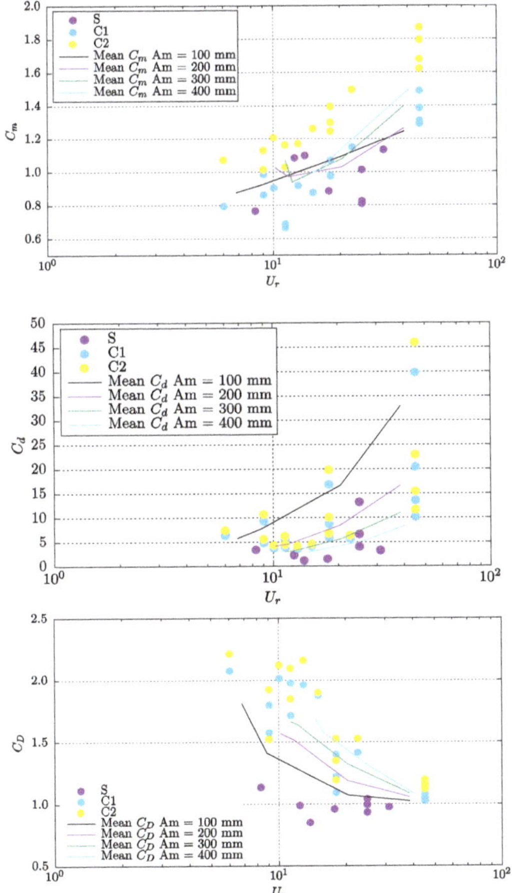

Figure 13. Evolution of C_m, C_d and C_D vs. U_r for the S, C1 and C2 cases.

5. Discussion and Conclusions

This study shows the impact of two realistic hard marine growth roughness *C1* and *C2* on the drag and inertia coefficients compared to a smooth case. The shape and organization of species are deduced from on-site observations of a full colonization by adult mussels that induce high relative roughness (0.09 for *C1* and 0.14 for *C2*) for mooring lines and power cables of Floating Offshore Wind Turbines. These results show that the surface roughness has an influence on the drag coefficient, the r.m.s. values of the lift fluctuations and the Strouhal number. The results from this experimental campaign highlight significant differences concerning the forces on the two rough cylinders. For instance, the effect of roughness is shown to be significant especially for low KC ($\approx$3) of oscillating motion, where $C_d \approx 1.3$ for *C1* and 2.1 for *C2*, leading to a 62% increase of drag forces. The results show a constant difference between the inertia coefficient of 0.3 between the two rough cylinders with a higher value for the higher relative roughness (*C2*). The vortices are shed into the wake with different frequencies and different amplitudes, the amplitude peaks of the vortex shedding frequencies are much higher for the smooth configuration than the rough configuration with a difference of about 90%. The r.m.s. values are always lower for the rough circular cylinders. A difference of about 10% between cases on the drag coefficient is observed for $R_e < 2 \times 10^5$. For the oscillating cases, the inertia coefficients for the rough cases present less dispersion than for the smooth cylinder. For $U_r < 10$, the mean drag coefficients are two times higher for the rough cases than for the smooth one. In this case, a strong dependency on the amplitude of the drag coefficients at fixed frequency for the rough cases has been highlighted, while they are stable in static. This shows that the commonly used approach of $C_d = \psi(KC)$. $C_D(R_e)$ is not legitimate. Moreover, while Morison's linearization for static drag force is justified, it means that it is not for oscillating cases, the KC defined only with the amplitude is not representative of the flow variety, this number should also depend on the frequency. These results highlight the fact that the behavior of the rough cylinders is mainly governed by the flow and not by their motions, contrary to the smooth cylinder for which its behavior is mainly governed by its motions. Moreover, the results have been compared with similar studies carried out for high relative roughness (0.1); significant differences have been observed due to the fact that the shape of the rough cylinders is not well described in these studies: key information about the organization and the areal density of peaks are usually not given.

For now, assumptions are strong: the marine growth is considered to be of a homogeneous circumferential and length volume. Due to internal and inter-species competition, it has been observed that mussels may be arranged in a bulbous manner. This phenomenon has not been studied here but the roughness variations must be studied to be compared to homogenous cover, which is considered in the engineering design phase. Pure current, regular forced oscillations and superimposed loadings must be tested in order to conclude on the validity of extracted coefficients but also on the standard hydrodynamic loading's formulation commonly used.

Author Contributions: Conceptualization, F.S.; methodology, A.M., G.G., C.B., T.S., F.S.; tests conception, A.M., J.-V.F., B.G., G.G., C.B., T.S., F.S.; tests realization and treatment, A.M., J.-V.F., B.G., G.G.; test analysis, A.M., J.-V.F., B.G., G.G., C.B., T.S., F.S.; writing—original draft preparation, F.S., A.M., G.G.; writing—review and editing, A.M., G.G., C.B., T.S., F.S. All authors have read and agreed to the published version of the manuscript.

Funding: Authors are grateful to Christian Berhault (senior Consultant) and Françoise Dubois (naval Energies), for their help in LEHERO-MG project. This work was carried out within the project LEHERO-MG (Load Effect of HEterogeneous ROughness of Marine Growth) granted by WEAMEC, West Atlantic Marine Energy Community with the support of Région Pays de la Loire and in partnership with Naval Energies. This work benefits from funding from France Energies Marines as well as the French National Research Agency under the Investments for the Future program bearing the reference ANR-10-IEED-0006-28 and within the project OMDYN-2. The authors would like to thank Guillaume Damblans from France Energies Marines. This project was partly financially supported by the European Union (FEDER), the French government, IFREMER and the region Hauts-de-France in the framework of the project CPER 2015–2020 MARCO. We are most grateful to Thomas Bacchetti for his participation conducting all the test campaign corresponding to about 800 cases.

Institutional Review Board Statement: Not applicable.

Informed Consent Statement: Not applicable.

Data Availability Statement: Data are shared in a consortium and will be published as soon as the agreement will permit it. They can be obtained under this permission by asking to gregogy.germain@ifremer.fr or franck.schoefs@univ-nantes.fr.

Conflicts of Interest: The authors declare no conflict of interest.

References

1. IEA. *World Energy Outlook 2019*; Energy Outlook Report; IEA: Paris, France, 2019. Available online: https://www.iea.org/reports/world-energy-outlook-2019 (accessed on 1 November 2019).
2. Spraul, C.; Arnal, V.; Cartraud, P.; Berhault, C. Parameter Calibration in Dynamic Simulations of Power Cables in Shallow Water to Improve Fatigue Damage Estimation. In *OMAE International Conference*; The American Society of Mechanical Engineers: Trondheim, Norway, 2017. [CrossRef]
3. Decurey, B.; Schoefs, F.; Barillé, A.L.; Soulard, T. Model of Bio-Colonisation on Mooring Lines: Updating Strategy based on a Static Qualifying Sea State for Floating Wind Turbine. *J. Mar. Sci. Eng.* **2020**, *8*, 108. [CrossRef]
4. Ma, K.T.; Duggal, A.; Smedley, P.; L'hostis, D.; Shu, H. A Historical Review on Integrity Issues of Permanent Mooring Systems. In Proceedings of the Offshore Technology Conference, Houston, TX, USA, 6 May 2013. [CrossRef]
5. Fontaine, E.; Kilner, A.; Carra, C.; Washington, D.; Ma, K.T.; Phadke, A.; Laskowski, D.; Kusinski, G. Industry Survey of Past Failures, Pre-emptive Replacements and Reported Degradations for Mooring Systems of Floating Production Units. In Proceedings of the Offshore Technology Conference, Houston, TX, USA, 5 May 2014. [CrossRef]
6. James, R.; Weng, W.Y.; Spradbery, C.; Jones, J.; Matha, D.; Mitzlaff, A.; Ahilan, R.V.; Frampton, M.; Lopes, M. Floating Wind Joint Industry Project—Phase I Summary Report. *Carbon Trust Tech. Rep.* **2018**, *19*, 2–20.
7. Braithwaite, R.A. McEvoy, L.A. Marine biofouling on fish farms and its remediation. *Adv. Mar. Biol.* **2005**, *47*, 215–252. [CrossRef]
8. Morison, J.R.; Johnson, J.W.; Schaaf, S.A. The Force Exerted by Surface Waves on Piles. *J. Pet. Technol.* **1950**, *2*, 149–154. [CrossRef]
9. Heaf, N.J. The Effect of Marine Growth on The Performance of Fixed Offshore Platforms in the North Sea. In Proceedings of the Offshore Technology Conference, Houston, TX, USA, 30 April 1979; p. 14. [CrossRef]
10. Jusoh, I.; Wolfram, J. Effects of marine growth and hydrodynamic loading on offshore structures. *J. Mek.* **1996**, *1*, 77–98.
11. Spraul, C.; Pham, H.D. Effect of Marine Growth on Floating Wind Turbines Mooring Lines Responses. In Proceedings of the 23ème Congrès Français de Mécanique, Lille, France, 1 September 2017.
12. Schoefs, F. Sensitivity approach for modelling the environmental loading of marine structures through a matrix response surface. *Reliab. Eng. Syst. Saf.* **2008**, *93*, 1004–1017. [CrossRef]
13. Schoefs, F.; Boukinda, M.L. Sensitivity Approach for Modeling Stochastic Field of Keulegan–Carpenter and Reynolds Numbers Through a Matrix Response Surface. *J. Offshore Mech. Arct. Eng.* **2010**, *132*, 011602. [CrossRef]
14. Wolfram, J.; Jusoh, I.; Sell, D. Uncertainty in the Estimation of Fluid Loading Due to the Effects of Marine Growth. In Proceedings of the 12th International Conference on Offshore Mechanical and Arctic Engineering, Glasgow, UK, 20–24 June 1993; Volume 2, pp. 219–228.
15. Zeinoddini, M.; Bakhtiari, A.; Schoefs, F.; Zandi, A.P. Towards an Understanding of the Marine Fouling Effects on VIV of Circular Cylinders: Partial Coverage Issue. *Biofouling* **2017**, *33*, 268–280. [CrossRef]
16. Sarpkaya, T. On the Effect of Roughness on Cylinders. *J. Offshore Mech. Arct. Eng.* **1990**, *112*, 334. [CrossRef]
17. Ameryoun, H.; Schoefs, F.; Barillé, L.; Thomas, Y. Stochastic Modeling of Forces on Jacket-Type Offshore Structures Colonized by Marine Growth. *J. Mar. Sci. Eng.* **2019**, *7*, 158. [CrossRef]
18. Theophanatos, A. Marine Growth and the Hydrodynamic Loading of Offshore Structures. Ph.D. Thesis, University of Strathclyde, Glasgow, UK, 1988. Available online: https://ethos.bl.uk/OrderDetails.do?uin=uk.bl.ethos.382439 (accessed on 1 February 1988).
19. API RP 2A WSD. *Recommended Practice for Planning, Designing, and Constructing Fixed Offshore Platforms*; The American Petroleum Institute: Washington, DC, USA, 2005; Volume 2.
20. DNV. *Recommended Practice DNV-RP-C205: Environmental Conditions and Environmental Loads*; Det Norske Veritas: Bærum, Norway, 2010.
21. Page, H.M.; Hubbard, D.M. Temporal and spatial patterns of growth in mussels Mytilus edulis on an offshore platform: Relationships to water temperature and food availability. *J. Exp. Mar. Biol. Ecol.* **1987**, *111*, 159–179. [CrossRef]
22. Parks, T.; Sell, D.; Picken, G. *Marine Growth Assessment of Alwyn North A in 1994*; Technical Report 820; Auris Environmental LTD: Coventry, UK, 1995.
23. Boukinda Mbadinga, M.L.; Quiniou Ramus, V.; Schoefs, F. Marine Growth Colonization Process in Guinea Gulf: Data Analysis. *J. Offshore Mech. Arctic Eng.* **2007**, *129*, 97–106. [CrossRef]

24. Handå, A.; Alver, M.; Edvardsen, C.V.; Halstensen, S.; Olsen, A.J.; Øie, G.; Reinertsen, H. Growth of farmed blue mussels (*Mytilus edulis* L.) in a Norwegian coastal area: Comparison of food proxies by DEB modeling. *J. Sea Res.* **2011**, *66*, 297–307. [CrossRef]
25. O'Byrne, M.; Schoefs, F.; Pakrashi, V.; Ghosh, B. An underwater lighting and turbidity image repository for analysing the performance of image based non-destructive techniques. *Struct. Infrastruct. Eng. Maint. Manag. Life Cycle Des. Perform.* **2018**, *14*, 104–123. [CrossRef]
26. O'Byrne, M.; Schoefs, F.; Pakrashi, V.; Ghosh, B. A Stereo-Matching Technique for Recovering 3D Information from Underwater Inspection Imagery. *Comput. Aided Civ. Infrastruct. Eng.* **2018**, *33*, 193–208. [CrossRef]
27. O'Byrne, M.; Pakrashi, V.; Schoefs, F.; Ghosh, B. Semantic Segmentation of Underwater Imagery Using Deep Networks. *J. Mar. Sci. Eng. Sect. Ocean Eng.* **2018**, *6*, 93. [CrossRef]
28. O'Byrne, M.; Pakrashi, V.; Schoefs, F.; Ghosh, B. Applications of Virtual Data in Subsea Inspections. *J. Mar. Sci. Eng. Sect. Ocean Eng.* **2020**, *8*, 328. [CrossRef]
29. Picken, G.B. Review of marine fouling organisms in the North Sea on offshore structures. Discussion Forum and Exhibition on Offshore Engineering with Elastomers. *Plast. Rubber Inst.* **1985**, *5*, 1–5.
30. Schoefs, F.; O'Byrne, M.; Pakrashi, V.; Ghosh, B.; Oumouni, M.; Soulard, T.; Reynaud, M. Feature-Driven Modelling and Non-Destructive Assessment of Hard Marine Growth Roughness from Underwater Image Processing. *J. Mar. Sci. Eng.* **2021**, accepted.
31. Achenbach, E.; Heinecke, E. On vortex shedding from smooth and rough cylinders in the range of Reynolds numbers 6e3 to 5e6. *J. Fluid Mech.* **1981**, *109*, 239–251. [CrossRef]
32. Gaurier, B.; Germain, G.; Facq, J.; Baudet, L.; Birades, M.; Schoefs, F. Marine growth effects on the hydrodynamical behaviour of circular structures. In Proceedings of the 14th Journées de l'Hydrodynamique, Val de Reuil, France, 18–20 November 2014.
33. Sarpkaya, T. *Moirison's Equation and the Wave Forces on Offshore Structures*; Technical Report; Naval Civil Engineering Laboratory: Carmel, CA, USA, 1981.
34. Marty, A.; Germain, G.; Facq, J.V.; Gaurier, B.; Bacchetti, T. Experimental investigation of the marine growth effect on the hydrodynamical behavior of a submarine cable under current and wave conditions. In Proceedings of the JH2020-17èmes Journées de l'Hydrodynamique, Cherbourg, France, 4–26 November 2020. [CrossRef]
35. Marty, A.; Berhault, C.; Damblans, G.; Facq, J.-V.; Gaurier, B.; Germain, G.; Soulard, T.; Schoefs, F. Experimental study of marine growth effect on the hydrod-ynamical behaviour of a submarine cable. *Appl. Ocean. Res.* **2021**. in Press.
36. Schlichting, H. *Boundary Layer Theory*; McGraw-Hill: New York, NY, USA, 1979.
37. Verley, R.L.P. Oscillations of Cylinders in Waves and Currents. Ph.D. Thesis, Loughborough University of Technology, Loughborough, UK, 1980.
38. Molin, B. *Hydrodynamique des Structures Offshore*; Technip Editions: Paris, France, 2002.
39. Melbourne, W.H.; Blackburn, H.M. The effect of free-stream turbulence on sectional lift forces on a circular cylinder. *J. Fluid. Mech.* **1996**, *11*, 267–292.
40. Sarpkaya, T. *In-Line and Transverse Forces on Smooth and Sand Roughened Circular Cylinders in Oscillating Flow at High Reynolds Numbers*; Technical Report No. NPS-69SL; Naval Postgraduate School: Monterey CA, USA, 1976.
41. Rodenbusch, G.; Gutierrez, C.A. *Forces on Cylinders in Two Dimensional Flow*; Technical Report No. BRC 123-83; Bellaire Research Center: Houston, TX, USA, 1983; Volume 1.
42. Nath, J.H. Heavily roughened horizontal cylinders in waves. *Proc. Conf. Behav. Offshore Struct.* **1982**, *1*, 387–407.

Journal of
Marine Science and Engineering

Article

Cavity Detachment from a Wedge with Rounded Edges and the Surface Tension Effect

Yuriy N. Savchenko, Georgiy Y. Savchenko and Yuriy A. Semenov *

Institute of Hydromechanics of the National Academy of Sciences of Ukraine, 8/4 Marii Kapnist Street,
03680 Kiev, Ukraine; hydro.ua@gmail.com (Y.N.S.); lenchik123@ukr.net (G.Y.S.)
* Correspondence: semenov@nas.gov.ua

Abstract: Cavity flow around a wedge with rounded edges was studied, taking into account the surface tension effect and the Brillouin–Villat criterion of cavity detachment. The liquid compressibility and viscosity were ignored. An analytical solution was obtained in parametric form by applying the integral hodograph method. This method gives the possibility of deriving analytical expressions for complex velocity and for potential, both defined in a parameter plane. An expression for the curvature of the cavity boundary was obtained analytically. By using the dynamic boundary condition on the cavity boundary, an integral equation in the velocity modulus was derived. The particular case of zero surface tension is a special case of the solution. The surface tension effect was computed over a wide range of the Weber number for various degrees of cavitation development. Numerical results are presented for the flow configuration, the drag force coefficient, and the position of cavity detachment. It was found that for each radius of the edges, there exists a critical Weber number, below which the iterative solution process fails to converge, so a steady flow solution cannot be computed. This critical Weber number increases as the radius of the edge decreases. As the edge radius tends to zero, the critical Weber number tends to infinity, or a steady cavity flow cannot be computed at any finite Weber number in the case of sharp wedge edges. This shows some limitations of the model based on the Brillouin–Villat criterion of cavity detachment.

Keywords: cavity detachment; free streamlines; Brillouin criterion

Citation: Savchenko, Y.N.;
Savchenko, G.Y.; Semenov, Y.A.
Cavity Detachment from a Wedge
with Rounded Edges and the Surface
Tension Effect. *J. Mar. Sci. Eng.* **2021**,
9, 1253. https://doi.org/10.3390/
jmse9111253

Academic Editors: Kamal Djidjeli and
Md Jahir Rizvi

Received: 21 October 2021
Accepted: 9 November 2021
Published: 11 November 2021

Publisher's Note: MDPI stays neutral
with regard to jurisdictional claims in
published maps and institutional affiliations.

1. Introduction

Surface tension arises at an air–liquid interface as a result of a reversible isothermoki-netic process on a free boundary. According to the Young–Laplace equation, surface tension results in a pressure jump, which is governed by the curvature of the cavity boundary. The dynamic boundary condition in the model of the ideal fluid (Bernoulli equation) includes this jump; therefore, the velocity modulus on the cavity boundary, the pressure on the liquid side of the interface, and the curvature of the interface are related to one another. Cavity detachment for smooth-shaped bodies is determined by the Brillouin–Villat crite-rion [1,2], which states that the curvature of the cavity boundary should be equal to the curvature of the body at the point of detachment. This criterion comes from geometrical restrictions on the cavity curve and, therefore, it should be valid for flows both with and without surface tension. As the surface tension tends to zero, the solution of the problem gradually approaches its limiting case without surface tension, as shown in [3], for cavity flow past a circular cylinder.

Cavity detachments for bodies whose shapes form corner trailing edges are different, and it is not fully understood yet when the surface tension is nonzero. Although the position of cavity detachment is predetermined at the edge, the curvature of the cavity boundary is infinite for the case of zero surface tension [4,5]. Even small surface tension might result in an infinite negative pressure at the cavity detachment point. The same infi-nite negative pressure occurs for flow around a corner without flow detachment. Therefore, flow detachment may not occur at all.

219

Attempts to take into account surface tension in problems of cavity flows were done in [6–9]. The case of small surface tension values was considered in [8] using the method of matched asymptotic expansions. For this case, the curvature of the cavity boundary equals the curvature of the plate at the point of detachment. Such a formulation of the problem leads to waves on the free surface. It was shown in [9] that these waves have no physical basis since they require an energy input from infinity. In [6,7], it is assumed that at the flow detachment point, the tangent to the flow boundary is a discontinuous function. According to the theory of jets of an ideal liquid [4,5], this assumption leads to zero velocity at the point, while the curvature of the cavity boundary remains undetermined. Alternative models of cavity detachment were proposed in the works [10,11]. These models account for the physical properties of the liquid/structure interaction [12,13]. The work [10] considers cavity detachment as bubbles growing within the boundary layer starting from the stagnation point; the coalition and merging of the bubbles form a cavity. The work [11] considers adhesion between the liquid and the material of the body, which determines the contact angle between the body and the free surface. However, these models are not used widely in practice since further experimental validation of these models is required.

The surface tension force becomes dominant on the scale of mechanical elements far below millimeters, which is the typical size of microelectromechanical systems (MEMS) and microfluidic systems [14,15]. The study of cavitation phenomena in such systems is of interest when developing micropumps, microvalves, microcoolers, etc., which are fabricated with a micro-orifice of a smaller diameter than the main microchannel. Cavitation in MEMS appears in the same way as cavitation in macroscale devices, when the local pressure drops below the vapor pressure. However, on microscale, the surface tension force dominates other forces and may affect flow characteristics, which may be treated as a scale effect [16,17].

In this paper, we consider classical two-dimensional potential free surface flow around a wedge with rounded edges and apply the Brillouin–Villat criterion to determine the cavity detachment for the case of nonzero surface tension. A sharp trailing edge is considered as the limiting case of a rounded edge when the radius of the edge tends to zero. An analytical solution to the problem is obtained using the integral hodograph method [18]. The complex velocity and the derivative of the complex potential are derived analytically in a parameter plane. A function mapping the parameter plane onto the physical plane is also obtained. The method is based on the integral formula [19], which gives a solution of a mixed boundary-value problem for a complex function (here, the complex velocity) defined in the first quadrant. The derivative of the complex potential is obtained by using Chaplygin's singular point method [5,20].

The complex velocity includes the slope to the body and the velocity modulus along the cavity boundary, which are determined from a system of integral equations derived by using the dynamic and the kinematic boundary conditions. By using an analytical expression for the curvature of the free boundary, it is possible to directly apply the Brillouin–Villat condition and obtain an equation to determine the point of cavity detachment.

The results are presented over a wide range of Weber numbers and the radius of the wedge edges. It is shown that the effect of surface tension becomes more pronounced as the Weber number or the radius of the edge decrease. The surface tension results in some reduction of the cavity length and the drag force coefficient and delays cavity detachment. The point of detachment moves downward, increasing the wetted part of the rounded edge. It is found that, for each edge radius, there exists a critical Weber number below which a solution cannot be obtained due to the iteration process failing to converge. This critical Weber number increases as the radius of the edge decreases. As the edge radius tends to zero, the Weber number tends to infinity, or a steady cavity flow cannot be computed for any finite Weber number in the case of a sharp wedge edge. This shows some limitation of the model based only on the Brillouin–Villat criterion of detachment.

2. Complex Potential of the Flow

We consider cavity flow past a wedge of width $2H$ with rounded edges of radius R sketched in Figure 1a. The wedge angle α lies in the range $0 < \alpha < \pi$. For $\alpha = 90°$, the wedge becomes a flat plate. We introduce Cartesian coordinates xy. The origin is located at point O. The liquid is inviscid and incompressible. The flow is symmetric about the x-axis, so we consider only half of the flow region. An assumption as to the cavity closure is necessary due to the Brillouin paradox at the point where the upper and the lower cavity boundaries are merged [4]. We introduce an implicit model of cavity closure according to which, along the streamline $OCDB$, the velocity modulus changes in such a manner that the portion OC corresponds to the cavity boundary, and on the portion CD the velocity modulus changes from the value V_0 at the end of the cavity (point C) to the value U at point D. The velocity U is the velocity at infinity. Along the contour DB, the velocity remains constant and equals U. As $s \to \infty$, the y-coordinate of the closing line DB tends to zero. Such a cavity closure model mimics a turbulent wake in real cavity flows [21,22].

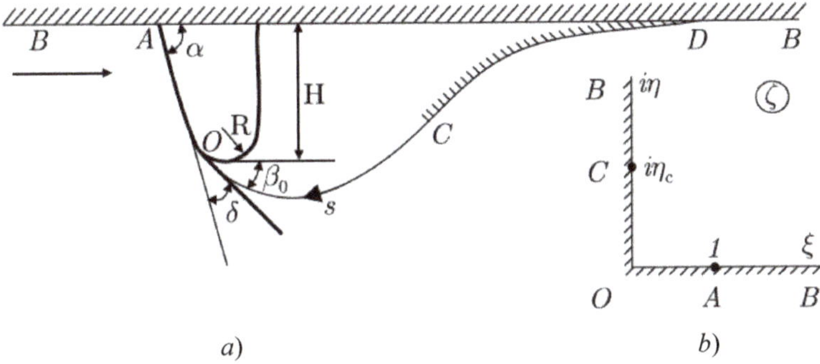

Figure 1. Flow sketch: (a) physical plane (b) parameter, or ζ-plane.

The pressure jump across the cavity boundary is determined by the Laplace–Young formula

$$p - p_c = \tau Y, \tag{1}$$

where τ is the coefficient of surface tension, p_c is the pressure in the cavity, and Y is the curvature. The pressure in the cavity may be equal to the vapor pressure, $p_c = p_v$, or be higher for gaseous liquids.

We choose H and U as the characteristic length and velocity, respectively. From the equation

$$\frac{V^2}{2} + \frac{p}{\rho} = \frac{U^2}{2} + \frac{p_\infty}{\rho} \tag{2}$$

and Equation (1) the velocity modulus on the cavity boundary is obtained

$$v = \frac{V}{U} = \sqrt{1 + \sigma - \frac{2\chi}{We}}, \tag{3}$$

where $\chi = YH$. Here, We is the Weber number that characterizes the ratio of the inertia force to the surface tension force, and σ is the cavitation number indicating the degree of cavitation development.

$$\sigma = \frac{p_\infty - p_c}{1/2\rho U^2}, \qquad We = \frac{\rho U^2 H}{\tau}, \tag{4}$$

where ρ is the density of the liquid, and p_∞ is the pressure at infinity.

We can introduce a complex potential, $w(z) = \phi(x,y) + i\psi(x,y)$, where $\phi(x,y)$ and $\psi(x,y)$ are the harmonic conjugate function of the potential and the stream function, respectively. The determination of the function $w(z)$ is a challenging problem. Instead, Michell [23] and Joukovskii [24] introduced an auxiliary ζ-plane and determined the complex conjugate velocity, dw/dz, and the derivative of the complex potential, $dw/d\zeta$, as functions of the parameter variable ζ. If these functions are known, then the solution of the problem is obtained in parametric form:

$$w(\zeta) = w_0 + \int_0^\zeta \frac{dw}{d\zeta'}d\zeta', \qquad z(\zeta) = z_0 + \int_0^\zeta \frac{dw}{d\zeta'}\Big/\frac{dw}{dz}\,d\zeta', \tag{5}$$

where w_0 is the value of the complex potential at point O, and z_0 is the coordinate of point O.

We choose the first quadrant of the ζ-plane in Figure 1b as the region corresponding to the liquid region in the physical plane (Figure 1a). According to the conformal mapping theorem, we can chose the position of three points. They are points O ($\zeta = 0$), A ($\zeta = 1$), and B ($\zeta = \infty$), as shown in Figure 1b. The interval $0 < \xi < 1$ corresponds to the wetted part of the wedge, and the interval $1 < \xi < \infty$ corresponds to the symmetry line AB. The imaginary η-axis corresponds to the boundary $OCDB$, which includes the cavity contour OC ($0 < \eta < \eta_c$) and the closure contour CB ($\eta_c < \eta < \infty$). The position of point C ($\zeta = i\eta_c$) has to be determined from the solution of the problem and physical considerations. In order to determine these functions, we assume that the velocity modulus v and the tangent to the wetted part of the body β are known functions of the variables η and ξ, respectively.

2.1. Expressions for the Complex Velocity and for the Derivative of the Complex Potential

The body is considered to be fixed; therefore, the velocity direction and the slope of the body coincide. Besides, at this stage we assume that the velocity magnitude on the free surface is a known function of the parameter variable, $v(\eta)$. Then the boundary-value problem for the complex velocity in the first quadrant of the parameter plane can be written as follows

$$\chi(\xi) = \arg\left(\frac{dw}{dz}\right) = \begin{cases} -\beta_b(\xi), & 0 \le \xi \le 1, \\ 0, & 1 \le \xi < \infty. \end{cases} \tag{6}$$

$$v(\eta) = \left|\frac{dw}{dz}\right|_{\zeta=i\eta}, \qquad 0 \le \eta < \infty. \tag{7}$$

Here, $\beta_b(\xi)_{\xi=0} = \beta_0$ at point O, and $\beta_b(\xi) = -\alpha$ as $\xi \to 1$. Equation (6) satisfies the conditions: $\chi(\xi) = 0$ along the interval AB on the symmetry line and $\chi(\xi) = -\beta_b(\xi)$ along the body. The argument of the complex velocity exhibits a jump $\Delta = \alpha$ at point A when we move along the boundary in the physical plane from point O to point B. This boundary-value problem can be solved by applying the following integral formula [19]:

$$\frac{dw}{dz} = v_\infty \exp\left[\frac{1}{\pi}\int_0^\infty \frac{d\chi}{d\xi}\ln\left(\frac{\zeta+\xi}{\zeta-\xi}\right)d\xi - \frac{i}{\pi}\int_0^\infty \frac{d\ln v}{d\eta}\ln\left(\frac{\zeta-i\eta}{\zeta+i\eta}\right)d\eta + i\chi_\infty\right], \tag{8}$$

where $v_\infty = v(\eta)_{\eta\to\infty}$ and $\chi_\infty = \chi(\xi)_{\xi\to\infty}$. Substituting Equations (6) and (7) into (8) and evaluating the first integral over the step change at the point $\zeta = 1$, we obtain

$$\frac{dw}{dz} = \left(\frac{\zeta-1}{\zeta+1}\right)^{\frac{\alpha}{\pi}}\exp\left[\frac{1}{\pi}\int_0^1 \frac{d\beta_b}{d\xi}\ln\left(\frac{\zeta-\xi}{\zeta+\xi}\right)d\xi - \frac{i}{\pi}\int_0^\infty \frac{d\ln v}{d\eta}\ln\left(\frac{\zeta-i\eta}{\zeta+i\eta}\right)d\eta\right]. \tag{9}$$

Here, we used $v_\infty = 1$, which follows from the definition. It can be easily verified that for $\zeta = \xi$ the argument of the right-hand side of (9) is the function $-\beta_b(\xi)$, while for $\zeta = i\eta$ the modulus of (9) is the function $v(\eta)$, i.e., the boundary conditions (6) and (7) are satisfied. It can also be seen that the complex velocity function has zeros of order α/π, which correspond to flow around a corner of angle $\pi - \alpha$ at point A.

The velocity is tangent to the boundary for steady flows and, therefore, $\Im(w) = 0$ both along the body and along the cavity boundary. The real part of the complex potential increases from $-\infty$ at point B to $+\infty$ at point B'. Thus, the region of the complex potential is a half-plane. The first quadrant and the half-plane of the w-plane are related as $w = K\zeta^2 + w_0$ where K is a positive real number. Then one can obtain

$$\frac{dw}{d\zeta} = K\zeta. \tag{10}$$

The derivative of the mapping function is obtained by dividing (10) by (9)

$$\frac{dz}{d\zeta} = K\zeta\left(\frac{\zeta+1}{\zeta-1}\right)^{\frac{\alpha}{\pi}} \exp\left[-\frac{1}{\pi}\int_0^1 \frac{d\beta_b}{d\xi}\ln\left(\frac{\zeta-\xi}{\zeta+\xi}\right)d\xi + \frac{i}{\pi}\int_0^\infty \frac{d\ln v}{d\eta}\ln\left(\frac{\zeta-i\eta}{\zeta+i\eta}\right)d\eta\right]. \tag{11}$$

The integration of this equation yields the mapping function $z = z(\zeta)$ relating the parameter and the physical planes. Equations (9) and (10) include the parameter K, and the functions $v(\eta)$ and $\beta_b(\xi)$, which are determined from the boundary conditions and from physical considerations.

The arc length coordinates along the body, $s_b(\xi)$, and along the cavity boundary, $s(\eta)$, are obtained as follows:

$$s_b(\xi) = \int_0^\xi \frac{ds_b}{d\xi}d\xi, \quad s(\eta) = -\int_0^\eta \left|\frac{dz}{d\zeta}\right|_{\zeta=\eta'} d\eta' = -K\int_0^\eta \frac{\eta'}{v(\eta')}d\eta', \tag{12}$$

where

$$\frac{ds_b}{d\xi} = \left|\frac{dz}{d\zeta}\right|_{\zeta=\xi} = K\xi\left(\frac{\xi+1}{1-\xi}\right)^{\alpha/\pi} \exp\left(-\frac{1}{\pi}\int_0^1 \frac{d\beta_b}{d\xi'}\ln\left|\frac{\xi-\xi'}{\xi+\xi'}\right|d\xi' + \frac{1}{\pi}\int_0^\infty \frac{d\ln v}{d\eta}2\tan^{-1}\frac{\eta}{\xi}d\eta\right) \tag{13}$$

The real factor K is determined from the condition for the length of the wetted part of the wedge, including the trailing edge S_w. By substituting Equation (13) into (12), we obtain

$$S_w = K\int_0^1 \xi\left(\frac{\xi+1}{1-\xi}\right)^{\alpha/\pi} \exp\left(-\frac{1}{\pi}\int_0^1 \frac{d\beta_b}{d\xi'}\ln\left|\frac{\xi-\xi'}{\xi+\xi'}\right|d\xi' + \frac{1}{\pi}\int_0^\infty \frac{d\ln v}{d\eta}2\tan^{-1}\frac{\eta}{\xi}d\eta\right)d\xi \tag{14}$$

2.2. Cavity Closure Model

A solution of the problem of cavity flow around a body is not unique in the framework of the model of ideal liquid due to a paradox that arises at the point of cavity closure. On the one hand, the velocity at this point must be equal to the velocity on the cavity boundary since this point belongs to it. On the other hand, the velocity at this point should be equal to zero since the upper and the lower contours of the cavity merge into one streamline. This is known as the Brillouin paradox [4].

Various assumptions as to the flow in the cavity closure region were proposed by Roshko, Riabouchinsky, Efros, in the re-entrant jet model, Tulin, etc., to resolve this paradox. They are now well known as classical models of cavity flows [5]. These models give similar results for developed cavity flows, for which the size of the cavity is larger than the size of the body. However, when the cavity size is smaller, the cavity contour ends on the body surface (partial cavitation), and the assumptions made may affect the results [10,25].

An implicit cavity closure model modeling effects of the liquid viscosity in the cavity closure region was proposed in [21,22,26]. It is based on dividing the flow region into viscous and inviscid subregions and the application of viscous/inviscid interaction conditions along the interface. In this work, we simplify the model by dropping the viscous wake and using an assumption as to the velocity distribution along the closing contour.

We assume that the velocity magnitude, $v(s)$, linearly decreases from the value v_0 at point C determined by Equation (3) to the value U at point D and then remains constant along BD

$$v^*(s') = \begin{cases} v_0 + (v_\infty - v_0)/2[1 - \cos(\pi s')], & 0 < s' < 1, \\ 1, & 1 < s' < \infty, \end{cases} \tag{15}$$

where $s' = (s_c - s)/(s_c - s_d)$, s_c and s_d are the arc length coordinates of points C and D, respectively.

From the balance of the flow rate upstream and downstream, the y-coordinates of the boundary CDB at right infinity (point B) should be equal to zero; that is

$$Im\left(\int_{\zeta=\infty} \frac{dz}{d\zeta} d\zeta\right) = -Im\left(\int_{\zeta'=0} \frac{dz}{d\zeta'}\frac{d\zeta'}{\zeta'^2}\right) = -\frac{i\pi}{4} \mathop{Res}_{\zeta'=0} \frac{d^2}{d\zeta'^2}\left(\frac{dz}{d\zeta'}\zeta'\right) = 0, \tag{16}$$

where $\zeta' = 1/\zeta$.

By using the theorem of residues, we evaluate the integral and get the following equation

$$\int_0^1 \frac{d\beta_b}{d\xi}\xi d\xi + \int_0^\infty \frac{d\ln v}{d\eta}\eta d\eta + \alpha = 0. \tag{17}$$

This is an implicit equation in the cavity length, s_c. The magnitude of the velocity $v(\eta) = v[s(\eta)]$, $\eta_c < \eta < \infty$, given by Equation (15) and the cavity length $s_c = s(\eta_c)$ affect the second integral in Equation (17). The unique value $s_c = s(\eta_c)$, satisfies Equation (17).

2.3. Brillouin–Villat Condition of Cavity Detachment

The Brillouin–Villat criterion [1,2] comes from the consideration of possible configurations of the cavity boundary near the detachment point. It states that the curvature of the cavity boundary at the point of cavity detachment should be finite and equal to the curvature of the body. This conclusion is based on the following considerations. In order to determine the curvature of the cavity boundary, we first find the slope of the cavity boundary by taking the argument of the complex velocity from Equation (9)

$$\delta(\eta) = -\arg\left(\frac{dw}{dz}\bigg|_{\zeta=i\eta}\right) = \frac{\alpha}{\pi}\left(2\tan^{-1}\eta - \pi\right) - \frac{1}{\pi}\int_0^1 \frac{d\beta_b}{d\xi}\left(\pi - 2\tan^{-1}\frac{\eta}{\xi}\right)d\xi + \frac{1}{\pi}\int_0^1 \frac{d\ln v}{d\eta}\ln\left|\frac{\eta'-\eta}{\eta'+\eta}\right|d\eta' \tag{18}$$

Then, differentiating the above equation, we find the curvature of the free surface

$$\chi(\eta) = \frac{d\delta}{ds} = \frac{d\delta/d\eta}{ds/d\eta} = -\frac{v(\eta)}{K\eta}\left(\frac{\alpha}{\pi}\frac{2}{1+\eta^2} + \frac{2}{\pi}\int_0^1 \frac{d\beta_b}{d\xi}\frac{\xi d\xi}{\xi^2+\eta^2} - \frac{2}{\pi}\int_0^\infty \frac{d\ln v}{d\eta'}\frac{\eta' d\eta'}{\eta'^2-\eta^2}\right) \tag{19}$$

At the detachment point $\eta = 0$; therefore, the denominator of the above equation equals zero. The curvature may take a finite value if the nominator at $\eta = 0$ also equals zero, or

$$\int_0^1 \frac{d\beta_b}{d\xi}\frac{d\xi}{\xi} - \int_0^\infty \frac{d\ln v}{d\eta'}\frac{d\eta'}{\eta'} + \alpha = 0 \tag{20}$$

This is an equation in the unknown length of the wetted part of the body, S_w, which affects the function $\beta_b(\xi) = \beta_b[s(\xi)]$, $0 < s < S_w$.

2.4. Integro-Differential Equations in the Functions $\beta_b(\zeta)$ and $v(\eta)$

The function $\beta_b(s)$ is known due to the given shape of the body. Then the function $\beta_b(\zeta)$ is determined from the integro-differential equation

$$\frac{d\beta_b}{d\zeta} = \frac{d\beta_b}{ds}\frac{ds}{d\zeta} = \chi[s(\zeta)]K\zeta\left(\frac{\zeta+1}{1-\zeta}\right)^{\alpha/\pi}\exp\left(-\frac{1}{\pi}\int_0^1\frac{d\beta_b}{d\zeta'}\ln\left|\frac{\zeta-\zeta'}{\zeta+\zeta'}\right|d\zeta' + \frac{1}{\pi}\int_0^\infty\frac{d\ln v}{d\eta}2\tan^{-1}\frac{\eta}{\zeta}d\eta\right) \tag{21}$$

where $\chi(s)$ is the curvature of the body given as a function of the arc length s.

In order to derive an integral equation in the function $v(\eta)$, we substitute (19) into the dynamic boundary condition (3)

$$\int_0^\infty\frac{d\ln v}{d\eta'}\frac{\eta'd\eta'}{\eta'^2-\eta^2} = \frac{\pi WeK\eta}{4Rv(\eta)}(1+\sigma-v^2(\eta)) + \frac{\alpha}{1+\eta^2} + \int_0^1\frac{d\beta_b}{d\zeta}\frac{\zeta d\zeta}{\zeta^2+\eta^2} \tag{22}$$

where

$$v(\eta) = v_0\exp\left(\int_0^\eta\frac{d\ln v}{d\eta'}d\eta'\right).$$

According to the Brillouin–Villat criterion, the curvature of the free surface at the cavity detachment point equals the curvature of the body, or $\chi_0 = H/R = 1/r$. Then, the velocity magnitude at the point of cavity detachment is obtained from Equation (3)

$$v_0 = \sqrt{1+\sigma+\frac{2}{rWe}}. \tag{23}$$

From the above equation, it follows that the velocity magnitude at the cavity detachment point tends to infinity as the radius of the edge tends to zero or the Weber number tends to zero.

3. Numerical Method and Results

3.1. Numerical Approach

In discrete form, the solution is sought on a fixed set of points ζ_j, $j = 1, ..., M$ distributed along the real axis of the parameter region and on a fixed set of points η_i, $i = 1, ..., N$ distributed along its imaginary axis. The points $0 < \eta_i \le \eta_C$, $i = 1, ..., K$, where $\eta_K = \eta_C$, correspond to the end of the cavity, and the points $\eta_K < \eta_i \le \eta^*$, $i = K+1, ..., N$ correspond to the cavity closure contour CD. The total number of the points η_i was chosen in the range from $N = 150$ to 450, and the total number of the points ζ_j was chosen as $M = 3N$ to check the convergence of the solution procedure. The points ζ_j are distributed to provide a higher density of the points $s_j = s(\zeta_j)$ near the rounded edges of the wedge and point A, at which the derivative $ds/d\zeta$ has a singularity. The distribution of the points η_i is chosen to provide a higher density of the points $s_i = s(\eta_i)$ on the free surface near point O where the curvature of the cavity boundary may vary quickly. The length of the contour CD is chosen as $s_{CD} = 10$.

The solution of Equation (22) can be found by the method of successive approximations, determining the $(k+1)^{th}$ approximation using the Hilbert transform

$$\left(\frac{d\ln v}{d\eta}\right)_i^{(k+1)} = \frac{8}{\pi}\int_0^\infty F^{(k)}(\eta')\frac{\eta'd\eta'}{\eta'^2-\eta_i^2}, \qquad i = 1, ..., K. \tag{24}$$

where

$$F^{(k)}(\eta) = \frac{\alpha}{1+\eta^2} + \int_0^1\left(\frac{d\beta}{d\zeta}\right)^{(k)}\frac{\zeta d\zeta}{\zeta^2+\eta^2} + \frac{\pi WeK\eta}{4Rv^{(k)}}(\eta)\left(1+\sigma-v^{2(k)}\right),$$

$$v^{(k)}(\eta) = v_0\exp\left(\int_0^\eta\left(\frac{d\ln v}{d\eta'}\right)^{(k)}d\eta'\right).$$

is the right-hand side of Equation (22).

At each iteration step k, the integro-differential Equation (21) in the function $d\beta/d\xi$ was solved using the inner iteration procedure

$$
\left(\frac{d\beta}{d\xi}\right)^{(l+1)} = \chi[s^{(l)}(\xi)]\frac{K^{(l)}\xi}{v_0}\left(\frac{1+\xi}{1-\xi}\right)^{\alpha/\pi}\exp\left[-\frac{1}{\pi}\int_0^1\left(\frac{d\beta}{d\xi}\right)^{(l)}\ln\left(\frac{\xi'-\xi}{\xi'+\xi}\right)d\xi' \right.
$$
$$
\left. -\frac{1}{\pi}\int_0^\infty\left(\frac{d\ln v}{d\eta}\right)^{(k)}\left(\pi-2\tan^{-1}\frac{\eta}{\xi}\right)d\eta\right] \tag{25}
$$

At each iteration step l, the parameters $K^{(l)}$ and η_C were calculated from Equations (14) and (17), respectively. The integrals appearing in the system of equations were evaluated using the linear interpolation of the functions $\beta(\xi)$ and $\ln v(\eta)$ on the intervals (ξ_{j-1},ξ_j) and η_{i-1},η_i, respectively. At the first iteration, the functions $\beta(\xi) \equiv -\alpha$ and $v(\eta) \equiv 1$.

The convergence of the inner iteration procedure required from 5 to 10 iterations to reach the condition $|\beta_j^{(l+1)} - \beta_j^{(l)}| < \varepsilon$, $j = 1,...,M$ for the chosen value $\varepsilon = 10^{-7}$. The convergence of the outer iteration procedure required from several hundreds to several thousands of iterations, and it was obtained applying the under-relaxation method. In the paper [3], a similar approach was applied to solving the problem of cavity flow past a circular cylinder. It was shown that the numerical procedure can provide an accuracy about 1%, which is satisfactory for this kind of problems. The results presented below were obtained for a larger number of nodes; therefore, we can expect a similar accuracy of the results.

An alternative numerical approach is based on the collocation method used earlier for solving boundary-value and related problems [27–29]. Instead of solving the integro-differential Equation (22), we solve the system of nonlinear equations

$$
G_k(\mathbf{V}) = 1 + \sigma - \frac{2\chi_k}{We} - v_k^2 = 0, \qquad k = 2,...,K. \tag{26}
$$

which is obtained from the dynamic boundary condition (3) at the collocation points $\bar{\eta}_k = (\eta_{k-1} + \eta_k)/2$, $k = 2,...,K$. Here, $\mathbf{V} = \{v_j\}, j = 2,...,K$ is the vector of unknown velocities at the points η_i, $i = 2,...,K$. For each set of values $\{v_j\}$, the system of Equations (17), (20) and (21) is solved, and the curvature of the cavity boundary, $\chi_k = \chi(\eta_k)$, at the points $\bar{\eta}_k$ can be computed using Equation (19). The system of nonlinear Equation (24) is solved using the Newton–Raphson method.

3.2. Numerical Results

The cavity contours for edge radius $r = 0.05$ and cavitation numbers $\sigma = 1.0$ and 0.5 are shown in Figure 2 for different Weber numbers. The cavity contours are shown as thin lines, and the closure contours are shown as thick lines. It can be seen that surface tension reduces the cavity size and affects the slope of the cavity near the detachment point. The latter is more pronounced for the larger cavitation number. The values $We = 70$ and $We = 50$ are the minimal values of the Weber number for which the solution is obtained for cavitation number $\sigma = 1.0$ and $\sigma = 0.5$, respectively. For smaller Weber numbers, the iteration process fails to converge. As can be seen in Figure 2a, the angle of cavity detachment, β_0, decreases, and the cavity surface becomes flatter. This tendency may lead to a positive curvature of the free surface and make the flow unstable.

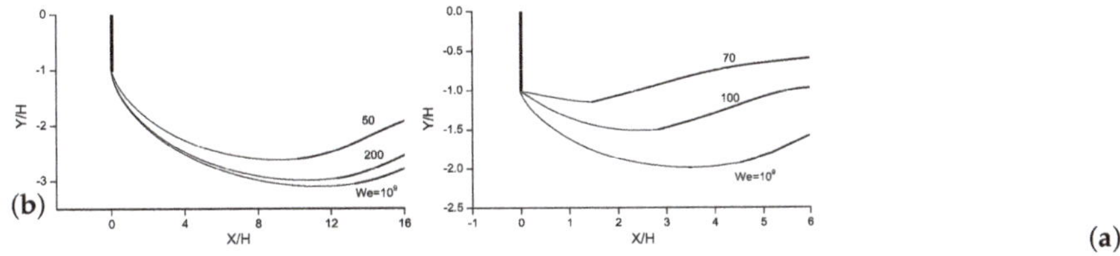

Figure 2. Cavity contours for various Weber numbers for edge radius r = 0.05 and cavitation number: (**a**) σ = 1.0 and (**b**) σ = 0.5.

The angle $\delta = \alpha - \beta_0$ shown in Figure 1a determines the wetted length of the rounded edge, $s^* = r\delta$. Figure 3 shows the angle δ versus the Weber number for different edge radii. At large Weber numbers, when the effect of surface tension on the flow is negligible, the angle δ decreases as the radius of the edge decreases. In the limiting case of $r \to 0$ (sharp edge) and $We \to \infty$ so that $rWe \to \infty$, the velocity at the detachment point (Equation (23)) becomes $v_0 = \sqrt{1 + \sigma}$ as in the case without surface tension. The angle δ tends to zero, and we can expect that the flow leaves the wedge tangentially to the flat part of the wedge. As the Weber number decreases, the angle δ increases, and the surface tension starts to affect the cavity contour. As can be seen in Figure 3, the smaller the wedge radius, the larger the Weber number at which the angle δ starts to increase. If the angle δ becomes $\delta = \pi - \alpha$, this corresponds to cavitation-free flow around the edge of radius r. In the limiting case of $r \to 0$, which corresponds to a sharp edge, and $rWe \to 0$, the velocity at the cavity detachment point $v_0 \to \infty$. This is the same as for flow around a corner without a cavity.

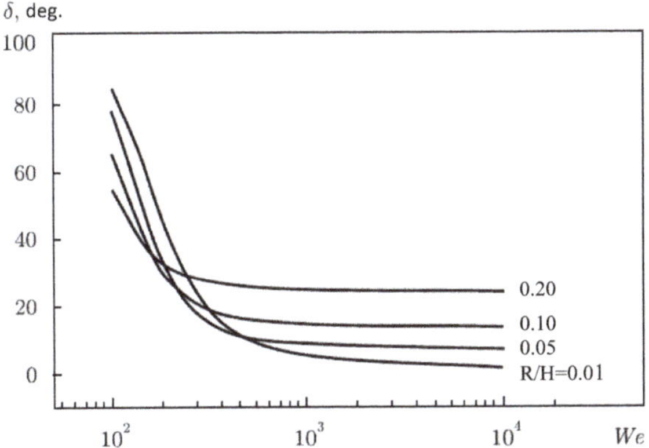

Figure 3. Angle of the wetted part of the edge versus the Weber number for cavitation number σ = 1.0 and various edge radii.

The drag force coefficient the versus cavitation number is shown in Figure 4 for edge radius r = 0.01. The relationships are almost linear, which is consistent with the theory of cavity flows [4,5]. The surface tension affects the drag force coefficient slightly: the smaller the Weber number, the smaller the drag force coefficient. This reduction occurs due to a higher velocity and, correspondingly, a lower pressure near the detachment point, as can be seen from Equation (23). The pressure near the edge may even become negative if the radius is very small, which decreases the total force acting on the wedge.

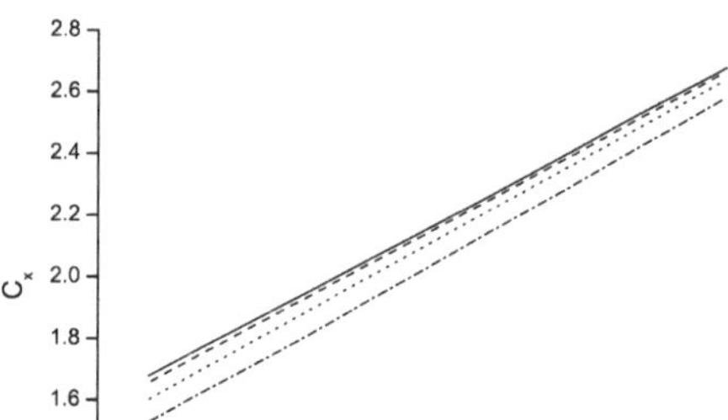

Figure 4. Drag force coefficient versus the cavitation number for radius $r = 0.01$ and Weber numbers: $We = 1000$ (solid line); $We = 250$ (dashed line); $We = 150$ (dotted line); and $We = 100$ (dot-dashed line).

The relative difference of the cavity flow parameters for Weber numbers $We = 2000$ and $We = 200$ versus the cavitation number is shown in Figure 5 for edge radius $r = 0.02$. Here, $\Delta h_c/h_c$, $\Delta l_c/l_c$ and $\Delta C_x/C_x$ are the relative difference of the cavity width, length and force coefficient, respectively. It can be seen that, for a smaller cavitation number, the effect of surface tension is larger.

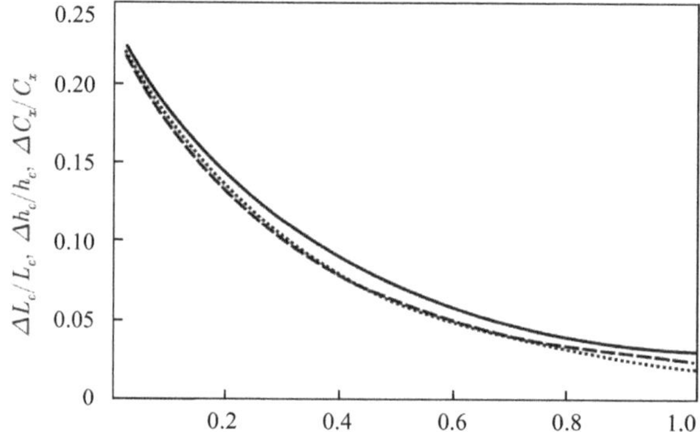

Figure 5. Relative difference of the cavity width (solid line), length (dashed line), and force coefficient (dotted line) for Weber numbers $We = 2000$ and $We = 200$, versus the cavitation number. The edge radius is $r = 0.02$.

4. Conclusions

The effect of surface tension on cavity flow past a wedge with rounded edges was studied theoretically based on an analytical solution that satisfies the Brillouin–Villat criterion for cavity detachment. The integral hodograph method was employed to derive analytical expressions for the complex velocity potential of the flow, the complex velocity, and the mapping function in integral form. These expressions contain the velocity magnitude along the cavity boundary and the slope of the wetted part of the wedge, both as functions of a parameter variable. The solution is obtained in the form of a system of integral equations using the dynamic and kinematic boundary conditions.

A case study is presented for a flat plate with rounded edges. The solution made it possible to determine the range of applicability of the model, which depends on the radius of the edges and the Weber number. For moderate Weber numbers, the surface tension slightly decreases the size of the cavity and the drag force. As the Weber number decreases further, the velocity at the point of cavity detachment increases, and the angle of flow detachment changes in such a manner that the wetting part of the edge becomes larger. The tendency seems to be towards to wetting the whole of the edge and making the flow attached to the wedge. However, this situation is not observed in the results due to the iteration process failing to converge for Weber numbers below some critical value. In these cases, a steady flow solution cannot be obtained. A similar situation occurs when the wedge radius tends to zero at some fixed value of the Weber number. The results obtained suggest that at Weber numbers below the critical value, the flow may become unsteady, and the point of cavity detachment may oscillate about some position on the edge. To verify this mechanism of cavity detachment, an advanced formulation of the problem, including flow unsteadiness and a more sophisticated flow model, are needed.

Author Contributions: Conceptualization, Y.N.S. and Y.A.S.; methodology, G.Y.S. and Y.A.S.; software, Y.A.S. and G.Y.S.; validation, Y.N.S., G.Y.S., and Y.A.S.; formal analysis, Y.A.S.; investigation, Y.N.S., G.Y.S., and Y.A.S.; resources, Y.N.S.; data curation, Y.N.S.; writing—original draft preparation, G.Y.S. and Y.A.S.; writing—review and editing, Y.A.S.; visualization, Y.N.S. and G.Y.S.; supervision, Y.N.S.; project administration, Y.N.S.; funding acquisition, Y.N.S. All authors have read and agreed to the published version of the manuscript.

Funding: This research received no external funding

Institutional Review Board Statement: Not applicable.

Informed Consent Statement: Not applicable.

Data Availability Statement: The data presented in this study are available on request from the corresponding author.

Conflicts of Interest: The authors declare no conflict of interest.

Nomenclature

x, y	Cartesian coordinates
$z = x + iy$	physical plane
$\zeta = \xi + i\eta$	parametric plane/parametric variable
w	complex potential
dw/dz	complex velocity
dw/ζ	derivative of the complex potential
ϕ	flow potential
ψ	stream function
s	arc length coordinate
H	half-width of the wedge
R	radius of the edge
p_∞	pressure at infinity
p_c	pressure in the cavity
U	inflow velocity
v	velocity magnitude
α	wedge angle
β	slope of the wedge
β_0	angle of cavity detachment
σ	cavitation number
δ	angle of detachment

References

1. Brillouin, M. Les surfaces de glissement de Helmholtz et la resistance des fluids. *Ann. Chim. Phys.* **1911**, *23*, 145–230.
2. Villat, H. Sur la validite des solutions de certains problemes d'hydrodynamique. *J. Math Pure Appl.* **1914**, *20*, 231–290.
3. Yoon, B.S.; Semenov, Y.A. Capillary cavity flow past a circular cylinder. *Eur. J. Mech. B/Fluids* **2009**, *28*, 670–676. [CrossRef]
4. Birkhoff, G.; Zarantonello, E.H. *Jets, Wakes and Cavities*; Academic Press: New York, NY, USA, 1957.
5. Gurevich, M.I. *Theory of Jets in Ideal Fluids*; Academic Press: New York, NY, USA, 1965.
6. Vanden-Broeck, J.-M. The influence of surface tension on cavitating flow past a curved obstacle. *J. Fluid Mech.* **1983**, *133*, 255–264. [CrossRef]
7. Vanden-Broeck, J.-M. Nonlinear capillary free-surface flows. *J. Eng. Math.* **2004**, *50*, 415–426. [CrossRef]
8. Ackerberg, R.C. The effects of capillarity on free-stream line separation. *IBID* **1975**, *70*, 333–352.
9. Cumberbatch, E.; Norbury, J. Capillarity modification of the singularity at a free-streamline separation point. *J. Mech. Appl. Maths.* **1979**, *32*, 303–312. [CrossRef]

10. Yoon, B.S.; Semenov, Y.A. Cavity detachment on a hydrofoil with the inclusion of surface tension effects. *Eur. J. Mech. -B/Fluids* **2011**, *30*, 17–25. [CrossRef]
11. Bouwhuis, W.; Snoeijer, J.H. The effect of surface wettability on inertial pouring flows. *arXiv* **2015**, arXiv:1507.05931.
12. Ni, B.Y.; Pan, Y.T.; Yuan, G.Y.; Xue, Y.Z. An experimental study on the interaction between a bubble and an ice floe with a hole. *Cold Reg. Sci. Technol.* **2021**, *187*, 103281. [CrossRef]
13. Yuan, G.Y.; Ni, B.Y.; Wu, Q.G.; Xue, Y.Z.; Zhang, A.M. An experimental study on the dynamics and damage capabilities of a bubble collapsing in the neighborhood of a floating ice cake. *J. Fluids Struct.* **2020**, *92*, 102833. [CrossRef]
14. Ho, C.-M.; Tai, Y.-C. Micro-electro-mechanical-systems (MEMS) and fluid flows. *Annu. Rev. Fluid Mech.* **1998**, *30*, 579. [CrossRef]
15. Gravesen, P.; Branegjerg, J.; Jensen, O.S. Microfluidics—A review. *J. Micromech. Microeng.* **1993**, *3*, 168. [CrossRef]
16. Mishra, C.; Pelesa, Y. Cavitation in flow through a micro-orifice inside a silicon microchannel. *Phisics Fluids* **2005**, *17*, 013601. [CrossRef]
17. Ghorbani, M.; Sadaghiani, A.K.; Villanueva, L.G.; Kosar, A. Hydrodynamic cavitation in microfluidic devices with roughened surfaces. *J. Micromech. Microeng.* **2018**, *28*, 075016. [CrossRef]
18. Semenov, Y.A.; Iafrati, A. On the nonlinear water entry problem of asymmetric wedges. *J. Fluid Mech.* **2006**, *547* 231–256. [CrossRef]
19. Semenov, Y.A.; Yoon, B.S. Onset of flow separation at oblique water impact of a wedge. *Phys. Fluids* **2009**, *21*, 112103-1. [CrossRef]
20. Chaplygin, S.A. *On the Pressure of a Plane Flow on Obstructing Bodies (To the Theory of an Airplane)*; Moscow University: Moscow, Russia, 1910; 49p.
21. Pilipenko, V.V.; Semenov, Y.A.; Pilipenko, O.V. Study of hydrodynamic cavitation in inducer centrifugal pumps. In Proceedings of the Third International Symposium on Cavitation, Grenoble, France, 7–10 April 1998; Volume 1, pp. 323–328.
22. Semenov, Y.A.; Tsujimoto, Y. A cavity wake model based on the viscous/inviscid interaction approach and its application to non-symmetric cavity flows in inducers. *Trans. ASME J. Fluids Eng.* **2003**, *125*, 758–766. [CrossRef]
23. Michell, J.H. On the theory of free stream lines. *Phil. Trans. R. Soc. Lond. A* **1890** *181*, 389–431.
24. Joukovskii, N.E. Modification of Kirchhof's method for determination of a fluid motion in two directions at a fixed velocity given on the unknown streamline. *Math. Sbornik.* **1890**, *15*, 121–278.
25. Anevlavi, D.; Belibassakis, K. An adjoint optimization prediction method for partially cavitating hydrofoils. *J. Mar. Sci. Eng.* **2021**, *9*, 976. [CrossRef]
26. Voskoboinick,V.A.; Grinchenko, V.T.; Makarenkov, A.P. Pseudo-Sound Behind an Obstacle on a Cylinder in Axial Flow. *Int. J. Fluid Mech. Res.* **2005**, *32*, 488–510. [CrossRef]
27. Andreev, M.V.; Drobakhin, O.O.; Saltykov, D.; Gorev, N.B.; Kodzhespirova, I.F. Simple technique for biconical cavity eigenfrequency determination. *Radioelectron. Commun. Syst.* **2017**, *60*, 555–561. [CrossRef]
28. Andreev, M.V.; Drobakhin, O.O.; Saltykov, D.; Gorev, N.B.; Kodzhespirova, I.F. Determination of biconical cavity eigenfrequencies using method of partial intersecting regions and approximation by rational fractions. *Radioelectron. Commun. Syst.* **2019**, *62*, 630–641. [CrossRef]
29. Semenov, Y.A. Nonlinear flexural-gravity waves due to a body submerged in the uniform stream. *Phys. Fluids* **2021**, *33*, 052115. [CrossRef]

MDPI
St. Alban-Anlage 66
4052 Basel
Switzerland
Tel. +41 61 683 77 34
Fax +41 61 302 89 18
www.mdpi.com

Journal of Marine Science and Engineering Editorial Office
E-mail: jmse@mdpi.com
www.mdpi.com/journal/jmse